ASK $y' \rightarrow y''$

ASC h SIMPLE

ASP $d12$

ENGINEERING
HEAT TRANSFER

Engineering Heat Transfer

JAMES R. WELTY

**Department of Mechanical Engineering,
Oregon State University, Corvallis**

John Wiley & Sons

New York   London   Sydney   Toronto

Library of Congress Cataloging in Publication Data:

Welty, James R.
  Engineering heat transfer.

  1. Heat–Transmission. 2. Electronic data processing–Heat–Transmission. I. Title.

TJ265.W38   621.4′022   73-22315
ISBN 0-471-93340-6

PREFACE

Numerous textbooks have been published, in the past twenty years, concerned with the subject of heat transfer. In each case the author or authors have undoubtedly believed their new entry into this field offered something new or unique to students engaged in the study of this important field. The field of heat transfer itself has advanced manyfold during this time span. Much in the way of new knowledge and new techniques for problem solving has been developed and published.

Every teacher at the university level develops approaches to the teaching of certain material and a way of organizing the subject matter which in his experience, work best. This text represents the treatment and chronology which this author has found to be the best of several alternate approaches used over a period of fifteen years.

It is believed that a student completing a first or second level course in heat transfer should have an understanding of the physical phenomena involved and certainly an ability to formulate and solve typical problems encountered in the field. Both basic understanding and problem solving are stressed in this text.

This text stresses fundamental aspects of heat transfer. No attempt has been made to be exhaustive concerning the advanced developments and up-to-date research in the field. Modern applications of heat transfer analyses are numerous. The nuclear, electronics, and aerospace industries have motivated many developments in the heat transfer field; heat transfer is of vital concern in many environmental applications; the heat pipe promises to be one of the most important technological developments of the past half century. Any author delving into these areas of application has trouble knowing when to stop. Allowing for the possibility of a "cop-out" label, this author has decided not to start. The basic ideas presented in this text along with an understanding of, and an ability to apply, the solution techniques which are treated should provide an engineer with enough background and a sufficient "tool kit" to approach new, and sometimes complex, areas of analysis.

This text includes an entire chapter on numerical formulation of heat transfer problems. The numerical work developed, and the application of numerical analysis to heat transfer is, perhaps, the most distinguishing feature of this text compared to others in print. Numerical solutions, including flow charts, example FORTRAN programs, and computer output are given for several problems where such techniques are appropriate. The digital computer is being used extensively to solve problems that were previously done by hand. Long involved solutions that are traditionally included in heat transfer texts are omitted here; an example of such an omission is the "relaxation" technique which becomes a routine application of numerical analysis. The chapter on numerical formulation may be omitted, as may the example problems solved numerically, in a particular course, but these techniques will surely be used by engineers on the job. The introduction to numerical procedures included in this text is intended to develop a vocabulary and enough understanding for an engineer to write simple programs and to converse with those who will write or use more sophisticated ones.

Numerical solutions are most easily employed with heat conduction problems. The majority of example problems solved numerically are included in Chapter 4, which deals entirely with conduction heat transfer. One example problem is solved in Chapter 6 concerning radiation. No convection problems are solved numerically. The reason for this apparent lack of balanced treatment lies in the complexity of the formulation and the computer programs necessary to solve realistic problems in these cases. Numerical solutions are being used in these areas and are extremely important, however such solutions are beyond the scope of this text.

The writing of any textbook requires some selection of material and a reasonable compromise concerning level of treatment. The choices made in this book are those thought to be most reasonable by this author. Students using this book are expected to have a command of mathematics through differential equations. Some background in thermodynamics will be most helpful as will an acquaintance with basic concepts in fluid mechanics.

Some material on fluid flow is presented in the early part of Chapter 5 dealing with convection heat transfer.

Many people have provided assistance to the author in preparing this book. Numerous discussions with colleagues, including graduate students, have helped in developing ideas concerning proper material to be included and the best way of developing certain subjects. Special appreciation is extended to Dr. Donald S. Trent and Mr. David P. Slack for their assistance and ideas concerning numerical analyses.

I am particularly indebted to my wife, Janet, and my children, Mark, Stephen, Dana, Jim, and Tracey for their encouragement and good nature during the period of time I was writing.

Finally, I wish to thank the editorial staff of John Wiley & Sons, in particular Mr. Al Beckett, Mr. Gene Davenport, and Mr. Gary Brahms for their support and continued confidence in me.

Corvallis, Oregon   James R. Welty

CONTENTS _____

CHAPTER 1

Basic concepts in heat transfer

In this chapter the modes of heat transfer will be introduced along with the basic quantitative relationships for determining rates of heat transfer by each mode.

In simplest form we may state that the rate of heat transfer is equal to the product of a driving force and a thermal conductance. Driving forces and thermal conductances (the reciprocal of thermal resistances) vary for each mode; these quantities will be discussed in detail in the sections to follow.

Certain terminology pertinent to the subject of heat transfer will also be introduced in this first chapter. These terms and symbols will be used throughout the succeeding chapters, so it is necessary that these fundamental concepts and terms be understood before proceeding.

The heat transfer modes of conduction, convection, and radiation will now be considered in some detail.

1.1 CONDUCTION

Heat transfer by conduction is accomplished via two mechanisms. The first is that of molecular interaction whereby molecules at relatively higher energy levels (indicated by their temperature) impart energy to adjacent molecules at lower energy levels. This type of transfer will occur in systems

1

where molecules of solid, liquid, or gas are present and in which a temperature gradient exists.

The second conduction heat transfer mechanism is via "free" electrons which are present primarily in pure metallic solids. The concentration of free electrons varies considerably for metallic alloys and is very low for nonmetals. The ability of solids to conduct heat varies directly with the free-electron concentration; thus we would expect pure metals to be the best heat conductors, and our experience has proven this to be so.

It has been mentioned that conduction is primarily a molecular phenomenon requiring a temperature gradient as a driving force. A quantitative expression relating a temperature gradient, the nature of the conducting medium, and the rate of heat transfer is attributed to Fourier,[1] who in 1822 presented the relation

$$\frac{q_x}{A} = -k\frac{dT}{dx} \tag{1-1}$$

where q_x is the x-directional heat flow rate in Btu/hr; A is the area normal to the direction of heat flow in ft²; dT/dx is the temperature gradient in the x direction in °F/ft; and k is the thermal conductivity, having units of Btu/hr-°F-ft²/ft. The ratio q_x/A, having the units Btu/hr-ft², is referred to as the x-directional *heat flux*. The complete expression for the heat flux is

$$\frac{\mathbf{q}}{A} = -k\,\boldsymbol{\nabla}T \tag{1-2}$$

where $\mathbf{q}$ is the heat flow vector and $\boldsymbol{\nabla}T$ is the temperature gradient in vector form. In both equations (1-1) and (1-2) the negative sign is necessary to account for the fact that heat flow by conduction occurs in the direction of a decreasing temperature gradient. These equations are scalar and vector forms, respectively, of the Fourier rate equation, sometimes referred to as Fourier's first "law" of heat conduction.

According to the Fourier rate equation, heat flux is proportional to the temperature gradient, this proportionality being represented by k, the thermal conductivity. Thermal conductivity is a property of a given medium, and equations (1-1) and (1-2) are the defining relationships for this quantity.

Thermal conductivity is an extremely important property of a material or medium. The value of thermal conductivity determines, in large part, the suitability of a material for a given application.

Values of thermal conductivity are shown in Figure 1.1 for several common materials. In this figure the temperature dependence of thermal conductivity can be seen; certain general conclusions can be made in this regard.

[1] J. B. J. Fourier, "Théorie Analytique de la Chaleur," Gouthier-Villars, 1822; English translation by Freeman, Cambridge, 1878.

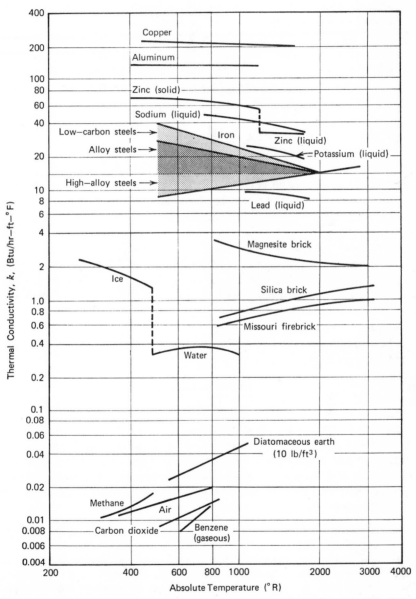

Figure 1.1 Thermal conductivity variation with temperature for various materials. (From M. Jakob and G. A. Hawkins, *Elements of Heat Transfer* (New York: McGraw-Hill Book Co., 1958), p. 23. By permission of the publishers.)

For gases, thermal conductivity values show an increase with an increase in temperature. This is due to the increased agitation of gaseous molecules at elevated temperatures resulting in greater frequency of contact and an attendant increase in molecular exchange rates.

Some considerable analytical work can be done in predicting the thermal conductivity of dilute monatomic gases. For a rough kinetic theory approach to the monatomic gas case the reader may consult Bird, Stewart, and Light-foot[2] or Welty, Wicks, and Wilson.[3] Taking the gas molecule to be a rigid sphere, the resulting equation for k is

$$k = \frac{1}{\pi^{3/2}d^2} \sqrt{\frac{\kappa^3 T}{m}} \qquad (1\text{-}3)$$

where d is the molecular diameter, κ is the Boltzmann constant, T is the absolute temperature, and m is the mass per molecule of the gas medium.

This equation predicts that thermal conductivity is a function of temperature to the $1/2$ power and is independent of pressure. The temperature dependence is a bit weak compared to experimental results; however, the pressure independence has been found to be correct up to about 10 atmospheres for most gases. Equation (1-3) and the simple analysis leading to it, although crude, should not be considered useless in that the result is qualitatively correct and provides a basis for predicting variations in k with temperature and pressure.

A more sophisticated intermolecular force model for a monatomic gas has been used in the Chapman-Enskog theory[4] for thermal conductivity. The Chapman-Enskog equation is

$$k = \frac{1.9891 \times 10^{-4}\sqrt{T/M}}{\sigma^2 \Omega_k} \qquad (1\text{-}4)$$

where k is thermal conductivity in cal/cm-sec, T is absolute temperature in $^\circ K$, M is the molecular weight, and σ and Ω_k are Lennard-Jones parameters associated with the Lennard-Jones intermolecular force potential model. Values of σ and Ω_k may be obtained in the references cited and in Hirschfelder, Curtiss, and Bird.[5] Equation (1-4) again indicates k to be independent of pressure and a function of T to the $1/2$ power.

[2] R. B. Bird, W. E. Stewart, and E. N. Lightfoot, *Transport Phenomena* (New York: John Wiley and Sons, Inc., 1960), chap. 8.
[3] J. R. Welty, C. E. Wicks, and R. E. Wilson, *Fundamentals of Momentum, Heat and Mass Transfer* (New York: John Wiley and Sons, Inc., 1969), chap. 15.
[4] S. Chapman and T. G. Cowling, *Mathematical Theory of Non-uniform Gases* 2nd ed. (Cambridge: Cambridge University Press, 1951).
[5] J. O. Hirschfelder, C. F. Curtiss, and R. B. Bird, *Molecular Theory of Gases and Liquids* (New York: John Wiley and Sons, Inc., 1954).

$$\frac{cal}{sec\ cm\ ^\circ K} \times 242.08 = \frac{Btu}{hr\ ft\ ^\circ R}$$

Table 1.1 Critical Constants for Gases

Substance	Molecular Weight	T_c (°K)	P_c (atm)	k_c (cal/sec-cm-°K)
Air	28.97	132	36.4	90.8
O_2	32.00	154.4	49.7	105.3
N_2	28.02	126.2	33.5	86.8
CO	28.01	133.0	34.5	86.5
CO_2	44.01	304.2	72.9	122.0
NO	30.01	180 00	64.0	118.2
N_2O	44.02	309.7	71.7	131.0
Cl_2	70.91	417.0	76.1	97.0
Ne	20.18	44.5	26.9	79.2
Ar	39.94	151.0	48.0	71.0
Kr	83.80	209.4	54.3	49.4
CH_4	16.04	190.7	45.8	158.0

Much helpful information is included in Figures 1.2 and 1.3 regarding thermal conductivity at a given temperature and pressure. Figure 1.2 shows the reduced thermal conductivity, $k_r = k/k_c$, which is the ratio of the thermal conductivity at given conditions to the value at the critical point, as a function of reduced temperature, $T_r = T/T_c$, and reduced pressure, $P_r = P/P_c$. This plot was developed for predicting thermal conductivities of monatomic gases but may be used in determining approximate k values for polyatomic gases as well. Values of k_c, T_c, and P_c are given in Table 1.1 for some of the more common gases.

In Figure 1.3 the reduced thermal conductivity function k^* is the ratio k/k^o, the ratio of the thermal conductivity at a given temperature and pressure to that at atmospheric pressure and the same temperature. This figure may be the more useful of the two since k^o is more readily available than k_c. As with Figure 1.2 any k values obtained for polyatomic gases should be regarded as approximations only.

Values of thermal conductivity for many gases of interest are tabulated versus temperature in Appendix A-3.

In liquid and solid materials the thermal conductivity is essentially independent of pressure, and much less a function of temperature than in the case of gases. Figure 1.1 shows the temperature dependence of various liquid and solid materials. Tables of k values for liquids and solids are contained in Appendix A-1.

A special mention might be made of the thermal conductivity of pure metals. In these materials free electrons exist and greatly enhance the heat-carrying and electric current-carrying capacities. We are familiar with the

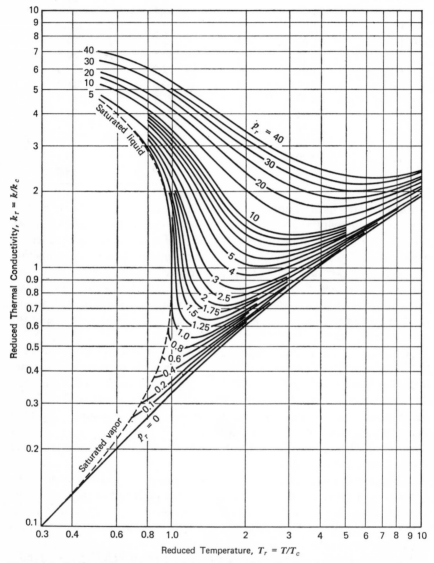

Figure 1.2 Reduced thermal conductivity as a function of reduced temperature and pressure for monatomic gases. (From E. J. Owens and G. Thodos, *AIChE Journal* 3 (1958): 461. By permission of the publishers.)

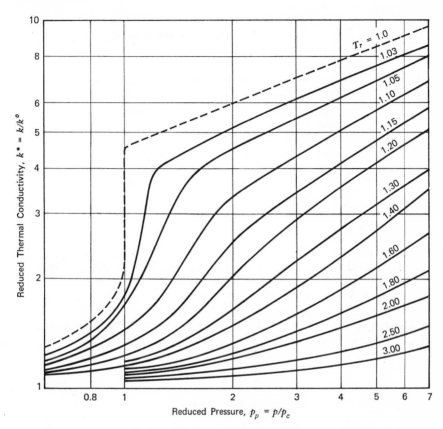

Figure 1.3 Reduced thermal conductivity as a function of reduced temperature and pressure. (From J. N. Lenoir, W. A. Junk, and E. W. Comings, *Chem. Engr. Progress* 49 (1953): 539. By permission of the publishers.)

superior electrical conducting properties of pure metals; the same physical traits which cause this to be so are also responsible for these materials being the best conductors of heat. Table 1.2 lists, in descending order, the general range in thermal conductivity for various categories of conducting media.

The thermal conductivities of numerous materials are tabulated in Appendix A-1. Special mention should be made in the case of wood. Note that values of k parallel to and normal to the grain differ by a substantial amount, for oak by a factor of 2. Wood is a good example of an *anisotropic* material, one whose properties vary in different directions. A medium whose properties do not vary with direction is designated *isotropic*.

The following examples illustrate the application of the Fourier rate equation in solving simple heat conduction problems.

**Table 1.2 Thermal Conductivity Values
for Various Material Categories**

Medium	k (Btu/hr-ft-°F)
Pure metals	20–250
Metal alloys	10–100
Liquid metals	5–50
Liquids (nonmetallic)	0.1–1.0
Nonmetallic solids	0.01–10
Insulating materials	0.01–0.2
Gases	0.001–0.1

Example 1.1

Figure 1.4 illustrates the situation in which steam is transported through a 1-1/2-inch schedule-80 mild steel pipe. The inside and outside pipe wall temperatures are 205°F and 195°F, respectively. Find

(a) the heat loss from 10 ft of pipe.

(b) the heat flux based upon inside and outside surface areas.

In this situation heat is transferred radially, hence the applicable scalar form of the Fourier rate equation is

$$q_r = -kA\frac{dT}{dr}$$

Noting that $A = 2\pi rL$, this equation becomes

$$q_r = -k(2\pi rL)\frac{dT}{dr}$$

For the steady-state case (no time dependence), q_r is constant. We may now

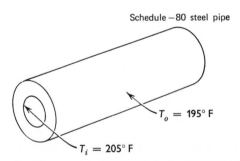

Schedule −80 steel pipe

$T_o = 195°$ F

$T_i = 205°$ F

Figure 1.4 Steady radial heat conduction through a pipe wall with uniform surface temperatures.

separate the variables in the above equation and solve as follows:

$$q_r \int_{r_i}^{r_o} \frac{dr}{r} = -2\pi kL \int_{T_i}^{T_o} dT$$

$$q_r \ln \frac{r_o}{r_i} = -2\pi kL(T_o - T_i) = 2\pi kL(T_i - T_o)$$

$$q_r = \frac{2\pi kL}{\ln \dfrac{r_o}{r_i}} (T_i - T_o) \qquad\qquad (1\text{-}5)$$

Equation (1-5) expresses the steady, radial heat flow rate in terms of geometrical and material properties, and in terms of the driving force, the temperature difference.

From Appendix A-1 we obtain k for mild steel to be 24.8 Btu/hr-ft-°F and from Appendix F-1 for 1-1/2-inch schedule-80 pipe we read OD = 1.900 in., ID = 1.500 in., wall thickness = 0.200 in. Substituting these values appropriately and solving, we obtain

heat flow → $$q_r = \frac{2\pi(24.8 \text{ Btu/hr-ft-°F})(10 \text{ ft})(10°F)}{\ln \dfrac{1.900}{1.500}}$$

$$= 65,000 \text{ Btu/hr}$$

The inside and outside surface areas of the pipe are

$$A_i = \pi \left(\frac{1.500}{12} \text{ ft} \right)(10 \text{ ft}) = 3.93 \text{ ft}^2$$

$$A_o = \pi \left(\frac{1.900}{12} \text{ ft} \right)(10 \text{ ft}) = 4.98 \text{ ft}^2$$

The fluxes evaluated at both surfaces are thus

heat flux → $$\frac{q_r}{A_i} = \frac{65,500}{3.93} = 16,700 \text{ Btu/hr-ft}^2$$

$$\frac{q_r}{A_o} = \frac{65,500}{4.98} = 13,150 \text{ Btu/hr-ft}^2$$

We have obtained heat flux values differing by over 25% for the same heat flow rate. The reader should note the importance of specifying the area upon which a given heat flux is based.

Example 1.2

For steady-state heat conduction through a plane wall with dimensions and surface temperatures as shown in Figure 1.5, express
(a) the heat flow rate for a constant thermal conductivity k.

(b) the heat flow rate when the thermal conductivity of the wall material varies linearly with temperature according to the expression

$$k = k_o(1 + \beta T)$$

Compare the result in (b) with that in (a) if a single value of k, evaluated at an arithmetic mean temperature, were used.

For k a constant the x-directional form of the Fourier rate equation is

$$q_x = -kA \frac{dT}{dx}$$

This expression may be separated and solved easily as follows:

$$q_x \int_0^L dx = -kA \int_{T_0}^{T_L} dT$$

$$q_x L = -kA(T_L - T_0) = kA(T_0 - T_L)$$

$$q_x = \frac{kA}{L}(T_0 - T_L) \tag{1-6}$$

Equation (1-6) applies to the case of a plane wall. As in equation (1-5), the steady-state heat flow rate is expressed in terms of geometric effects, material properties,

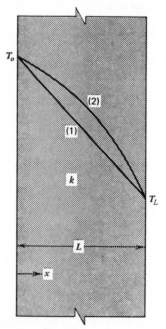

Figure 1.5 Conditions for steady heat conduction through a plane wall with (1) k constant and (2) k a linear function of temperature.

and the driving force, i.e., the temperature difference. Notice how the expression for a plane wall is relatively simple compared to that for a hollow cylinder.

For the variable thermal conductivity case the given k expression substituted into the Fourier rate equation is

$$q_x = -k_0(1 + \beta T)A\frac{dT}{dx}$$

Separating and solving, we obtain

$$q_x\int_0^L dx = -k_0 A\int_{T_0}^{T_L}(1 + \beta T)\,dT$$

$$q_x L = -k_0 A\left[T_L - T_0 + \frac{\beta}{2}(T_L{}^2 - T_0{}^2)\right]$$

$$q_x = \frac{k_0 A}{L}\left[(T_0 - T_L) + \frac{\beta}{2}(T_0{}^2 - T_L{}^2)\right]$$

$$q_x = \frac{A}{L}\left\{k_0\left[1 + \frac{\beta}{2}(T_0 + T_L)\right]\right\}(T_0 - T_L) \qquad (1\text{-}7)$$

In equation (1-7) we observe that the term in braces is the thermal conductivity calculated at the arithmetic mean temperature. Thus an equivalent form of equation (1-7) is

$$q_x = \frac{k_{\mathrm{Avg}}A}{L}(T_0 - T_L)$$

which completes the example.

1.2 CONVECTION

Convection heat transfer, the second basic mode to be considered, involves energy exchange between a bulk fluid and a surface or interface. Two kinds of convective processes exist, these being *forced convection* in which fluid motion past a surface is caused by an external agency such as a fan or pump, and *natural* or *free convection* in which density changes in the fluid resulting from the energy exchange cause a natural fluid motion to occur.

Sir Isaac Newton, in 1701, first expressed the basic rate equation for convective heat transfer. This very simple expression known as the Newton rate equation or Newton's "law" of cooling is

$$q = hA(T_{\mathrm{surf}} - T_{\mathrm{fluid}}) \qquad (1\text{-}8)$$

where q is the rate of convective heat transfer in Btu/hr, A is the area normal to the direction of heat flow in ft², $T_{\mathrm{surf}} - T_{\mathrm{fluid}}$ is the temperature driving force in °F, and h is the convective heat transfer coefficient in Btu/hr-ft²-°F.

The temperature difference may be written as in equation (1-7) or as $T_{\mathrm{fluid}} - T_{\mathrm{surf}}$. This temperature driving force determines whether heat transfer is to or from a given surface. The Newton rate equation is seldom written in vector form as was the Fourier rate equation. The orientation of the surface, to or from which heat is exchanged with an adjacent fluid, determines the direction of heat transfer.

Convective exchange will be considered in depth in later chapters. In addition to basic heat transfer and first law concepts, the consideration of momentum effects as described by Newton's second law of motion will be necessary. Regardless of the flow phenomena involved, it is known that directly adjacent to a surface the energy transfer mechanism is that of conduction. It is these surface conductive layers of fluid or fluid "film" which control the heat transfer rate and thus determine a given value for h. The coefficient h is often designated the "film coefficient" for this reason. Many people insist that there should be no separate distinction made for convection inasmuch as the controlling factor is conduction. We shall continue to make this distinction, however, along the lines described above.

Quantitative work involving equation (1-8) is extremely simple as the equation itself is simple. The difficulty in describing convective phenomena lies in the evaluation of the film coefficient. A large portion of our time in convection considerations will be devoted to the determination of h.

Energy transfer associated with changes in phase, particularly between liquid and vapor phases, is also evaluated by equation (1-8). The processes of boiling and condensation are both associated with relatively high h values. Table 1.3 gives some approximate limits for ranges in h values which characterize free and forced convection heat transfer in air and water.

Equation (1-8) may be compared with both equations (1-6) and (1-7), which are similar in form. The coefficients of ΔT on the right-hand sides of each play similar roles, namely, the *conductance* of each situation and geometry for heat transfer. The conductances for convection, conduction through a hollow cylindrical wall, and conduction through a plane wall are

$$K_{\text{convection}} = hA \quad \text{Btu/hr-}^\circ\text{F}$$

$$K_{\substack{\text{conduction} \\ \text{hollow cylinder}}} = \frac{2\pi kL}{\ln \dfrac{r_o}{r_i}} \text{ Btu/hr-}^\circ\text{F}$$

$$K_{\substack{\text{conduction} \\ \text{plane wall}}} = \frac{kA}{L} \quad \text{Btu/hr-}^\circ\text{F}$$

The reciprocals of these quantities may be thought of as *thermal resistances* offered by each mechanism or geometry.

Table 1.3 Approximate Values for h in Various Convective Situations

Mechanism	h (Btu/hr-ft^2-$^{\circ}$F)
Condensing water vapor	1000–20,000
Boiling water	500–5,000
Forced convection, water	50–3,000
Forced convection, air	5–100
Free convection, air	1–10

Example 1.3

Given the steel pipe with conditions as described in Example 1.1, with 10°F air surrounding the pipe and 210°F steam flowing on the inside, evaluate the convective heat transfer coefficients on each of the pipe surfaces, and indicate the convective mechanism which occurs on the inside and outside surfaces.

The calculated heat flux at the inside surface was 16,700 Btu/hr-ft^2-$^{\circ}$F. Solving for h, using equation (1-8), we obtain

$$h = \frac{q/A}{(T_{\text{stm}} - T_{\text{surf}})}$$
$$= \frac{16,700 \text{ Btu/hr-ft}^2}{5^{\circ}\text{F}}$$
$$= 3,340 \text{ Btu/hr-ft}^2\text{-}^{\circ}\text{F}$$

From Table 1.3 it is apparent that steam is condensing on the inside pipe surface.

Using the same approach for the outside surface, the film coefficient on the air side is seen to be

$$h = \frac{q/A}{T_{\text{surf}} - T_{\text{air}}}$$
$$= \frac{13,150 \text{ Btu/hr-}^{\circ}\text{F}}{(195 - 10)^{\circ}\text{F}}$$
$$= 71 \text{ Btu/hr-ft}^2\text{-}^{\circ}\text{F}$$

The outside air must definitely be in forced convection to achieve an h value of this magnitude.

1.3 THERMAL RADIATION

Heat transfer by radiation requires no medium for propagation. Radiant exchange between surfaces is, in fact, a maximum when no material occupies the intervening space. Radiant energy exchange can occur between two surfaces, between a surface and a gas or participating medium, or it may involve a complex interaction between several surfaces and intervening fluid constituents. Energy transfer by radiation is an electromagnetic phenomenon and the exact nature of this transfer is not known. It is possible, however, to treat this complex subject with reasonable accuracy.

A perfectly emitting or absorbing body is designated a *black* body. The rate at which a black body emits radiant energy is given by

$$\frac{q}{A} = \sigma T^4 \qquad (1\text{-}9)$$

$\sigma = .1714 \times 10^{-8} \, \frac{Btu}{hr \, ft^2 \, {}^{\circ}R^4}$

where q is the radiant emission in Btu/hr; A is the area of the emitting surface in ft²; T is the absolute temperature in °R; and σ is the Stefan-Boltzmann constant, numerically equal to 0.1714×10^{-8} Btu/hr-ft²-°R⁴. Equation (1-9) is the basic rate equation for radiant energy emission and is known as the Stefan-Boltzmann rate equation or as the Stefan-Boltzmann law of thermal radiation. Equation (1-9) was postulated empirically by Stefan in 1879, and derived from thermodynamic principles by Boltzmann in 1884.

The Stefan-Boltzmann rate equation as presented is extremely simple in that no consideration is given to geometry, to the interaction between the emitting surface and other surfaces or media which it sees, or to any non-black surface behavior. Each of these factors will be considered at some length in Chapter 6.

1.4 COMBINED MECHANISMS OF HEAT TRANSFER

Although conduction, convection, and radiation can be separated for discussion purposes, it is unusual when a real engineering situation involving heat transfer does not include at least two, or perhaps all three, mechanisms.

Figure 1.6 shows the situation where a composite plane wall, consisting of two different materials, separates two gases at different temperatures. This situation might be thought of physically as a furnace wall where one material is placed next to the hot furnace gas for its insulating qualities and the other is placed on the outside, adjacent to the air, for its appearance, or perhaps its structural qualities, or both. Denoting the hot and cool gas temperatures as T_h and T_c, respectively, the surface temperatures of the materials by T_1 and T_3, and the interface temperature between the two wall materials as T_2,

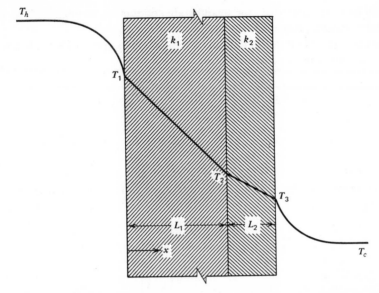

Figure 1.6 Heat transfer through a composite plane wall with convection at the surfaces.

we may write, for the heat transfer rate for each part of this process,

$$q_{h \to 1} = h_h A (T_h - T_1)$$

$$q_{1 \to 2} = -k_1 A \frac{dT}{dx}$$

$$q_{2 \to 3} = -k_2 A \frac{dT}{dx}$$

$$q_{3 \to c} = h_c A (T_3 - T_c)$$

For the steady-state case, the two conduction expressions will take the same form, the difference being in the limits of integration. Solving for $q_{1 \to 2}$, we obtain

$$q_{1 \to 2} \int_0^{L_1} dx = -k_1 A \int_T^{T_2} dT$$

$$q_{1 \to 2} = \frac{k_1 A}{L_1} (T_1 - T_2) \qquad (1\text{-}10)$$

A similar analysis for $q_{2 \to 3}$ will yield

$$q_{2 \to 3} = \frac{k_2 A}{L_2} (T_2 - T_3) \qquad (1\text{-}11)$$

It may be noted here that, for steady-state conduction through a plane wall, the temperature profile is linear.

Steady state also requires that all of the q's be the same; thus the following string of equalities may be written:

$$q = h_h A(T_h - T_1) = \frac{k_1 A}{L_1}(T_1 - T_2)$$

$$= \frac{k_2 A}{L_2}(T_2 - T_3) = h_c A(T_3 - T_c)$$

Each of these forms is sufficient to calculate the heat flow. Additional expressions may be obtained if each temperature difference is written in terms of q as

$$T_h - T_1 = q\left(\frac{1}{h_h A}\right)$$

$$T_1 - T_2 = q\left(\frac{L_1}{k_1 A}\right)$$

$$T_2 - T_3 = q\left(\frac{L_2}{k_2 A}\right)$$

$$T_3 - T_c = q\left(\frac{1}{h_c A}\right)$$

The addition of any two or more of these will yield another expression for q. If, for instance, the two conduction expressions are added, we obtain

$$q = \frac{T_1 - T_3}{\dfrac{L_1}{k_1 A} + \dfrac{L_2}{k_2 A}} \tag{1-12}$$

which involves the difference in surface temperatures as the driving force. This enables q to be evaluated without a knowledge of the interface temperature T_2.

If all of the temperature difference equations are added, the resulting expression for q is

$$q = \frac{T_h - T_c}{\dfrac{1}{h_h A} + \dfrac{L_1}{k_1 A} + \dfrac{L_2}{k_2 A} + \dfrac{1}{h_c A}} \tag{1-13}$$

which involves only the total or overall temperature difference between the hot and cold gases.

Equation (1-13) may be thought of as relating the quantity being transferred (heat) to the ratio of the driving force (temperature difference) to the combined effect of each part of the heat flow path. These latter effects individually constitute resistances offered by each part of the path to the flow of heat. This concept is analogous to Ohm's law, the heat transfer counterpart being simply

$$q = \frac{\Delta T}{\sum R_t} \tag{1-14}$$

where $\sum R_t$ represents an appropriate sum (series, parallel, or combinations) of the various thermal resistances involved.

The thermal resistance of a convective film is seen to be $1/hA$; that of conduction through a plane wall is L/kA; and for a hollow cylinder the thermal resistance is $[\ln (r_o/r_i)]/2\pi kL$. These terms are the reciprocals of the thermal conductances mentioned in Section 1.2.

Another common way of expressing the heat transfer rate when combined modes are involved is

$$q = UA\Delta T \tag{1-15}$$

where U is the *overall heat transfer coefficient*, having units of Btu/hr-ft²-°F, the same as h. Quite obviously, the overall heat transfer coefficient and combined thermal resistance are related according to

$$U = \frac{1}{A \sum R_t} \tag{1-16}$$

Not infrequently, the portion of the heat flow expression involving system geometry is separated from the other terms and designated the *shape factor*. The shape factor is defined by the expression

$$q = kS\Delta T \tag{1-17}$$

Finally, the shape factor, the overall heat transfer coefficient, and the combined thermal resistance may be related as

$$kS = UA = \frac{1}{\sum R_t} \tag{1-18}$$

The treatment of combined heat transfer mechanisms is illustrated in the following example.

Example 1.4

A one-inch nominal diameter steel pipe having its outside surface at 400°F is placed in still air at 90°F, with the convective heat transfer coefficient between the pipe surface and air equal to 1.5 Btu/hr-ft²-°F. It is proposed to add 85% magnesia insulation to the outside surface of the pipe to reduce the heat loss. What thickness of insulation will be required to reduce the heat loss by one-half if the pipe surface temperature and the convective heat transfer coefficient remain unchanged?

For 1-inch steel pipe, OD = 1.315 in. = 0.1095 ft.

For bare pipe,

$$q_0 = hA\Delta T = (1.5\ \text{Btu/hr-ft}^2\text{-}°\text{F})(\pi)(0.1095\ \text{ft})(400 - 90)°\text{F}$$
$$= 160\ \text{Btu/hr} \qquad \text{(per ft)}$$

For insulated pipe, letting the outside diameter of the insulation be D_2,

$$q = \frac{q_0}{2} = \frac{T_{\text{stm}} - T_{\text{air}}}{\dfrac{\ln \dfrac{D_2}{D_1}}{2\pi k} + \dfrac{1}{\pi D_2 h}}$$

Putting all terms involving D_2 on one side of the equation and simplifying as much as possible, we have

$$\frac{h}{2k} \ln \frac{D_2}{D_1} + \frac{1}{D_2} = \frac{2}{q_0}\ \Delta T \pi h$$

$$18.3 \ln \frac{D_2}{0.1095} + \frac{1}{D_2} = 18.25$$

The value of D_2 which satisfies this equality is 0.235 ft; thus the outside diameter of the insulated pipe is 2.82 in., and the thickness of insulation required is slightly over 0.75 inch.

1.5   CLOSURE

In this introductory chapter concerning the basic ideas and rate equations regarding heat transfer analysis, the basic modes of heat transfer—conduction, convection, and radiation—have been discussed. The basic rate equations for each mode are

$$\text{conduction:} \qquad q_x = -k_x A \frac{dT}{dx} \qquad\qquad (1\text{-}1)$$

$$\text{convection:} \qquad q_x = hA(T_{\text{surf}} - T_{\text{fluid}}) \qquad\qquad (1\text{-}8)$$

$$\text{radiation:} \qquad q_x = \sigma A T^4 \qquad\qquad (1\text{-}9)$$

Considerable discussion and useful information have been presented regarding thermal conductivity values for gas, liquid, and solid materials.

Concepts and solution techniques associated with combined modes of heat transfer have been discussed. Useful terms in this regard are the total thermal resistance $\sum R_t$ of a heat transfer path represented by

$$q_x = \frac{\Delta T}{\sum R_t} \qquad (1\text{-}14)$$

the overall heat transfer coefficient U, which is defined according to

$$q_x = UA\Delta T \qquad (1\text{-}15)$$

and the shape factor S, the defining relationship for which is

$$q_x = kS\Delta T \qquad (1\text{-}17)$$

CHAPTER 2

THE EQUATIONS FOR HEAT
TRANSFER

In this chapter we shall develop most of the equations which are basic to heat transfer analysis. Fundamental to all analyses is the first law of thermodynamics, or the energy equation; this will be the starting point for any problem solution. In convection problems one must also use the law of conservation of mass and Newton's second law of motion (the momentum theorem) to evaluate a given situation adequately.

The basic rate equations discussed in the previous chapter, used in conjunction with the fundamental laws, will allow the pertinent heat transfer quantities to be determined. We are usually interested in finding the temperature at a point, the temperature distribution along a boundary or throughout a region, or the heat transfer rate.

The manner of expressing the basic laws and associated rate equations leads to different sorts of solutions. The type of information desired or the capabilities for solution will often dictate the form to be used. For instance, the availability of a large, high-speed computer will make a numerical solution the logical type to be sought, and the governing equations will be best written in difference form. Other choices will be better in different circumstances.

2.1 BASIC LAWS OF HEAT TRANSFER ANALYSIS

Those physical laws which are basic to all heat transfer analysis are

1. First law of thermodynamics.
2. Conservation of mass.
3. Newton's second law of motion.

We will now consider lumped, integral, and differential formulations of these laws. Some special difference forms will be developed and solved in subsequent chapters but will not be considered at this time.

2.1-1 Lumped Formulation of the Basic Laws

A law expressed in lumped form is simply an expression whose terms are independent of space variables. This may be true totally in the case of the sole plate of an iron, which may be considered to be at constant temperature throughout at any time; or in part, as in the case of a fin extending from a hot wall surrounded by a cooler gas. For such a fin, which is very thin, it is usual to consider the temperature to be constant at any given distance from the wall or "lumped" in the direction transverse to the fin axis.

A lumped formulation has the desirable feature of reducing the number of independent variables pertinent to a differential equation. Thus a partial differential equation in three independent variables may be reduced to one in two variables, a partial differential equation in two independent variables will become an ordinary differential equation, and an ordinary differential equation becomes a simple algebraic equation.

Consider the system and control volume shown in Figure 2.1. The *control volume*, or fixed region in space, is depicted by the solid line at time t and $t + \Delta t$. The system is shown by dashed lines at the same two times. A *system*, as we shall consider the term, is a fixed collection of particles having constant mass. Notice that at time $t + \Delta t$ the system lies entirely within the control volume. At time t, at location i, a portion of the system, Δm_i, having volume ΔV_i, and a specific value—i.e., on a unit mass basis—Δp_i, of the property $\mathscr{P}$ lies outside the control volume. This amount of the system crosses the control volume boundary in the time interval Δt.

The change in value of the property $\mathscr{P}$ during the time interval Δt is

$$\Delta \mathscr{P} = \mathscr{P}\big|_{t+\Delta t} - \mathscr{P}\big|_t \tag{2-1}$$

At each time, the total property value includes the contribution of the portions lying within and without the control volume. At each instant in

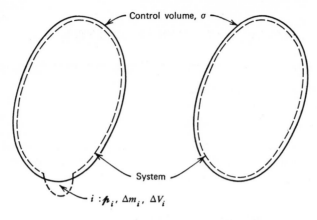

Figure 2.1   System and control volume for lumped formulation of basic laws.

time we may write

$$\mathscr{P}\big|_{t+\Delta t} = \mathscr{P}_\sigma\big|_{t+\Delta t} \tag{2-2}$$

$$\mathscr{P}\big|_t = \mathscr{P}_\sigma\big|_t + \mathit{p}_i\,\Delta m_i \tag{2-3}$$

Substituting these relations into equation (2.1) and denoting changes within the control volume by $\Delta\mathscr{P}_\sigma$, we obtain

$$\begin{aligned}
\Delta\mathscr{P} &= \mathscr{P}_\sigma\big|_{t+\Delta t} - \mathscr{P}_\sigma\big|_t - \mathit{p}_i\,\Delta m_i \\
&= \Delta\mathscr{P}_\sigma - \mathit{p}_i\,\Delta m_i
\end{aligned} \tag{2-4}$$

If portions of the system cross the control surface at more than one place, the more general relation

$$\Delta\mathscr{P} = \Delta\mathscr{P}_\sigma - \sum_{i=1}^{N} \mathit{p}_i\,\Delta m_i \tag{2-5}$$

is obtained, where N is the number of crossings.

Finally, if the time interval Δt between property evaluations is introduced into the denominator of equation (2-5), the expression which results involves the *rate* of change. Performing this operation on equation (2-5), we have

$$\frac{\mathscr{P}\big|_{t+\Delta t} - \mathscr{P}\big|_t}{\Delta t} = \frac{\mathscr{P}_\sigma\big|_{t+\Delta t} - \mathscr{P}_\sigma\big|_t}{\Delta t} - \sum_{i=1}^{N} \mathit{p}_i\,\frac{\Delta m_i}{\Delta t}$$

Evaluating in the limit as $\Delta t \to 0$, we obtain

$$\frac{d\mathscr{P}}{dt} = \frac{d\mathscr{P}_\sigma}{dt} - \sum_{i=1}^{N} \mathit{p}_i\,\frac{dm_i}{dt} \tag{2-6}$$

Equation (2-6) is a general expression for the rate of change of a system property, equating it to the rate of change of the property for the control volume less the net rate at which the property is changed by the effect of mass flow across the control surface. The mass flow rate dm_i/dt is positive for flow into the control volume.

2.1-1.1 Lumped Form of the Law of Conservation of Mass

In the case of the law of conservation of mass, the property under consideration is simply the mass, m; and the specific form is 1. Thus, using the terminology of equation (2-6), $\mathscr{P} = m$, $p_i = 1$, and substituting into this equation, we obtain

$$\frac{dm}{dt} = \frac{dm_\sigma}{dt} - \sum_{i=1}^{N} \frac{dm_i}{dt}$$

Since a system has fixed mass by definition, the derivative

$$\frac{dm}{dt} = 0$$

and the lumped form of the law of conservation of mass is written finally as

$$\frac{dm_\sigma}{dt} - \sum_{i=1}^{N} \frac{dm_i}{dt} = 0 \qquad (2\text{-}7)$$

2.1-1.2 Lumped Form of the First Law of Thermodynamics

Written for a system the first law, in difference form, is

$$\delta Q - \delta W = \Delta E \qquad (2\text{-}8)$$

and in rate form this expression becomes

$$\frac{\delta Q}{dt} - \frac{\delta W}{dt} = \frac{dE}{dt} \qquad (2\text{-}9)$$

In order, these terms represent the rate of heat addition to a system, the rate of work done by a system on its surroundings, and the rate of energy accumulation within the system. The sign conventions used with each term should be noted and used with care in any subsequent discussion or problem solution.

The property of interest in this case is E, the total energy. This single term includes all forms of energy related to the system, including potential, kinetic, internal, nuclear, chemical, and some others. Unless specifically noted, we will include in E only kinetic, potential, and internal energy components. In more detail E can be written

$$E = \frac{mv^2}{2} + mgY + U \tag{2-10}$$

where $mv^2/2$ is the total kinetic energy, mgY is the total potential energy, and U is the total internal energy of the system.

If each term in equation (2-10) is divided by m, the specific form for total energy

$$e = \frac{v^2}{2} + gY + u \tag{2-11}$$

is obtained, where each term is the specific form of its counterpart in the preceding relation.

Using the total energy and substituting into equation (2-6) where $\mathscr{P} = E$, and $\rho_i = e_i$, we obtain

$$\frac{dE}{dt} = \frac{dE_\sigma}{dt} - \sum_{i=1}^{N} e_i \frac{dm_i}{dt} \tag{2-12}$$

With this expression for dE/dt, the lumped form of the first law of thermodynamics is written

$$\frac{\delta Q}{dt} - \frac{\delta W^{\cdot}}{dt} = \frac{dE_\sigma}{dt} - \sum_{i=1}^{N} e_i \frac{dm_i}{dt} \tag{2-13}$$

Equation (2-13) still must be modified to be in useful form. The problem is in properly defining the work rate term, $\delta W/dt$. Without going into great detail, let us consider the work done to be composed of (1) shaft work, that done to produce an effect external to the system which can be made to turn a shaft or raise a weight through a distance; (2) flow work, that done in overcoming pressure effects at any place on the boundary where mass flow occurs; and (3) viscous work, that done in overcoming fluid friction effects at locations on the boundary where mass flow occurs.

The shaft work rate will be simply denoted by the subscript s and written as $\delta W_s/dt$. In similar fashion the viscous work rate will be written $\delta W_\mu/dt$, where the subscript μ symbolizes the viscous nature of this work rate component.

The flow work rate will be written $d(PV)/dt$ or, considering the pressure to be constant, $P\,dV/dt$. Introducing the mass flow rate in place of the volume flow rate, this term finally becomes

$$\frac{P}{\rho}\frac{d}{dt}(\rho V) = \frac{P}{\rho}\frac{dm}{dt}$$

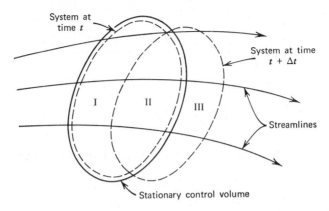

Figure 2.2 System and control volume for integral formulation of basic laws.

Introducing these work rate components into equation (2-13), we have the usable lumped form of the first law of thermodynamics

$$\frac{\delta Q}{dt} - \frac{\delta W_s}{dt} - \frac{\delta W_\mu}{dt} = \frac{dE_\sigma}{dt} - \sum_{i=1}^{N}\left(e_i + \frac{P}{\rho}\right)\frac{dm_i}{dt} \qquad (2\text{-}14)$$

2.1-2 Integral Formulation of the Basic Laws

In a fashion similar to that for the lumped case we shall consider a general relation or transformation formula involving a transport of a general property $\mathscr{P}$, then use this formula to express each of the basic laws in integral form.

Consider the flow field illustrated in Figure 2.2. The control volume, or fixed region, to be considered is shown by the solid line. The system is depicted by dashed lines and is shown at time t, when the system and control boundaries coincide, and at $t + \Delta t$, when a portion of the system has moved outside the control volume. Referring to the figure, we see that

region I is occupied by the system at time t only,

region II is common to the system at both t and $t + \Delta t$,

region III is occupied by the system at $t + \Delta t$ only.

If we once again consider the disposition of the property $\mathscr{P}$, whose specific value is ρ_i, we may write

$$\frac{d\mathscr{P}}{dt} = \lim_{\Delta t \to 0} \frac{\mathscr{P}\big|_{t+\Delta t} - \mathscr{P}\big|_t}{\Delta t} \qquad (2\text{-}15)$$

and in terms of regions I, II, and III,

$$\frac{d\mathscr{P}}{dt} = \lim_{\Delta t \to 0} \frac{\mathscr{P}_{II}|_{t+\Delta t} + \mathscr{P}_{III}|_{t+\Delta t} - \mathscr{P}_{I}|_t - \mathscr{P}_{II}|_t}{\Delta t} \tag{2-16}$$

Equation (2-16) may be rearranged in the form

$$\frac{d\mathscr{P}}{dt} = \lim_{t \to 0} \left[\frac{\mathscr{P}_{II}|_{t+\Delta t} - \mathscr{P}_{II}|_t}{\Delta t} + \frac{\mathscr{P}_{III}|_{t+\Delta t}}{\Delta t} - \frac{\mathscr{P}_{I}|_t}{\Delta t} \right] \tag{2-17}$$

The first term on the right side of equation (2-17) may be written as

$$\lim_{\Delta t \to 0} \frac{\mathscr{P}_{II}|_{t+\Delta t} - \mathscr{P}_{II}|_t}{\Delta t} = \frac{d}{dt} \mathscr{P}_{II}$$

which is the rate of change of the property $\mathscr{P}$ within the control volume itself since, at $\Delta t \to 0$, region II becomes coincident with the control volume. In integral form this term is

$$\lim_{\Delta t \to 0} \frac{\mathscr{P}_{II}|_{t+\Delta t} - \mathscr{P}_{II}|_t}{\Delta t} = \left(\frac{d}{dt} \mathscr{P} \right)_{cv} = \frac{\partial}{\partial t} \int_{cv} \not p \rho \, dV \tag{2-18}$$

where dV is a volume element, ρ is the mass density, and $\not p$ is the specific form of $\mathscr{P}$.

The second and third terms on the right side of equation (2-17) represent the leaving and entering amounts of $\mathscr{P}$ due to mass flow across the control volume boundary. A more compact form for these flux terms is obtained if we first consider the control volume as shown in Figure 2.3.

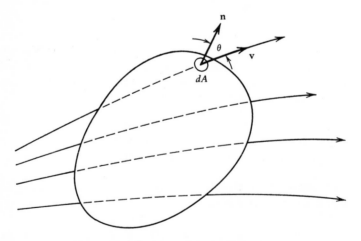

Figure 2.3 Control volume for integral analysis.

For the small area dA on the control volume boundary, the rate of movement of the property $\mathscr{P}$ may be written as

$$\text{flux of } \mathscr{P} = \not{p}(\rho v)(dA \cos \theta)$$

Note that the product $\rho v \, dA \cos \theta$ is the mass flow rate through dA, where the velocity vector $\mathbf{v}$ and outwardly directed unit normal vector $\mathbf{n}$ to dA are separated by the angle θ. From vector algebra the product of terms may be rewritten as

$$\not{p}(\rho v)(dA \cos \theta) = \not{p}(\rho \, dA)[|\mathbf{v}||\mathbf{n}| \cos \theta]$$

The term in square brackets is the scalar or dot product, $\mathbf{v} \cdot \mathbf{n}$, and the property flux term becomes

$$\not{p}\rho(\mathbf{v} \cdot \mathbf{n}) \, dA$$

Integrated over the entire control surface, the property flux due to mass flow is

$$\int_{cs} \not{p}\rho(\mathbf{v} \cdot \mathbf{n}) \, dA$$

It should be noted that this integral, when evaluated over that portion of the control surface where mass flow is outward, is positive since the angle θ will, in all cases, be less than 90°. Wherever mass flow is into the control volume, the angle θ is greater than 90°, and the dot product $\mathbf{v} \cdot \mathbf{n}$ is negative. Thus this single integral taken over the entire control surface includes flow out of and into the control volume; it is equivalent to the last two terms on the right of equation (2-17). The transformation formula in final form may now be written

$$\left.\frac{d\mathscr{P}}{dt}\right|_{\text{system}} = \frac{\partial}{\partial t} \int_{cv} \not{p}\rho \, dV + \int_{cs} \not{p}\rho(\mathbf{v} \cdot \mathbf{n}) \, dA \qquad (2\text{-}19)$$

In words, equation (2-19) states that the rate of change of a property $\mathscr{P}$ for a system is equivalent to the rate of change of $\mathscr{P}$ for a control volume plus the net rate of efflux of $\mathscr{P}$ by virtue of mass flow across the control volume boundaries.

The basic laws will next be put in integral form, using the transformation relation, equation (2-19).

2.1-2.1 Conservation of Mass; Integral Formulation

In the case of conservation of mass, the general and specific property forms are $\mathscr{P} = m$, $\not{p} = 1$. Applying equation (2-19), we obtain

$$\left.\frac{dm}{dt}\right|_{\text{system}} = \frac{\partial}{\partial t} \int_{cv} \rho \, dV + \int_{cs} \rho(\mathbf{v} \cdot \mathbf{n}) \, dA \qquad (2\text{-}20)$$

We again note that the mass of a system is constant by definition, requiring that $(dm/dt)_{\text{system}} = 0$. The resulting formulation for conservation of mass thus becomes

$$\frac{\partial}{\partial t} \int_{\text{cv}} \rho \, dV + \int_{\text{cs}} \rho(\mathbf{v} \cdot \mathbf{n}) \, dA = 0 \tag{2-21}$$

2.1-2.2 First Law of Thermodynamics; Integral Formulation

We have previously noted that the first law of thermodynamics stated in rate form for a system according to equation (2-9) is

$$\frac{\delta Q}{dt} - \frac{\delta W}{dt} = \frac{dE}{dt}\bigg|_{\text{system}} \tag{2-9}$$

In order to use the integral transformation relation, we must first express the property $\mathscr{P}$ appropriately. For this case,

$$\mathscr{P} = E, \qquad p = e$$

and, employing equation (2-19), we obtain

$$\frac{dE}{dt}\bigg|_{\text{system}} = \frac{\partial}{\partial t} \int_{\text{cv}} e\rho \, dV + \int_{\text{cs}} e\rho(\mathbf{v} \cdot \mathbf{n}) \, dA \tag{2-22}$$

An integral form for the first law may thus be written as

$$\frac{\delta Q}{dt} - \frac{\delta W}{dt} = \frac{\partial}{\partial t} \int_{\text{cv}} e\rho \, dV + \int_{\text{cs}} e\rho(\mathbf{v} \cdot \mathbf{n}) \, dA \tag{2-23}$$

As mentioned in Section 2.1-1.2, the work rate term includes shaft work rate, viscous work rate, and flow work rate components. We again write the shaft and viscous work rates as $\delta W_s/dt$ and $\delta W_\mu/dt$, respectively. The flow work rate, for steady pressure, is written

$$P \frac{dV}{dt} = \frac{P}{\rho} \frac{d}{dt}(\rho V) = \frac{P}{\rho} \frac{dm}{dt} \tag{2-24}$$

where dm/dt is the mass flow rate crossing the control surface. Using the integral form for net mass efflux across the control surface, the flow work rate term becomes

$$\frac{P}{\rho} \frac{dm}{dt} = \int_{\text{cs}} \frac{P}{\rho} \rho(\mathbf{v} \cdot \mathbf{n}) \, dA \tag{2-25}$$

Finally, all of the components of work rate may be included in equation (2-23), yielding the integral form of the first law of thermodynamics

$$\frac{\delta Q}{dt} - \frac{\delta W_s}{dt} - \frac{\delta W_\mu}{dt} = \frac{\partial}{\partial t} \int_{\text{cv}} e\rho \, dV + \int_{\text{cs}} \left(e + \frac{P}{\rho}\right)\rho(\mathbf{v} \cdot \mathbf{n}) \, dA \tag{2-26}$$

The reader is again reminded that heat flow rate is positive when directed into the control volume, and work rate is positive when done by the control volume on the surroundings.

In words, equation (2-26) states that the net heat added to a control volume less the rate of shaft and shear work is equal to the rate of energy increase within the control volume plus the net rate of energy efflux across the control volume boundary due to mass flow.

2.1-2.3 Newton's Second Law of Motion; Integral Formulation

Newton's second law of motion is written

$$\sum \mathbf{F} = \frac{d}{dt}(m\mathbf{v}) \qquad (2\text{-}27)$$

It may be stated in words as "the net force exerted on a system is equal to the time rate of change of momentum of the system." It should be noted that this equation involves vector quantities in contrast to the scalar equations that have been considered thus far.

Applying equation (2-19), the applicable values of $\mathscr{P}$ and $\not{p}$ are

$$\mathscr{P} = m\mathbf{v}, \qquad \not{p} = \mathbf{v}$$

and we obtain the integral form of Newton's second law

$$\sum \mathbf{F} = \frac{\partial}{\partial t} \int_{\mathrm{cv}} \mathbf{v}\rho \, dV + \int_{\mathrm{cs}} \mathbf{v}\rho(\mathbf{v} \cdot \mathbf{n}) \, dA \qquad (2\text{-}28)$$

Equation (2-28) states that the net force on a control volume is equivalent to the time rate of change of momentum of the control volume plus the net rate of efflux of momentum from the control volume by virtue of mass flow.

Equations (2-21), (2-26), and (2-28) are the integral formulations that will be referred to and applied in subsequent sections.

2.1-3 Differential Formulation of the Basic Laws

Having generated the integral forms of the basic laws in the previous section, it is now possible for us to obtain the differential forms quickly by applying the divergence theorem from vector calculus. This approach is mathematically sound, but it provides limited physical insight regarding the resulting differential equations. The reader who is interested in an alternate development of the differential forms of the basic laws is referred to Welty, Wicks, and Wilson.[1]

For a general vector quantity $\mathbf{A}$, which is continuously differentiable, and for a volume V, enclosed by a piecewise smooth surface S, the divergence

[1] J. R. Welty, C. E. Wicks, and R. E. Wilson, *Fundamentals of Momentum, Heat and Mass Transfer.* (New York: John Wiley and Sons, Inc., 1969), chaps. 9, 15, 25.

theorem states that

$$\int_S \mathbf{A} \cdot \mathbf{n} \, dS = \int_v \mathbf{\nabla} \cdot \mathbf{A} \, dV \tag{2-29}$$

This relation will now be applied to the integral forms of the basic laws to yield the desired differential equations.

2.1-3.1 Conservation of Mass; Differential Formulation

In Section 2.1-2.1 the integral form of the law of conservation of mass was stated as

$$\int_{\text{cv}} \frac{\partial}{\partial t} \rho \, dV + \int_{\text{cs}} \rho(\mathbf{v} \cdot \mathbf{n}) \, dA = 0 \tag{2-21}$$

The surface integral may be converted to a volume integral by application of equation (2-29) as follows:

$$\int_{\text{cs}} \rho(\mathbf{v} \cdot \mathbf{n}) \, dA = \int_{\text{cv}} \mathbf{\nabla} \cdot \rho \mathbf{v} \, dV \tag{2-30}$$

Including the entire left-hand side of equation (2-21) in a single volume integral, we may write

$$\int_{\text{cv}} \left(\frac{\partial \rho}{\partial t} + \mathbf{\nabla} \cdot \rho \mathbf{v} \right) dV = 0 \tag{2-31}$$

For equation (2-31) to be true in general, the integrand must be equal to zero, and the equation which results is

$$\frac{\partial \rho}{\partial t} + \mathbf{\nabla} \cdot \rho \mathbf{v} = 0 \tag{2-32}$$

which is the differential form of the law of conservation of mass. If the divergence term is separated, an equivalent form to this equation is

$$\frac{\partial \rho}{\partial t} + \mathbf{v} \cdot \mathbf{\nabla} \rho + \rho \mathbf{\nabla} \cdot \mathbf{v} = 0 \tag{2-33}$$

Those terms involving partial derivatives of density are in a form which is encountered sufficiently often that a special name and symbol are reserved for this grouping. The name is the *substantial derivative*, and the symbol for the substantial derivative operator is D/Dt. By definition, this operation is

$$\frac{D}{Dt} \equiv \frac{\partial}{\partial t} + \mathbf{v} \cdot \mathbf{\nabla} \tag{2-34}$$

so the continuity equation in differential form, equation (2-32), may also be written as

$$\frac{D\rho}{Dt} + \rho\nabla \cdot \mathbf{v} = 0 \tag{2-35}$$

At this point it is appropriate to discuss briefly the physical meaning of the substantial derivative. Consider first the atmospheric pressure P, considered in general to be a function of position and time, expressed as $P = P(x, y, z, t)$. To evaluate the change in atmospheric pressure, the differential of P, in Cartesian coordinates, may be written as

$$dP = \frac{\partial P}{\partial t}\, dt + \frac{\partial P}{\partial x}\, dx + \frac{\partial P}{\partial y}\, dy + \frac{\partial P}{\partial z}\, dz$$

where dx, dy, and dz are arbitrary displacements in the x, y, and z directions, respectively. If we divide through by dt, the rate of change of pressure is obtained as

$$\frac{dP}{dt} = \frac{\partial P}{\partial t} + \frac{\partial P}{\partial x}\frac{dx}{dt} + \frac{\partial P}{\partial y}\frac{dy}{dt} + \frac{\partial P}{\partial z}\frac{dz}{dt} \tag{2-36}$$

Consider three approaches for the evaluation of dP/dt. In the first case, the instrument to measure pressure is located in a weather station which, naturally, is fixed to the earth's surface. The coefficients dx/dt, dy/dt, and dz/dt are all zero in this case, and for a fixed point of observation the total derivative dP/dt is equal to the local time derivative $\partial P/\partial t$.

A second approach involves the pressure-measuring instrument housed in an aircraft which may be made to fly in any chosen direction, climb, or descend as the pilot may choose. In such a case, the derivatives dx/dt, dy/dt, and dz/dt are the x, y, and z components of the aircraft velocity, respectively. These components are arbitrarily chosen and bear only co-incidental relationship with the air currents.

A third situation is one in which the pressure-measuring instrument is in a balloon which drifts, rises, or falls with the air in which the balloon is suspended. Here the derivatives dx/dt, dy/dt, and dz/dt are the components of the velocity of the medium itself, v_x, v_y, and v_z, respectively. This third situation corresponds to the definition of the substantial derivative, and the terms in equation (2-36) may be grouped as follows:

$$\frac{dP}{dt} = \frac{DP}{Dt} = \underbrace{\frac{\partial P}{\partial t}}_{\substack{\text{local rate} \\ \text{of pressure} \\ \text{change}}} + \underbrace{v_x\frac{\partial P}{\partial x} + v_y\frac{\partial P}{\partial y} + v_z\frac{\partial P}{\partial z}}_{\substack{\text{rate of pressure} \\ \text{change due to fluid} \\ \text{motion}}} \tag{2-37}$$

It is a simple exercise to show that equation (2-37) is the Cartesian coordinate form of the more general substantial derivative operator given in equation (2-34). The substantial derivative is thus seen to be the *derivative following the flow* of the fluid. The operator, D/Dt, may be interpreted as the time rate of change of a given variable evaluated along the path of a fluid element in the flow field. This operator may be applied to both scalar and vector quantities, and we shall encounter and use it in both contexts.

A final observation regarding the continuity equation given by equations (2-32) and (2-35) is the very simple form which results for the case of incompressible flow. In this case, the equation becomes

$$\mathbf{\nabla} \cdot \mathbf{v} = 0 \qquad (2\text{-}38)$$

2.1-3.2   First Law of Thermodynamics; Differential Approach

Our starting place for the generation of the differential form of the first law of thermodynamics will be equation (2-26), repeated below for reference.

$$\frac{\delta Q}{dt} - \frac{\delta W_s}{dt} - \frac{\delta W_\mu}{dt} = \frac{\partial}{\partial t} \int_{\text{cv}} e\rho \, dV + \int_{\text{cs}} \left(e + \frac{P}{\rho} \right) \rho(\mathbf{v} \cdot \mathbf{n}) \, dA \qquad (2\text{-}26)$$

Working first with the right-hand side of equation (2-26), we interchange the order of differentiation and integration for the volume integral, and rewrite the surface integral as a volume integral. The result of these changes is

$$\frac{\delta Q}{dt} - \frac{\delta W_s}{dt} - \frac{\delta W_\mu}{dt} = \int_{\text{cv}} \left[\frac{\partial}{\partial t} (e\rho) + \mathbf{\nabla} \cdot (e\rho \mathbf{v} + P\mathbf{v}) \right] dV \qquad (2\text{-}39)$$

Considering next the terms on the left-hand side of these equations, we first encounter the heat flow rate $\delta Q/dt$. The heat flow is considered to be either that due to conduction across the control volume boundary or that from internal generation by electrical resistance heating, chemical reaction, or similar effects. Including both of these effects, the heat flow rate term may be written

$$\frac{\delta Q}{dt} = \int_{\text{cs}} -\frac{\mathbf{q}}{A} \cdot \mathbf{n} \, dA + \int_{\text{cv}} \dot{q} \, dV \qquad (2\text{-}40)$$

The first term, involving the surface integral, expresses conductive heat transfer across the boundary in terms of the heat flow vector $\mathbf{q}$, expressed in equation (1-2) as a function of thermal conductivity and temperature gradient. The minus sign with this term in equation (2-40) is due to the heat flow being defined positive *into* the control volume. The second term involves the total rate of heat generated within the control volume.

Equation (2-40) may be modified by writing the heat flow vector in terms of the temperature gradient as in equation (1-2). Equation (2-40) now becomes

$$\frac{\delta Q}{dt} = \int_{cs} k \, \nabla T \cdot \mathbf{n} \, dA + \int_{cv} \dot{q} \, dV \qquad (2\text{-}41)$$

and, applying the divergence theorem to the surface integral term, we have

$$\frac{\delta Q}{dt} = \int_{cv} (\nabla \cdot k \, \nabla T + \dot{q}) \, dV \qquad (2\text{-}42)$$

The two work rate terms are to be considered next. For a differential element, the shaft work rate—that done by some effect within the control volume—is taken to be zero; thus

$$\frac{\delta W_s}{dt} = 0 \qquad (2\text{-}43)$$

The viscous work rate is evaluated by taking the scalar product of the velocity and the viscous stress over the entire control surface. To simplify the evaluation of this quantity, we define the term Λ as the viscous work rate per unit volume and write the energy equation term as

$$\frac{\delta W_\mu}{dt} = \int_{cv} \Lambda \, dV \qquad (2\text{-}44)$$

With the substitutions indicated by equations (2-42), (2-43), and (2-44) made, the energy equation now becomes

$$\int_{cv} \left[\frac{\partial}{\partial t}(e\rho) + \nabla \cdot (e\rho \mathbf{v} + P\mathbf{v}) - \nabla \cdot k \nabla T - \dot{q} + \Lambda \right] dV = 0 \quad (2\text{-}45)$$

We again use the argument that, for equation (2-45) to be true in general, the integrand must be zero. The result is the general differential form of the first law of thermodynamics

$$\frac{\partial}{\partial t}(e\rho) + \nabla \cdot (e\rho \mathbf{v}) + \nabla \cdot P\mathbf{v} - \nabla \cdot k \nabla T - \dot{q} + \Lambda = 0 \qquad (2\text{-}46)$$

An alternate form of equation (2-46) is possible by writing it as

$$e \left[\frac{\partial \rho}{\partial t} + \nabla \cdot \rho \mathbf{v} \right] + \rho \left[\frac{\partial e}{\partial t} + \mathbf{v} \cdot \nabla e \right] + \nabla \cdot P\mathbf{v} - \nabla \cdot k \nabla T - \dot{q} + \Lambda = 0 \quad (2\text{-}47)$$

The first bracketed term in equation (2-47) is zero according to continuity, equation (2-32). The second bracketed term may be written more simply in substantial derivative form. The entire equation becomes

$$\rho \frac{De}{Dt} + \mathbf{\nabla} \cdot P\mathbf{v} - \mathbf{\nabla} \cdot k \mathbf{\nabla} T - \dot{q} + \Lambda = 0 \qquad (2\text{-}48)$$

Equation (2-48) is quite general in nature, including convective as well as conductive effects. For a stationary system, conduction is the only significant heat transfer mode, and equation (2-48) reduces to

$$\rho \frac{\partial e}{\partial t} - \mathbf{\nabla} \cdot k \mathbf{\nabla} T - \dot{q} = 0 \qquad (2\text{-}49)$$

In this case, the only significant energy form in e is the internal energy, which may be written as cT,[2] where c is the heat capacity. Equation (2-49) now becomes

$$\rho c \frac{\partial T}{\partial t} - \mathbf{\nabla} \cdot k \mathbf{\nabla} T - \dot{q} = 0 \qquad (2\text{-}50)$$

For an isotropic medium with material properties presumed independent of temperature, we may write

$$\frac{\partial T}{\partial t} - \frac{k}{\rho c} \nabla^2 T - \frac{\dot{q}}{\rho c} = 0 \qquad (2\text{-}51)$$

The ratio of physical properties, $k/\rho c$, is itself a physical property designated α, the *thermal diffusivity*. Values of α are tabulated versus temperature in Appendix A for various materials.

For a system without heat sources, equation (2-51) becomes

$$\frac{\partial T}{\partial t} = \alpha \nabla^2 T \qquad (2\text{-}52)$$

This equation is referred to as Fourier's second "law" of heat conduction, as the Fourier field equation, or simply the heat equation.

For a steady-state system with heat sources, equation (2-51) reduces to the Poisson equation

$$\nabla^2 T + \frac{\dot{q}}{k} = 0 \qquad (2\text{-}53)$$

In the very simple case of a conducting medium without heat sources and with no time dependence, the resulting relation is the Laplace equation

$$\nabla^2 T = 0 \qquad (2\text{-}54)$$

[2] Strictly speaking, the relationship $e = cT$ presumes simplistic physical behavior viz., all gases are ideal, solids are incompressible, etc. While this expression is not true in an absolute sense, it is an approximation which holds in practically all cases of engineering interest.

The Laplace equation written in Cartesian coordinates is

$$\frac{\partial^2 T}{\partial x^2} + \frac{\partial^2 T}{\partial y^2} + \frac{\partial^2 T}{\partial z^2} = 0 \qquad (2\text{-}55)$$

in cylindrical coordinates it is

$$\frac{\partial^2 T}{\partial r^2} + \frac{1}{r}\frac{\partial T}{\partial r} + \frac{1}{r^2}\frac{\partial^2 T}{\partial \theta^2} + \frac{\partial^2 T}{\partial z^2} = 0 \qquad (2\text{-}56)$$

and in spherical coordinates it is

$$\frac{1}{r^2}\frac{\partial}{\partial r}\left(r^2\frac{\partial T}{\partial r}\right) + \frac{1}{r^2 \sin\theta}\frac{\partial}{\partial \theta}\left(\sin\theta\,\frac{\partial T}{\partial \theta}\right) + \frac{1}{r^2 \sin^2\theta}\frac{\partial^2 T}{\partial \phi^2} = 0 \qquad (2\text{-}57)$$

2.1-3.3   Newton's Second Law of Motion; Differential Formulation

As in the previous cases, our starting point for generating the differential form of Newton's second law of motion will be the previously considered integral form. Equation (2-28) repeated below for reference is

$$\sum \mathbf{F} = \frac{\partial}{\partial t}\int_{cv} \mathbf{v}\rho \, dV + \int_{cs} \mathbf{v}\rho(\mathbf{v}\cdot\mathbf{n})\,dA \qquad (2\text{-}28)$$

The term on the left-hand side of equation (2-28), the net force exerted on the control volume, is the result of surface and body forces. Acting on the surface of a fluid control volume is the stress tensor, including both pressure and viscous effects. The only body force we will consider will be gravity. The net force, including these effects, may be written as

$$\sum \mathbf{F} = \int_{cs} \underset{\sim}{\boldsymbol{\tau}}\cdot\mathbf{n}\,dA + \int_{cv}\rho\mathbf{g}\,dV \qquad (2\text{-}58)$$

or, applying the divergence theorem to the surface integral term,

$$\sum \mathbf{F} = \int_{cv} (\boldsymbol{\nabla}\cdot\underset{\sim}{\boldsymbol{\tau}} + \rho\mathbf{g})\,dV \qquad (2\text{-}59)$$

The right-hand side of equation (2-28) may also be modified, using the divergence theorem to recast the surface integral term, giving

$$\frac{\partial}{\partial t}\int_{cv}\mathbf{v}\rho\,dV + \int_{cs}\mathbf{v}\rho(\mathbf{v}\cdot\mathbf{n})\,dA = \int_{cv}\left[\frac{\partial}{\partial t}(\mathbf{v}\rho) + \boldsymbol{\nabla}\cdot\rho\mathbf{v}\mathbf{v}\right]dV \qquad (2\text{-}60)$$

Using equations (2-59) and (2-60), the entire equation can be written as a volume integral

$$\int_{cv}\left[\boldsymbol{\nabla}\cdot\underset{\sim}{\boldsymbol{\tau}} + \rho\mathbf{g} - \frac{\partial}{\partial t}(\mathbf{v}\rho) - \boldsymbol{\nabla}\cdot\rho\mathbf{v}\mathbf{v}\right]dV = 0 \qquad (2\text{-}61)$$

And, with the usual arguments, setting the integrand equal to zero, we obtain the desired differential form

$$\nabla \cdot \underset{\sim}{\tau} + \rho \mathbf{g} = \frac{\partial}{\partial t} (\mathbf{v}\rho) + \nabla \cdot \rho \mathbf{v}\mathbf{v} \tag{2-62}$$

The right-hand side of equation (2-62) may be modified, giving

$$\nabla \cdot \underset{\sim}{\tau} + \rho \mathbf{g} = \mathbf{v}\left[\frac{\partial \rho}{\partial t} + \nabla \cdot \rho \mathbf{v}\right] + \rho\left[\frac{\partial \mathbf{v}}{\partial t} + \mathbf{v} \cdot \nabla \mathbf{v}\right] \tag{2-63}$$

The first bracketed term on the right vanishes as required by continuity, equation (2-32), and the second bracketed term may be written more compactly in terms of the substantial derivative. The result is

$$\rho \frac{D\mathbf{v}}{Dt} = \rho \mathbf{g} + \nabla \cdot \underset{\sim}{\tau} \tag{2-64}$$

The Cartesian scalar component forms of equation (2-64) are

$$\rho \frac{Dv_x}{Dt} = \rho g_x + \frac{\partial}{\partial x} \sigma_{xx} + \frac{\partial}{\partial y} \tau_{yx} + \frac{\partial}{\partial z} \tau_{zx} \tag{2-65a}$$

$$\rho \frac{Dv_y}{Dt} = \rho g_y + \frac{\partial}{\partial x} \tau_{xy} + \frac{\partial}{\partial y} \sigma_{yy} + \frac{\partial}{\partial z} \tau_{zy} \tag{2-65b}$$

$$\rho \frac{Dv_z}{Dt} = \rho g_z + \frac{\partial}{\partial x} \tau_{xz} + \frac{\partial}{\partial y} \tau_{yz} + \frac{\partial}{\partial z} \sigma_{zz} \tag{2-65c}$$

The above equations are valid for any type of fluid for any kind of stress rate-of-strain behavior. For the case of a Newtonian fluid (one with constant viscosity) in laminar, incompressible flow, the resulting vector form of equation (2-64) is the incompressible form of the *Navier-Stokes* equation

$$\rho \frac{D\mathbf{v}}{Dt} = \rho \mathbf{g} - \nabla P + \mu \nabla^2 \mathbf{v} \tag{2-66}$$

For a flow in which viscous effects are negligibly small, the governing relation is Euler's equation

$$\rho \frac{D\mathbf{v}}{Dt} = \rho \mathbf{g} - \nabla P \tag{2-67}$$

These equations, pertaining to fluid flow, allow the general form of the energy equation, equation (2-48), to be considered in more detail. This will be done in the convective heat transfer section, Chapter 5.

2.2 CLOSURE

The governing equations for heat transfer analysis have been developed in this chapter in lumped, integral, and differential form.

The equations which will be basic to all subsequent analyses are the following:

Law of conservation of mass
  lumped form

$$\frac{dm_\sigma}{dt} - \sum_{i=1}^{N} \frac{dm_i}{dt} = 0 \tag{2-7}$$

  integral form

$$\frac{\partial}{\partial t} \int_{cv} \rho \, dV + \int_{cs} \rho(\mathbf{v} \cdot \mathbf{n}) \, dA = 0 \tag{2-21}$$

  differential form

$$\frac{\partial \rho}{\partial t} + \nabla \cdot \rho \mathbf{v} = 0 \tag{2-32}$$

First law of thermodynamics
  lumped form

$$\frac{\delta Q}{dt} - \frac{\delta W_s}{dt} - \frac{\delta W_\mu}{dt} = \frac{dE_\sigma}{dt} - \sum_{i=1}^{N} \left(e_i + \frac{P}{\rho} \right) \frac{dm_i}{dt} \tag{2-14}$$

  integral form

$$\frac{\delta Q}{dt} - \frac{\delta W_s}{dt} - \frac{\delta W_\mu}{dt} = \frac{\partial}{\partial t} \int_{cv} e\rho \, dV + \int_{cs} \left(e + \frac{P}{\rho} \right) \rho(\mathbf{v} \cdot \mathbf{n}) \, dA \tag{2-26}$$

  differential form

$$\rho \frac{De}{Dt} + \nabla \cdot P\mathbf{v} - \nabla \cdot k \nabla T - \dot{q} + \Lambda = 0 \tag{2-48}$$

Newton's second law of motion
  integral form

$$\sum \mathbf{F} = \frac{\partial}{\partial t} \int_{cv} \mathbf{v}\rho \, dV + \int_{cs} \mathbf{v}\rho(\mathbf{v} \cdot \mathbf{n}) \, dA \tag{2-28}$$

  differential form

$$\rho \frac{D\mathbf{v}}{Dt} = \rho \mathbf{g} + \nabla \cdot \underset{\sim}{\tau} \tag{2-64}$$

CHAPTER 3

NUMERICAL FORMULATION OF
HEAT TRANSFER EQUATIONS

The traditional formulations of heat transfer problems into lumped, differential, and integral forms were presented in the previous chapter. While the results of these type formulations are of extreme importance, they are rapidly assuming a lesser role in the heat transfer hierarchy of methods as digital computers are more readily available and as heat transfer analysts become more skillful in computer programming and numerical analysis. This chapter is devoted, in its entirety, to the application of numerical analysis to the solution of heat transfer problems, with particular attention paid to those considerations which are necessary for digital computer use.

This treatment of numerical formulation and analysis is not intended to be exhaustive or to be complete in its mathematical rigor. We shall present techniques on the basis of their being of practical use. Special vocabulary will be explained when words are encountered for the first time. The reader is assumed to have had some previous experience with FORTRAN programming.

3.1 FUNDAMENTAL CONSIDERATIONS OF NUMERICAL FORMULATION

The presentation in this chapter will focus primarily on heat conduction; some convection and radiation problem solutions by numerical means will be presented in Chapters 5 and 6.

The differential equations which are of greatest interest in conduction heat transfer are the heat equation, Poisson's equation, and Laplace's equation. These were developed in Chapter 2 and presented as equations (2-52), (2-53), and (2-54), respectively. They are repeated below and renumbered for ready reference in this chapter.

$$\text{Heat equation} \qquad \frac{\partial T}{\partial t} = \alpha \nabla^2 T \qquad (3\text{-}1)$$

$$\text{Poisson's equation} \qquad \nabla^2 T + \frac{\dot{q}}{k} = 0 \qquad (3\text{-}2)$$

$$\text{Laplace's equation} \qquad \nabla^2 T = 0 \qquad (3\text{-}3)$$

Each of the three equations listed above may be thought of as a special case of equation (2-51), whose development will be repeated below for reference. Figure 3.1 shows the two-dimensional volume element and nomenclature for the development which follows. The volume of material will be referred to as "node i" and is one of many such volumes or "nodes" which comprise the total volume of material of interest. The heat balance (application of the first law of thermodynamics) on node i yields the following expression:

$$
\left[\frac{\rho c T|_{t+\Delta t} - \rho c T|_t}{\Delta t} \right] \Delta V = k_x \Delta y \left. \frac{\partial T}{\partial x} \right|_{\substack{x+\Delta x \\ t=\xi}} - k_x \Delta y \left. \frac{\partial T}{\partial x} \right|_{\substack{x \\ t=\xi}}
$$

$$
+ k_y \Delta x \left. \frac{\partial T}{\partial y} \right|_{\substack{y+\Delta y \\ t=\xi}} - k_y \Delta x \left. \frac{\partial T}{\partial y} \right|_{\substack{y \\ t=\xi}} + \dot{q}(x, y, t) \Delta V \qquad (3\text{-}4)
$$

where k_x and k_y represent the thermal conductivity values in the x and y directions, respectively; the time ξ at which these terms apply is in the interval $t \leq \xi \leq t + \Delta t$; and the node volume is given by $\Delta V = \Delta x \Delta y(1)$.

As Δx, Δy, and Δt approach zero in the limit, a general differential equation for a conducting medium

$$
\rho c \frac{\partial T}{\partial t} = \frac{\partial}{\partial x}\left(k_x \frac{\partial T}{\partial x} \right) + \frac{\partial}{\partial y}\left(k_y \frac{\partial T}{\partial y} \right) + \dot{q}
$$

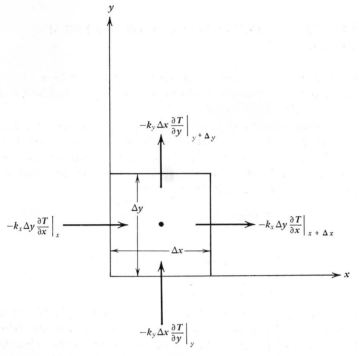

Figure 3.1   Volume element in a conducting medium.

is obtained, with the possibility that ρ, c, k_x, and k_y may be position dependent and that $\dot{q}$ may vary with both position and time. For an isotropic medium, ρ, c, and k may be taken as constant, and this expression reduces to the more familiar form

$$\nabla^2 T + \frac{\dot{q}}{k} = \frac{1}{\alpha}\frac{\partial T}{\partial t} \tag{3-5}$$

Equation (3-5) is seen to be equivalent to equation (2-51) developed in the previous chapter.

At this juncture two different approaches are possible for the formulation of a transient heat conduction problem in difference form; these are

1.   Application of finite difference methods to equation (3-5).
2.   Direct use of equation (3-4).

The former of these approaches will be designated the "difference method" and the latter the "heat balance method."

The two approaches to developing difference equations, although different, yield equivalent results. Differencing techniques applied to a known differential equation are compact and useful in the study of such theoretical considerations as numerical stability. Greater flexibility is possible, however, in the heat balance method; better physical insight is possible also. These methods will be explored separately in the sections which follow.

3.1-1 Finite Difference Representation of Derivatives

The replacement of time and space derivatives by finite differences involves expressing the derivatives in terms of a truncated Taylor series expansion. In one-dimensional space, T expanded about point x_i is given by

$$T(x_i + h) = T(x_i) + h\left(\frac{dT}{dx}\right)_i + \frac{h^2}{2}\left(\frac{d^2T}{dx^2}\right)_i + \frac{h^3}{6}\left(\frac{d^3T}{dx^3}\right)_i + \cdots \quad (3\text{-}6)$$

and, similarly,

$$T(x_i - h) = T(x_i) - h\left(\frac{dT}{dx}\right)_i + \frac{h^2}{2}\left(\frac{d^2T}{dx^2}\right) - \frac{h^3}{6}\left(\frac{d^3T}{dx^3}\right)_i + \cdots \quad (3\text{-}7)$$

Adding equations (3-6) and (3-7) through those terms involving h^3, we obtain

$$\left.\frac{d^2T}{dx^2}\right|_i = \frac{[T(x_i + h) + T(x_i - h) - 2T(x_i)]}{h^2} + O(h^2) \quad (3\text{-}8)$$

The additive term at the end of equation (3-8), $O(h^2)$, indicates that truncated terms cause the "error" in this representation of the second derivative to be *on the order of* h^2.

The result given by equation (3-8) is called a 3-*point central difference* operation. For 3 points, equally spaced, the middle one designated by the index i, the central difference representation for the second derivative may be written as

$$\left.\frac{d^2T}{dx^2}\right|_i = \frac{T_{i+1} + T_{i-1} - 2T_i}{\Delta x^2} + O(\Delta x^2) \quad (3\text{-}9)$$

Should equation (3-6) be truncated after the linear term in h, the result is

$$\left.\frac{dT}{dx}\right|_i = \frac{T(x_i + h) - T(x_i)}{h} + O(h) = \frac{T_{i+1} - T_i}{\Delta x} + O(\Delta x) \quad (3\text{-}10)$$

which is called the *first forward difference* representation for the first derivative. The *first backward difference* form comes from equation (3-7), which, when truncated past the linear term in h, yields

$$\left.\frac{dT}{dx}\right|_i = \frac{T(x_i) - T(x_i - h)}{h} + O(h) = \frac{T_i - T_{i-1}}{\Delta x} + O(\Delta x) \quad (3\text{-}11)$$

Another approach to achieving first derivatives is to subtract equation (3-6) from equation (3-7) and neglect terms involving h^3 and higher. The resulting expression for dT/dx is

$$\frac{dT}{dx}\bigg|_i = \frac{T(x_i + h) - T(x_i - h)}{2h} + 0(h^2) = \frac{T_{i+1} - T_{i-1}}{2\Delta x} + 0(\Delta x^2) \quad (3\text{-}12)$$

which is a *central difference* representation of the first derivative, dT/dx.

One should observe in these simple, one-dimensional representations for derivatives that each involves some error, that the index notation for equal node spacings is convenient and simple to use and understand, and that a simple extension to two and three dimensions is possible.

3.1-2   Difference Equation Formulation Using Difference Methods

Directing our attention now to equation (3-5), several differencing schemes could be used to replace the time and space derivatives. As an initial approach we shall write the term $\nabla^2 T$ in central difference form so that, in two dimensions, we have

$$\frac{T_{i-1,j} - 2T_{i,j} + T_{i+1,j}}{\Delta x^2} + \frac{T_{i,j-1} - 2T_{i,j} + T_{i,j+1}}{\Delta y^2} + \frac{\dot{q}}{k} = \frac{1}{\alpha}\frac{\partial T_{i,j}}{\partial t} \quad (3\text{-}13)$$

The temperature of node i, j, designated $T_{i,j}$, relates to the rectangle shown in Figure 3.2. The rectangle, outlined with dashed lines, for which i, j is the center, is characterized by properties at its center such as $T_{i,j}$, $\rho_{i,j}$, and $c_{i,j}$. The rectangular areas shown comprise a rectangular grid into which a total conducting medium is subdivided. The center points of such rectangles or subvolumes also form a rectangular grid with points of intersection designated as *nodes*. Property values at a node point are assumed representative of the subvolume having that node as center point. Properties are thus "lumped" as single values for each node in an array.

Figure 3.2 shows an array where spacings in the x direction are equal, with value Δx, and Δy is the constant spacing in the y direction; however, $\Delta x \neq \Delta y$. Additional simplification is possible in equation (3-13) if the array is square, i.e., $\Delta x = \Delta y$, in which case equation (3-13) becomes

$$T_{i-1,j} - 2T_{i,j} + T_{i+1,j} + T_{i,j-1} - 2T_{i,j} + T_{i,j+1} + \frac{\dot{q}}{k}\Delta x^2 = \frac{\Delta x^2}{\alpha}\frac{\partial T_{i,j}}{\partial t} \quad (3\text{-}14)$$

The completed difference formulation is achieved when the first derivative in time is expressed in forward, backward, or central difference fashion as indicated in equations (3-10), (3-11), and (3-12), respectively. The central difference form given by equation (3-12) leads to problems with numerical

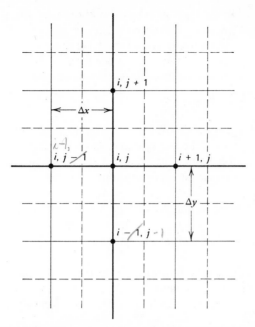

Figure 3.2   Finite-difference grid for node i, j in a conducting medium.

instability and will not be considered further. The other two forms will be used in the following discussion.

3.1-2.1   The Explicit Method

An *explicit* numerical scheme is one in which the unknown quantity may be solved for directly.

When the time derivation in equation (3-14) is written in first forward difference form, we obtain

$$T^n_{i-1,j} - 2T^n_{i,j} + T^n_{i+1,j} + T^n_{ij,-1} - 2T^n_{i,j} + T^n_{i,j+1} + \frac{\dot{q}}{k}\Delta x$$

$$= \frac{\Delta x^2}{\alpha\,\Delta t}(T^{n+1}_{i,j} - T^n_{i,j}) \quad (3\text{-}15)$$

where the superscript n denotes the nth time step.

Equation (3-15) is the classic explicit representation of the two-dimensional, transient conduction expression with internal generation of heat. The unknown $T^{n+1}_{i,j}$ appears only once, thus the designation "explicit." A solution for $T^{n+1}_{i,j}$ requires a known temperature distribution initially, $n = 0$.

This initial distribution may somehow be known, or the steady-state temperature distribution, with known boundary conditions, may be solved for by setting the time derivative in equation (3-14) equal to zero. This operation yields the two-dimensional, finite difference form of Poisson's equation

$$T_{i-1,j} - 2T_{i,j} + T_{i+1,j} + T_{i,j-1} - 2T_{i,j} + T_{i,j+1} + \frac{\dot{q}}{k}\Delta x^2 = 0 \quad (3\text{-}16)$$

The solution of equation (3-16) for $T_{i,j}$ may then be used as initial values for a problem involving the use of equation (3-15). Equation (3-16) may be more difficult to solve than the transient case; i.e., attaining valid initial conditions may be the most time-consuming part of a transient conduction problem solution.

For most problems, initial conditions are known or obvious. One technique in solving a steady-state problem is that of assuming an arbitrary initial temperature distribution and carrying out a transient solution until $T_{i,j}^{n+1} \to T_{i,j}^{n}$.

Wherever possible in the following developments we shall write difference equations in one space dimension rather than in the more cumbersome two- and three-dimensional forms. In one dimension, equations (3-14) and (3-15) become

$$T_{i-1} - 2T_i + T_{i+1} + \frac{\dot{q}}{k}\Delta x^2 = \frac{\Delta x^2}{\alpha}\frac{\partial T_i}{\partial t} \quad (3\text{-}17)$$

and

$$T_{i-1}^{n} - 2T_i^{n} + T_{i+1}^{n} + \frac{\dot{q}}{k}\Delta x^2 = \frac{\Delta x^2}{\alpha\,\Delta t}(T_i^{n+1} - T_i^{n}) \quad (3\text{-}18)$$

Solving for T_i^{n+1} from equation (3-18), we obtain

$$T_i^{n+1} = \frac{\alpha\,\Delta t}{\Delta x^2}\left[T_{i-1}^{n} + T_{i+1}^{n} + \frac{\dot{q}\,\Delta x^2}{k} + \left(\frac{\Delta x^2}{\alpha\,\Delta t} - 2\right)T_i^{n}\right] \quad (3\text{-}19)$$

3.1-2.2   The Implicit Method

In an *implicit* formulation the unknown appears in a series of equations which must be solved simultaneously.

Equation (3-14) is now written with the time derivative in first backward difference form; the resulting expression is

$$T_{i-1,j}^{n+1} - 2T_{i,j}^{n+1} + T_{i+1,j}^{n+1} + T_{i,j-1}^{n+1} - 2T_{i,j}^{n+1} + T_{i,j+1}^{n+1} + \frac{\dot{q}}{k}\Delta x^2$$

$$= \frac{\Delta x^2}{\alpha\,\Delta t}(T_{i,j}^{n+1} - T_{i,j}^{n}) \quad (3\text{-}20)$$

The unknown $T_{i,j}^{n+1}$ cannot be solved for directly from equation (3-20), thus the designation "implicit." The solution is achieved by solving simultaneously the set of algebraic equations resulting at time step $n + 1$.

Equation (3-20) is a "fully implicit" form for solving the transient heat conduction problem with generation. A "mixed" scheme will now be discussed.

3.1-2.2   A Mixed, or Weighted Average, Method

The explicit and implicit methods just considered may also be combined by a *weighted average*, or *mixed*, technique. The procedure is to multiply equation (3-20) by a factor F, where F has a value between 0 and 1; multiply equation (3-15) by $(1 - F)$; and add. Performing these steps, we obtain

$$F[T_{i-1,j}^{n+1} - 2T_{i,j}^{n+1} + T_{i+1,j}^{n+1} + T_{i,j-1}^{n+1} - 2T_{i,j}^{n+1} + T_{i,j+1}^{n+1}]$$

$$+ (1 - F)[T_{i-1,j}^{n} - 2T_{i,j}^{n} + T_{i+1,j}^{n} + T_{i,j-1}^{n} - 2T_{i,j}^{n} + T_{i,j+1}^{n}]$$

$$+ \frac{\dot{q}\,\Delta x^2}{k} = \frac{\Delta x^2}{\alpha\,\Delta t}(T_{i,j}^{n+1} - T_{i,j}^{n}) \quad (3\text{-}21)$$

Explicit and implicit forms are obtained for $F = 0$ and $F = 1$, respectively. The arithmetic average of the explicit and implicit schemes is obtained if we let $F = 1/2$; this value of F is associated with the Crank-Nicholson method.[1]

Thus far our examination of differencing schemes has involved two levels of time, t and $t + \Delta t$ (or n and $n + 1$), and spatial derivatives with, at most, 3 (or 5 in three dimensions) space points. The three methods of considering the time derivative have been the first forward difference (explicit), first backward difference (implicit), and weighted average of these two (implicit). These are the basic approaches for expressing the general heat conduction equation in difference form. Variations on these approaches are possible, however, and some of these variations will be discussed later in this chapter.

3.1-3   Difference Equation Formulation Using the Heat Balance Method

The differencing of differential equations as discussed in the previous section is very straightforward and easy to do. When certain types of complications arise, this method may prove unusually cumbersome. Complications which are common include

1.   The additional effects of radiation and convection.
2.   Spatially varying thermal properties.
3.   Heat generation which may vary with time and position.

[1] J. Crank and P. Nicholson, *Proc. Cambridge Phil. Soc.* **43** (1947): 50–64.

4. Unequal node sizes.
5. Varying node shapes.
6. Changes in phase.

These factors are considered in much simpler fashion with improved physical insight, using the *heat balance* method.

Equation (3-4) comprises the principal tool for numerical formulation, using the heat balance approach. Figure 3.3 shows node i surrounded by its four immediate neighboring nodes. The quantities designated δ_{ij} are the semi-thicknesses of each node in the directions shown; each may have a different value. The subscripts i, j designate the node of reference, i, and the node adjacent, j. The system of subscripting is apparent from the figure.

The heat flux at the boundary of node i may now be expressed as

$$q_{1 \to i} = \frac{1}{R_{1i} + R_{i1}} (T_1 - T_i) = K_{1i}(T_1 - T_i)$$

$$q_{2 \to i} = \frac{1}{R_{2i} + R_{i2}} (T_2 - T_i) = K_{2i}(T_2 - T_i)$$

$$q_{3 \to i} = \frac{1}{R_{3i} + R_{i3}} (T_3 - T_i) = K_{3i}(T_3 - T_i)$$

$$q_{4 \to i} = \frac{1}{R_{4i} + R_{i4}} (T_4 - T_i) = K_{4i}(T_4 - T_i)$$

Quantities designated R_{ij} are thermal resistances between node i and the boundary with node j; $R_{ij} = \delta_{ij}/k_i A_{ij}$. The quantities designated K_{ij} are the thermal conductances between nodes i and j; note that $K_{ij} = K_{ji} = 1/(R_{ij} + R_{ji})$. The quantities written as $q_{j \to i}$ are the heat fluxes across the boundaries between nodes j and node i.

We now apply equation (3-4) to node i, using the nomenclature just developed to obtain

$$V_i \left[\frac{\rho c T_i|_{t+\Delta t} - \rho c T_i|_t}{\Delta t} \right] = \sum_{j=1}^{N} K_{ij}{}^{\xi} T_j{}^{\xi} - T_i{}^{\xi} \sum_{j=1}^{N} K_{ij}{}^{\xi} + \dot{q}_i{}^{\xi} V_i \qquad (3\text{-}22)$$

where the summations are over j from 1 to N, N representing the total number of nodes in thermal communication with node i. The superscript ξ designates the level of time when all terms are evaluated.

There are in this case, as before, choices to make on the time levels which will make the resulting difference equations explicit or implicit.

Setting the superscript $\xi = t$, equation (3-22) reduces to the *explicit* form

$$T_i^{n+1} = T_i \left[1 - \sum_{j=1}^{N} \frac{K_{ij}}{C_i} \right] + \sum_{j=1}^{N} \frac{K_{ij}}{C_i} T_j + \dot{q}_i \frac{V_i}{C_i} \qquad (3\text{-}23)$$

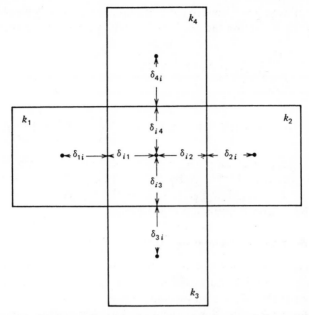

Figure 3.3 Node i and neighboring node points for heat balance analysis.

The quantity c_i includes a combination of terms; it is defined as $c_i \equiv \rho_i c_i V_i / \Delta t$. Those quantities without superscripts are evaluated at time level t. This formulation, that of setting $\xi = t$, is equivalent to a first forward differencing scheme for the time derivative in our previous development, and equation (3-23) is explicit in T_i^{n+1}.

By setting the superscript $\xi = t + \Delta t$ in equation (3-22), we obtain an *implicit* form

$$T_i^{n+1} = \frac{1}{\left[1 + \sum_{j=1}^{N} \dfrac{K_{ij}}{c_i} \right]^{n+1}} \left[T_i^n + \left(\frac{\dot{q}_i V_i}{c_i} \right)^{n+1} + \sum_{j=1}^{N} \left(\frac{K_{ij}}{c_i} T_j \right)^{n+1} \right] \qquad (3\text{-}24)$$

This form is equivalent to that obtained by a first backward difference for the time derivative in our earlier formulation. All physical properties in equation (3-24) are evaluated at time $n + 1$. A correction is often made, when properties change rapidly with time, to evaluate physical properties at the average temperature during the interval Δt, this temperature being designated $T_i^{n+1/2}$, and evaluated according to

$$T_i^{n+1/2} = \frac{T_i^n + T_i^{n+1}}{2} \qquad (3\text{-}25)$$

A weighted average of the explicit and implicit forms is again obtained by introducing the weighting factor F. The resulting form, which is implicit, is

$$T_i^{n+1} = \frac{F}{\left[1 + \sum_{j=1}^{N} \dfrac{K_{ij}}{C_i}\right]^{n+1}} \left[T_i^n + \left(\dot{q}_i \frac{V_i}{C_i}\right)^{n+1} + \sum_{j=1}^{N}\left(\frac{K_{ij}}{C_i} T_j\right)^{n+1}\right]$$

$$+ (1 - F)\left[T_i^n\left(1 - \sum_{j=1}^{N} K_{ij}{}^n\right) + \sum_{j=1}^{N}\left(\frac{K_{ij}}{C_i} T_j\right)^n + \left(\frac{\dot{q}_i V_i}{C_1}\right)^n\right] \quad (3\text{-}26)$$

The thermal properties in equation (3-26) may again be evaluated at the average temperature $T^{n+1/2}$, given by equation (3-25), if desired. With $F = 1/2$, i.e., equal weighting given to both the explicit and implicit forms, the Crank-Nicholson formulation is achieved.

3.1–3.1 Boundary Conditions and Internode Conductance

The complete formulation of any problem, as discussed in Chapter 2, requires stipulating initial and boundary conditions as well as expressing the governing equation. This requirement remains in the case of difference equations. Boundary conditions in heat conduction problems are in the form of specified temperatures or heat fluxes at the system boundaries. The boundary temperature or the temperature of the adjacent fluid medium is a common specification; radiant heat fluxes may also be encountered at boundaries.

Numerical solutions to difference equations are much more amenable to solving problems with complex boundary conditions than analytical solutions. The most direct approach to handling boundary conditions is that of defining boundary nodes and incorporating the boundary conditions into internode conductances. Some examples of these procedures are illustrated in Table 3.1 for some important cases.

The examples shown in Table 3.1 are some of the more common cases; others are possible. With the heat balance method there is no major problem in handling boundary conditions other than some tedious bookkeeping.

3.2 COMPUTATION SCHEMES FOR NUMERICALLY FORMULATED PROBLEMS

The development thus far has indicated some basic ways of reducing the temperature distribution in a conducting medium to a set of algebraic equations, one for each node in the system. The solutions to the set of equations will yield the temperature of each node.

Table 3.1 Internode Conductances for Some Important Boundary Conditions

Case   Internode Conductance

1. Adiabatic surface

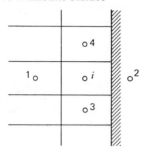

$$K_{i2} = 0$$

2. Convection at a surface

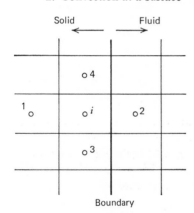

$$K_{i2} = \cfrac{1}{\cfrac{\delta_{i2}}{k_i A_{i2}} + \cfrac{1}{h_i A_{i2}}}$$

Cases: a) $T_{\text{wall}} = T_\infty : h_i \simeq \infty$
 b) $h_i = $ constant
 c) $h_i = h_i(T)$

Example, equation (5-166)

3. Radiation at a surface

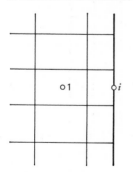

$$K_{is} = A_{is} F_{is} \sigma \frac{T_i^4 - T_s^4}{T_i - T_s}$$

s = surrounding at temperature T_s
F_{is} = shape factor

For a radiation condition at a surface, nodes centered at the surface are useful

Table 3.1   (*continued*)

        Case   Internode Conductance

4. A known contact resistance R_c between nodes

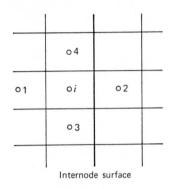

$$K_{i2} = \cfrac{1}{\dfrac{\delta_{i2}}{k_i A_{i2}} + \dfrac{\delta_{2i}}{k_2 A_{2i}} + R_c}$$

Internode surface

5. A specified heat flux

$$Q_{2-i} = q_2 A_{2i}$$

where q_2 is the boundary heat flux

For an *explicit* solution scheme, each equation in the set is independent of the rest, and the solution presents no problem.

Implicit schemes, on the other hand, require that the equations be solved simultaneously. For a set N of equations involving the unknown T_i^{n+1} implicitly, the matrix form of the set may be represented as

$$
\begin{Bmatrix}
A_{11} & A_{12} & \cdots & A_{1N} \\
 & & & \\
 & & & \\
 & & & \\
A_{21} & A_{22} & \cdots & A_{2N} \\
 & & & \\
 & & & \\
 & & & \\
A_{N1} & A_{N2} & \cdots & A_{NN}
\end{Bmatrix}
\begin{Bmatrix}
T_1^{n+1} \\
\\
\\
T_2^{n+1} \\
\\
\\
T_N^{n+1}
\end{Bmatrix}
=
\begin{Bmatrix}
B_1 \\
\\
\\
B_2 \\
\\
\\
B_N
\end{Bmatrix}
\tag{3-27}
$$

or, in simpler matrix notation,

$$\mathbf{A}T = \mathbf{B} \tag{3-28}$$

The terms in the coefficient matrix, the A_{ij}, include values of temperatures determined at earlier time steps. The number of nodes N may be extremely large, making the solution for T virtually impossible without the aid of a digital computer.

There are two commonly used solution techniques for sets of algebraic equations as given by equation (3-27); these are designated Gauss elimination and Gauss-Seidel iteration. These methods will be discussed briefly in the sections to follow.

3.2-1 The Gauss Elimination Technique

By formulating a heat conduction problem into finite-difference form, we generate sets of algebraic equations. The *Gauss elimination* technique is one method of solving sets of algebraic equations. It is a direct technique in that each value obtained in the solution is evaluated once only and is subsequently used to solve for the remaining unknowns. Gauss elimination is particularly attractive if the set of equations is linear; this would be the case where, in equation (3-27), the A_{ij} and B_i are constants (i.e., not temperature dependent).

The basic idea in the Gauss elimination scheme is the solution to a set of algebraic equations by substitution. We shall examine the following general set of three equations with three unknowns:

$$A_{11}T_1 + A_{12}T_2 + A_{13}T_3 = B_1$$
$$A_{21}T_1 + A_{22}T_2 + A_{23}T_3 = B_2$$
$$A_{31}T_1 + A_{32}T_2 + A_{33}T_3 = B_3$$

An initial requirement is that A_{11} be nonzero. Any arrangement involving a zero coefficient for the first term in the first equation must be altered by an interchange of rows.

With $A_{11} \neq 0$, a multiplier m_2 is defined as

$$m_2 = \frac{A_{21}}{A_{11}}$$

The first equation is multiplied by m_2 and subtracted from the second, yielding

$$(A_{21} - m_2A_{11})T_1 + (A_{22} - m_2A_{12})T_2 + (A_{23} - m_2A_{13})T_3 = B_2 - m_2B_1$$

The following coefficients are defined:

$$A_{21}' = A_{21} - m_2A_{11} \equiv 0$$
$$A_{22}' = A_{22} - m_2A_{12}$$
$$A_{23}' = A_{23} - m_2A_{13}$$
$$B_2' = B_2 - m_2B_1$$

Since $A_{21}' = 0$, the new set of equations becomes

$$A_{11}T_1 + A_{12}T_2 + A_{13}T_3 = B_1$$
$$A_{22}'T_2 + A_{23}'T_3 = B_2'$$
$$A_{31}T_1 + A_{32}T_2 + A_{33}T_3 = B_3$$

The third equation is now operated on in like manner as the second; each term in the first equation is multiplied by the factor m_3, which is defined as

$$m_3 = \frac{A_{31}}{A_{11}}$$

and subtracted from the third equation, yielding

$$A_{32}'T_2 + A_{33}'T_3 = B_3'$$

where

$$A_{31}' = A_{31} - m_3 A_{11} \equiv 0$$
$$A_{32}' = A_{32} - m_3 A_{12}$$
$$A_{33}' = A_{33} - m_3 A_{13}$$
$$B_3' = B_3 - m_3 B_1$$

The original set of three equations has now been reduced to

$$A_{11}T_1 + A_{12}T_2 + A_{13}T_3 = B_1$$
$$A_{22}'T_2 + A_{23}'T_3 = B_2'$$
$$A_{32}'T_2 + A_{33}'T_3 = B_3'$$

The last two equations are now operated on in like manner. The multiplier m_3' is defined as

$$m_3' \equiv \frac{A_{32}'}{A_{22}'}$$

The second equation in the modified set is multiplied by m_3' and subtracted from the third equation, yielding

$$A_{33}''T_3 = B_3''$$

where

$$A_{32}'' = A_{32}' - m_3' A_{22}' \equiv 0$$
$$A_{33}'' = A_{33}' - m_3' A_{23}'$$
$$B_3'' = B_3' - m_3' B_2'$$

The final set of algebraic equations is now

$$A_{11}T_1 + A_{12}T_2 + A_{13}T_3 = B_1$$
$$A_{22}'T_2 + A_{23}'T_3 = B_2'$$
$$A_{33}''T_3 = B_3''$$

where T_1 has a nonzero coefficient in the first equation alone; each of the subsequent equations has one less term than its predecessor.

Values of T_1, T_2, and T_3 are obtained by solving the third equation for T_3; substituting T_3 into the second equation and solving for T_2; and, finally, substituting values of T_3 and T_2 into the first equation and solving for T_1. The expressions for T_1, T_2, and T_3 are

$$T_3 = B_3''/A_{33}''$$
$$T_2 = (B_2' - A_{23}'T_3)/A_{22}'$$
$$T_1 = (B_1 - A_{12}T_2 - A_{13}T_3)/A_{11}$$

Should A_{11}, A_{22}', or A_{33}'' be zero, the original system of equations is singular and has no solution.

The Gauss elimination procedure is one that would normally be tried from an intuitive approach. It is a systematic way of using the process of substitution to solve a set of simultaneous algebraic equations.

The following short example problem should aid in understanding the Gauss elimination procedure.

Example 3.1

Solve the following set of equations for the unknowns T_1, T_2, and T_3:

$$2T_1 + 3T_2 + T_3 = 5$$
$$2T_1 - T_2 + 4T_3 = 3$$
$$T_1 + 4T_2 + 2T_3 = 6$$

A table will be helpful in keeping track of terms.

Coefficients			Constants
2	3	1	5
2	−1	4	3
1	4	2	6
	−4	3	−2
	5/2	3/2	7/2
		27/8	18/8

We have the following equations:

$$2T_1 + 3T_2 + T_3 = 5$$
$$- 4T_2 + 3T_3 = -2$$
$$27T_3 = 18$$

Back calculation next yields, for T_i, the values

$$T_3 = 18/27 = 2/3$$

$$T_2 = \frac{-2-2}{-4} = 1$$

$$T_1 = \frac{5-3-2/3}{2} = 2/3$$

The utility of any numerical solution scheme lies in its adaptability to the digital computer for solving large sets of equations. The Gauss elimination technique is a useful tool and will now be stated in a more general fashion.

Having N linear equations in N unknowns, we may write, in general,

$$A_{11}T_1 + A_{12}T_2 + \cdots + A_{1j}T_j + \cdots + A_{1N}T_N = B_1$$
$$A_{21}T_1 + A_{22}T_2 + \cdots + A_{2j}T_j + \cdots + A_{2N}T_N = B_2$$

$$A_{i1}T_1 + A_{i2}T_2 + \cdots + A_{ij}\dot{T}_j + \cdots + A_{iN}T_N = B_i$$

$$A_{N1}T_1 + A_{N2}T_2 + \cdots + A_{Nj}T_j + \cdots + A_{NN}T_N = B_N$$

We now define the multipliers $m_i \equiv A_{i1}/A_{11}$, and subtract m_i times the first equation from the ith equation to obtain

$$A_{11}T_1 + A_{12}T_2 + \cdots + A_{1j}T_j + \cdots + A_{1N}T_N = B_1$$

$$0 + A_{i2}'T_2 + \cdots + A_{ij}'T_j + \cdots + A_{iN}'T_N = B_i'$$

$$0 + A_{N2}'T_2 + \cdots + A_{Nj}'T_j + \cdots + A_{NN}'T_N = B_N'$$

This process is continued; each successive operation reduces the number of terms in subsequent equations by 1. The result of all operations of this type is the modified set

$$A_{11}T_1 + A_{12}T_2 + \cdots \quad A_{1j}T_j + \cdots + A_{1N}T_N = B_1$$

$$A_{22}'T_2 + \cdots + A_{2j}'T + \cdots + A_{2N}'T_N = B_2$$

$$\vdots \qquad\qquad \vdots \qquad\qquad \vdots$$

$$A_{jj}^{(i-1)}T_j + \cdots + A_{iN}^{(i-1)}T_N = B_i^{(i-1)}$$

$$\vdots \qquad\qquad \vdots$$

$$A_{NN}^{(N-1)}T_N = B_N^{(N-1)}$$

As before, the solution for T is obtained by solving for T_N and working back until T_1 is obtained.

The flow diagrams associated with Gauss elimination are shown in Figure 3.4. An example problem using a computer solution with Gauss elimination is given in the next chapter.

Some difficulty is encountered with Gauss elimination when nonlinearities are present. If the A_{ij} and B_i are temperature dependent, then they are estimated initially to solve for the T_i. New values are then determined for the coefficient terms, and the temperatures are calculated once more, using Gauss elimination. This process is continued until properties and T_i obtained from successive calculations differ by less than some prescribed amount.

As the number of unknowns becomes very large and when nonlinearities are present, an iterative or indirect solution scheme may be more attractive than the direct technique just considered.

3.2-2   The Gauss-Seidel Iteration Technique

An indirect or iterative technique is often used for solving sets of simultaneous algebraic equations when direct techniques become too burdensome. A popular method is designated *Gauss-Seidel iteration*.

The approach used in Gauss-Seidel iteration is illustrated by considering the general set of these algebraic equations:

$$A_{11}T_1 + A_{12}T_2 + A_{13}T_3 = B_1$$
$$A_{21}T_1 + A_{22}T_2 + A_{23}T_3 = B_2$$
$$A_{31}T_1 + A_{32}T_2 + A_{33}T_3 = B_3$$

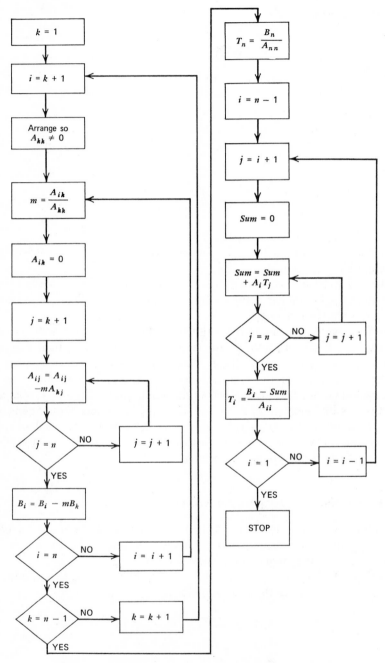

Figure 3.4 Flow diagram for Gauss elimination technique.

For $A_{11} \neq 0$, $A_{22} \neq 0$, $A_{33} \neq 0$, or, in general, $A_{ii} \neq 0$, each of the unknowns T_i is solved for as

$$T_1 = \frac{1}{A_{11}} (B_1 - A_{12}T_2 - A_{13}T_3) \qquad \text{(i)}$$

$$T_2 = \frac{1}{A_{22}} (B_2 - A_{21}T_1 - A_{23}T_3) \qquad \text{(ii)}$$

$$T_3 = \frac{1}{A_{33}} (B_3 - A_{31}T_1 - A_{32}T_2) \qquad \text{(iii)}$$

The procedure now is to guess initial values of the dependent variables and update these estimates by solving the above equations in the order shown, *always using the most recent information.* The procedure, using the three equations above, goes like this:

1. Assume approximate values for T_1^0, T_2^0, T_3^0
2. With T_2^0 and T_3^0, solve equation (i) for T^1
3. With T_1^1 and T_3^0, solve equation (ii) for T_2^1
4. With T_1^1 and T_2^1, solve equation (iii) for T_3^1
5. With T_2^1 and T_3^1 solve equation (i) for T_1^2

$$\cdot$$
$$\cdot$$
$$\cdot$$

etc.

The kth approximation for T_i can be expressed in general terms as

$$T_1^{(k)} = \frac{1}{A_{11}} [B_1 - A_{12}T_2^{(k-1)} - A_{13}T_3^{(k-1)}]$$

$$T_2^{(k)} = \frac{1}{A_{22}} [B_2 - A_{21}T_1^{(k)} - A_{23}T_3^{(k-1)}]$$

$$T_3^{(k)} = \frac{1}{A_{33}} [B_3 - A_{31}T_1^{(k)} - A_{32}T_2^{(k)}]$$

A numerical example of this technique is given in Example 3.2.

Example 3.2

Solve the following set of three simultaneous algebraic equations for T_1, T_2, and T_3:

$$5T_1 + T_2 + 2T_3 = 32$$
$$2T_1 + 8T_2 + T_3 = 29$$
$$T_1 + 2T_2 + 4T_3 = 28$$

Rewriting and solving for T_1, T_2, and T_3 in the prescribed manner, we have

$$T_1 = (1/5)(32 - T_2 - 2T_3)$$
$$T_2 = (1/8)(29 - 2T_1 - T_3)$$
$$T_3 = (1/4)(28 - T_1 - 2T_2)$$

Making a common assumption, that $T_2^0 = T_3^0 = 0$, we obtain

$$T_1 = (1/5)(32 - 0 - 0) = 6.40$$
$$T_2 = (1/8)(29 - 12.8 - 0) = 2.025$$
$$T_3 = (1/4)(28 - 6.4 - 4.05) = 4.388$$

The second iteration yields

$$T_1 = (1/5)(32 - 2.025 - 8.776) = 4.240$$
$$T_2 = (1/8)(29 - 8.48 - 4.388) = 2.017$$
$$T_3 = (1/4)(28 - 4.24 - 4.034) = 4.932$$

A tabular record of these results and of two additional iterations, carried four places, follows:

Iteration No.	T_1	T_2	T_3
0	0	0	0
1	6.400	2.025	4.388
2	4.240	2.017	4.932
3	4.024	2.003	4.993
4	4.002	2.000	5.000
5	4.000	2.000	5.000

The exact answer to the example problem is obviously $T_1 = 4$, $T_2 = 2$, $T_3 = 5$.

Any desired accuracy is attainable by the Gauss-Seidel technique, provided a solution exists, if the iteration is carried out a sufficient number of times. For purposes of computer programming, one of two methods is commonly used to stop the process. These two methods are

1. Establish some limit for the maximum difference between two successive values of T_i, i.e.,

$$\max |T_i^{(k)} - T_i^{(k-1)}| < \varepsilon \qquad (3\text{-}29)$$

    for all i.

2. Establish some limit ε on the relative difference between successive values of T_i

$$\max \left| \frac{T_i^{(k)} - T_i^{(k-1)}}{T_i^{(k)}} \right| < \varepsilon \qquad (3\text{-}30)$$

The flow diagram for a Gauss-Seidel iterative computation scheme is shown in Figure 3.5.

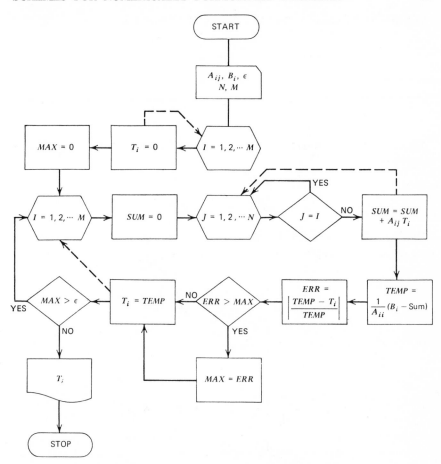

Figure 3.5 Flow diagram for Gauss-Seidel iteration.

Convergence of the Gauss-Seidel technique is a logical concern at this point. In Example 3.2 four-place accuracy was obtained in five iterations starting from zero. One might ask the questions: What would have made the set converge faster? slower? What conditions would make convergence impossible?

On intuitive grounds it seems helpful to have relatively large values for the A_{ii}, that is, to divide by large numbers each time. This can be demonstrated more formally if we consider the simplest possible case, that of two equations in two unknowns:

$$A_{11}T_1 + A_{12}T_2 = B_1$$
$$A_{21}T_1 + A_{22}T_2 = B_2$$

For the kth iteration, values of $T_1^{(k)}$ and $T_2^{(k)}$ are found from

$$T_1^{(k)} = \frac{1}{A_{11}} [B_1 - A_{12}T_2^{(k-1)}]$$

$$T_2^{(k)} = \frac{1}{A_{22}} [B_2 - A_{21}T_1^{(k)}]$$

After the kth iteration, the errors in T_1 and T_2, denoted ΔT_1 and ΔT_2, are expressed as

$$\Delta T_1^{(k)} = T_1 - T_1^{(k)}$$

$$\Delta T_2^{(k)} = T_2 - T_2^{(k)}$$

Making substitutions for T_i and $T_i^{(k)}$ from the original equations and from the iteration expressions, we may write $\Delta T_i^{(k)}$ as

$$\Delta T_1^{(k)} = \frac{1}{A_{11}} [B_1 - A_{12}T_2 - B_1 + A_{12}T_2^{(k-1)}]$$

$$= -\frac{A_{12}}{A_{11}} \Delta T_2^{(k-1)}$$

and

$$\Delta T_2^{(k)} = \frac{1}{A_{22}} [B_2 - A_{21}T_1 - B_2 + A_{21}T_1^{(k)}]$$

$$= -\frac{A_{21}}{A_{22}} \Delta T_1^{(k)}$$

These results may be combined to yield

$$\Delta T_1^{(k)} = \frac{A_{12}A_{21}}{A_{11}A_{22}} \Delta T_1^{(k-1)}$$

$$\Delta T_1^{(k-1)} = \frac{A_{12}A_{21}}{A_{11}A_{22}} \Delta T_1^{(k-2)}$$

.

.

.

etc.

or, in general,

$$\Delta T_1^{(k)} = \left(\frac{A_{12}A_{21}}{A_{11}A_{22}}\right)^k \Delta T_1^{(0)}$$

and

$$\Delta T_2^{(k)} = \left(\frac{A_{12}A_{21}}{A_{11}A_{22}}\right)^k \Delta T_2^{(0)}$$

For convergence to occur, it is quite obvious that, as k becomes large, $\Delta T_1^{(k)} \to 0$, and $\Delta T_2^{(k)} \to 0$; this makes it necessary that

$$\left(\frac{A_{12}A_{21}}{A_{11}A_{22}}\right) < 1$$

This condition will be met under the following conditions: either

$$\begin{cases} |A_{11}| > |A_{12}| \\ |A_{22}| \geq |A_{21}| \end{cases} \quad \text{or} \quad \begin{cases} |A_{11}| \geq |A_{12}| \\ |A_{22}| > |A_{21}| \end{cases}$$

i.e., diagonal terms must be at least as large as off-diagonal terms and larger in at least one case. The terminology used is that the coefficient matrix, A_{ij}, be *diagonally dominant*. Naturally, convergence is more rapid for increasingly larger values of diagonal terms.

A word is in order concerning the choice one should make between direct and iterative solutions. When both are possible, iterative schemes will often yield acceptable results with less work. Iterative schemes are definitely preferred when variable medium properties are involved in the coefficients and constants, the A_{ij} and B_i; these terms can be updated as newer values are determined for $T_i^{(k)}$. Gauss elimination may be necessary, however, if the required conditions for convergence of Gauss-Seidel iteration are not met, i.e., the coefficient matrix is not diagonally dominant.

Problem solutions involving both Gauss elimination and Gauss-Seidel iteration are included in later chapters.

3.3 ADDITIONAL CONSIDERATIONS INVOLVING NUMERICAL SOLUTIONS

An unfortunate aspect of results obtained with modern computing machinery is that the solution obtained may not be correct if, in fact, a solution is obtained at all. Questions to be considered in this context concern solution *stability*, *accuracy*, and *convergence*. These items will now be discussed briefly.

3.3-1   Errors in Numerical Solutions

The error in a numerical solution is generally considered to be one of two types: *numerical*, or *roundoff error* and *discretization*, or *truncation error.*

Roundoff error occurs when an insufficient number of significant figures is carried through the computation. Roundoff in the 8th or 12th place is insignificant when a single term is considered; however, after thousands or millions of calculations are accomplished, the accumulated error due to roundoff may be intolerable. Fortunately, modern computing machinery has sufficient capacity to make this type of error of little concern in a carefully programmed numerical solution scheme.

Truncation, or discretization, errors occur as a result of replacing derivatives by finite differences. The quantities designated $0(\Delta x)$ and $0(\Delta x^2)$ in equations (3-11) and (3-12) are of this type. This type of error is inherent in differencing and is, therefore, unavoidable. It can be reduced, however, by reducing the size of the time and space increments in some prescribed fashion.

If errors grow as a solution proceeds and this growth is unbounded, the solution is said to be *unstable*. Instability also results if errors grow at a faster rate than that at which convergence is approached. Figure 3.6 illustrates some of the ideas being discussed.

The unstable solution, shown with the dashed line, is a typical pattern for the case of unbounded error growth.

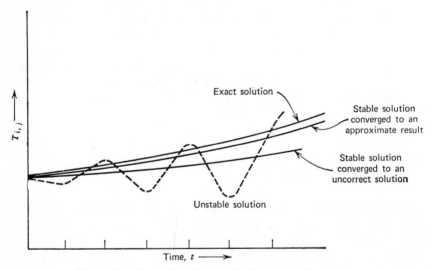

Figure 3.6   Types of results possible in numerical solutions.

The two lines showing stable solutions both yield inaccurate results: one is acceptable, the other is obviously not. The upper solution curve is sufficiently close to the exact solution to be acceptable; the error is due to discretization.

The lower line for a converged numerical solution is unacceptable because it is grossly in error, even though the solution is stable. Possible causes of this behavior are:

1. The governing differential equation is not adequately represented by the difference equation.
2. Iteration at each time step was halted before convergence to correct values was achieved.
3. All convergence criteria were satisfied, but convergence to the wrong solution was achieved. Reasons for this are often unknown.

This last category is not encountered frequently; such results may occur in solving sets of highly nonlinear equations.

Convergence criteria in heat conduction problems should be made on an overall heat balance basis, i.e., | heat in − heat out | should be less than some small amount ε. The criteria based upon the temperature change between successive iterations as specified by equations (3-29) and (3-30) are not sufficient.

3.3-1.1 Criteria for Stability

The explicit method of solving the heat equation, using a first forward difference representation for the time derivative, was given in equation (3-19). Computationally, this approach is desirable; however, it may be unstable in certain instances.

Equation (3-19) is stable when

$$\frac{\alpha \, \Delta t}{\Delta x^2} \leq 1/2 \tag{3-31}$$

or, in two dimensions, when

$$\alpha \, \Delta t \left(\frac{1}{\Delta x^2} + \frac{1}{\Delta y^2} \right) \leq 1/2 \tag{3-32}$$

The criterion for stability of the explicit heat balance expression, equation (3-23), is not so straightforward due to the influence of the temperature-dependent properties. Bray and MacCracken[2] suggest as the criterion in this case the relation

$$\Delta t_i < \frac{C_i}{\sum\limits_{j=1}^{N} K_{ij}} \tag{3-33}$$

[2] A. P. Bray and S. J. MacCracken, Knolls Atomic Power Laboratory Report, KAPL 2044 (AEC), May 1959.

Each of the three preceding equations indicates that a decrease in size of a space increment must be accompanied by an associated decrease in the size of the time step. This frequently results in extremely small time increments.

Implicit solution schemes are unconditionally stable for the most part. The implicit form for solving the heat equation, given as equation (3-20), and the Crank-Nicholson method, equation (3-21) with $F = 1/2$, are inherently stable.

A conclusion from this discussion is that implicit schemes may be more desirable than explicit schemes since stability problems need cause no concern. This is true even though large sets of simultaneous equations may need solution. A helpful effect is that larger time steps may be used in implicit schemes; thus the time to achieve a solution may not be as large in comparison to an explicit method as one would initially suppose. One must still be concerned with discretization errors when employing implicit methods so that acceptable accuracy will not be sacrificed by using too large increments in space and time.

Certain developments have been made in numerical schemes which incorporate the computational ease of explicit methods with the accuracy and numerical stability of implicit schemes. Some of the more successful approaches are termed multilevel and alternating-direction procedures. A discussion of these approaches is beyond the scope of the present treatment. The interested reader may consult Carnahan, Luther, and Wilkes[3] or Trent and Welty[4] for details of these more advanced schemes.

3.4 CLOSURE

This chapter has been devoted entirely to aspects of numerical formulation of heat transfer equations. In future years heat transfer calculations will be performed more and more by computers, hence the importance attached to this subject in the present text.

Fundamental formulations based on differencing the governing differential equations and in generating difference equations from a heat balance over a small element were discussed; the heat balance method was generally recommended due, in part, to its greater flexibility in accepting complex boundary conditions.

Solution techniques involving explicit and implicit methods were discussed. Gauss elimination and Gauss-Seidel iteration techniques were discussed in detail with flow diagrams presented for each.

[3] B. Carnahan, H. A. Luther, and J. O. Wilkes, *Applied Numerical Methods* (New York: John Wiley and Sons, Inc., 1969).
[4] D. S. Trent and J. R. Welty, *A Summary of Numerical Methods for Solving Transient Heat Conduction Problems*, Oregon State Univ. Engr. Expt. Station Bulletin (in press).

A brief discussion of error, convergence, and stability problems associated with numerical solutions was presented. Implicit schemes are generally numerically stable, and are frequently more desirable than are explicit schemes which must satisfy certain criteria for insured stability.

Example problems incorporating numerical solutions will be included in the chapters to follow.

CHAPTER 4

CONDUCTION HEAT TRANSFER

The conduction mode of heat transfer was introduced in Chapter 1. Some basic terms and concepts have already been discussed. In this chapter conduction heat transfer will be considered in some detail and numerous example problems will be worked, including several by numerical means.

Steady-state systems will be considered first, beginning with one-dimensional cases, then going to two- and three-dimensional systems. Special cases of practical interest will be considered where appropriate, such as the section on heat transfer from fins and extended surfaces. Analytical and numerical solution techniques will be used. Unsteady-state situations will be considered next in essentially the same order as for the steady-state systems. In the unsteady cases, graphical solution techniques will be employed as well as analytical and numerical approaches to problem solutions.

4.1 STEADY-STATE CONDUCTION

A heat transfer situation in which time is not a factor is designated *steady state*. The consideration of heat transfer where time is not considered affords some simplification in the analysis. As discussed in Chapter 2, the governing equation for steady-state conduction with internal generation is

$$\nabla^2 T + \frac{\dot{q}}{k} = 0 \qquad (2\text{-}53)$$

which is known as the Poisson equation; and for steady-state conduction without internal generation of heat, the Laplace equation applies:

$$\nabla^2 T = 0 \tag{2-54}$$

Both of the above equations apply to an isotropic medium, i.e., one whose properties do not vary with direction; physical properties are also presumed independent of temperature.

Our initial consideration will be one-dimensional steady-state conduction without internal generation of energy.

4.1-1 One-Dimensional Systems without Generation

As just discussed, the Laplace equation applies to this case. A general form of the Laplace equation in one dimension is

$$\frac{d}{dx}\left(x^i \frac{dT}{dx}\right) = 0 \tag{4-1}$$

where $i = 0$, 1, or 2 in rectangular, cylindrical, and spherical coordinates, respectively.

4.1-1.1 Plane Walls

In the case of a plane wall as shown in Figure 4.1, equation (4-1) with $i = 0$ applies.

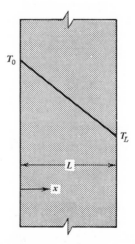

Figure 4.1 Steady-state conduction in a plane wall.

The equation and boundary conditions to be satisfied are

$$\frac{d^2T}{dx^2} = 0 \qquad\qquad (4\text{-}2)$$

$$T(x) = T(0) = T_0 \qquad \text{at } x = 0$$
$$T(x) = T(L) = T_L \qquad \text{at } x = L$$

Equation (4-2) may be separated and integrated twice to yield

$$T(x) = c_1 x + c_2 \qquad\qquad (4\text{-}3)$$

and the constants of integration c_1 and c_2 evaluated, by applying the boundary equations, to be

$$c_1 = \frac{T_L - T_0}{L} \qquad \text{and} \qquad c_2 = T_0$$

When c_1 and c_2 are substituted into equation (4-3), the final expression for the temperature profile becomes

$$T(x) = \frac{T_L - T_0}{L} x + T_0$$

or

$$T(x) = T_0 - \frac{T_0 - T_L}{L} x \qquad\qquad (4\text{-}4)$$

According to equation (4-4), the temperature variation in a plane wall under the conditions specified is linear as shown in Figure 4.1.

The Fourier rate equation may be used to determine the heat flux or heat flow rate in this case. The rate equation, presented in Chapter 1, is repeated below, in scalar form, for reference.

$$q_x = -kA \frac{dT}{dx} \qquad\qquad (1\text{-}1)$$

Since, in the steady-state case, q_x is constant, this equation may be separated and integrated directly as

$$q_x \int_0^L dx = -kA \int_{T_0}^T dT$$

giving

$$q_x = \frac{kA}{L}(T_0 - T_L) \qquad\qquad (4\text{-}5)$$

Alternately, the temperature gradient dT/dx could have been evaluated from equation (4-4) and substituted into equation (1-1) to achieve the identical result. These two alternate means of evaluating heat flux, either by direct integration of the Fourier rate equation or by solving for the temperature profile and substituting the temperature gradient expression into the rate equation, will both be employed in subsequent examples. One approach may be simpler than another in certain cases, but no general statement can be made in this regard.

The quantity kA/L, in equation (4-5), is the *thermal conductance* for a flat plate or wall. The reciprocal of this quantity, L/kA, is designated the *thermal resistance*.

The following two examples illustrate the use of equations (4-4) and (4-5) as well as the concept of thermal resistance.

Example 4.1

Pressurized water is to be transported through a pipe imbedded in a 4-foot-thick wall. It is desired to locate the pipe in the wall where the temperature is 250°F. One surface of the wall is held at 425°F and the other remains at 160°F, as shown in Figure 4.2. If the wall material has a thermal conductivity which varies with temperature according to

$$k = 0.5(1 + 0.06T)$$

where T is in °F and k is in Btu/hr-ft-°F, how far from the hot wall should the pipe be located?

In this problem the linear temperature profile is not valid due to the temperature dependence of k. With k written as a linear function of T in the form $k(T) = \alpha(1 + \beta T)$, the rate equation becomes

$$q_x = -\alpha(1 + \beta T)A\frac{dT}{dx}$$

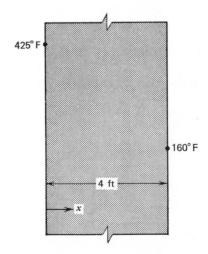

Figure 4.2 Plane wall system for Example 4.1.

Separating variables and integrating, we obtain

$$\frac{q_x}{A} \int_0^x dx = -\alpha \int_{T_0}^{T_x} (1 + \beta T)\, dT$$

or

$$\frac{q_x}{A} x = \alpha \left[T_0 - T_x + \frac{\beta}{2}(T_0{}^2 - T_x{}^2) \right]$$

when integrating from $x = 0$ to a general position x. If we integrate from x to $x = L$, we achieve the relation

$$\frac{q_x}{A}(L - x) = \alpha \left[T_x - T_L + \frac{\beta}{2}(T_x{}^2 - T_L{}^2) \right]$$

For the problem under consideration, q_x/A is constant, $L = 4$ ft, $T = 250°F$, $T_0 = 425°F$, $T_L = 160°F$, and α and β are the coefficients in the expression for k. Equating expressions for q_x/A, we have

$$T_0 - T_x + \frac{\beta}{2}(T_0{}^2 - T_x{}^2) = \frac{x}{L - x} \left[T_x - T_L + \frac{\beta}{2}(T_x{}^2 - T_L{}^2) \right]$$

and inserting the known values for T_x, T_0, T_L, and β, x is 2.88 ft. It is interesting to note that, had the temperature dependence of the thermal conductivity been ignored and the linear temperature variation assumed, the pipe would have been located 2.64 ft from the hot wall. At this location the actual temperature is 294°F. The saturation pressure of water at 250°F is 29.82 psi, and at 294°F the value is 61.201 psi, a factor of over 2 to 1! Obviously, the inclusion of the temperature dependence of thermal conductivity was worthwhile in this case.

Example 4.2

A furnace wall is constructed with 3 in. of fire clay brick ($k = 0.65$ Btu/hr-ft-°F) next to the fire box and 1/4 in. of mild steel ($k = 24$ Btu/hr-ft-°F) on the outside. The inside surface of the brick is at 1200°F, and the steel is surrounded by air at 80°F with an outside surface coefficient of 12 Btu/hr-ft²-°F. Find

(a) the heat flux through each square foot of furnace wall.
(b) the outside surface temperature of the steel.
(c) the percent increase in heat flux if, in addition to the conditions specified, two 3/4-inch-diameter steel bolts extend through the composite wall per square foot of wall area.

For the composite wall, the analogous electrical circuit is

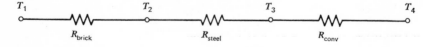

and the flux may be evaluated, using equation (1-14):

$$q = \frac{\Delta T}{\sum R_t} \qquad (1\text{-}14)$$

The thermal resistances are, in turn,

$$R_{\text{brick}} = \frac{L}{kA}\bigg|_{\text{brick}} = \frac{3/12 \text{ ft}}{(0.65 \text{ Btu/hr-ft-}^\circ\text{F})(A, \text{ft}^2)} = \frac{0.385}{A} \frac{\text{hr-}^\circ\text{F}}{\text{Btu}}$$

$$R_{\text{steel}} = \frac{L}{kA}\bigg|_{\text{steel}} = \frac{0.25/12 \text{ ft}}{(24 \text{ Btu/hr-ft-}^\circ\text{F})(A, \text{ft}^2)} = \frac{0.00087}{A} \frac{\text{hr-}^\circ\text{F}}{\text{Btu}}$$

$$R_{\text{conv}} = \frac{1}{hA} = \frac{1}{(12 \text{ Btu/hr-ft}^2\text{-}^\circ\text{F})(A, \text{ft}^2)} = \frac{0.0833}{A} \frac{\text{hr-}^\circ\text{F}}{\text{Btu}}$$

The combined thermal resistance is

$$R = \frac{1}{A}(0.385 + 0.00087 + 0.0833) = \frac{0.469}{A} \frac{\text{hr-}^\circ\text{F}}{\text{Btu}}$$

and the heat flux is

$$\frac{q}{A} = \frac{1200 - 80}{0.469} = \frac{1120}{0.469} = 2390 \text{ Btu/hr-ft}^2$$

With the heat flux now known, the steel surface temperature may be found, using either of the following equations:

$$1200 - T_3 = q(R_{\text{brick}} + R_{\text{steel}})$$

$$T_3 - 80 = qR_{\text{conv}}$$

The resulting value for T_3 is

$$T_3 = 1200 - (2390 \text{ Btu/hr-ft}^2)(0.386 \text{ hr-}^\circ\text{F/Btu})$$

$$= 80 + (2390 \text{ Btu/hr-ft}^2)(0.0833 \text{ hr-}^\circ\text{F/Btu}) = 278^\circ\text{F}$$

In the case of the additional steel bolts through the wall, there are now two paths whereby heat may flow from the inside of the furnace wall to the outside air. The equivalent electrical circuit in this case is shown below.

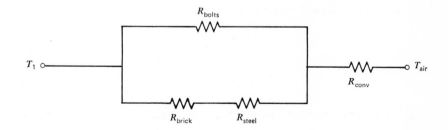

R_{brick}, R_{steel}, and R_{conv} are all known. The thermal resistance for the steel bolts is calculated as

$$R_{bolts} = \frac{L}{kA} = \frac{3.25/12 \text{ ft}}{(24 \text{ Btu/hr-ft-}°\text{F})} \bigg/ \left[2 \times \frac{\pi}{4} \left(\frac{0.75}{12}\right)^2 \right] \frac{\text{ft}^2}{\text{ft}^2} A$$

$$= \frac{1.84}{A} \frac{\text{hr-}°\text{F}}{\text{Btu}}$$

The equivalent resistance of the parallel portion of the circuit is

$$R_{equiv} = \frac{1}{\dfrac{1}{R_{bolts}} + \dfrac{1}{R_{brick} + R_{steel}}}$$

$$= \frac{1}{(0.543 + 2.59)A} = \frac{0.319}{A} \frac{\text{hr-}°\text{F}}{\text{Btu}}$$

The total thermal resistance for the wall and bolts is

$$R_{total} = R_{equiv} + R_{conv}$$

$$= \frac{0.319}{A} + \frac{0.0833}{A} = \frac{0.402}{A}$$

and the resulting heat flux is

$$\frac{q}{A} = \frac{\Delta T}{R} = \frac{1200 - 80}{0.402} = \frac{1120}{0.402} = 2790 \text{ Btu/hr-ft}^2$$

which is an increase of 400 Btu/hr-ft², or 16.7%, over the wall without bolts.

4.1-1.2 Hollow Cylinders

For steady-state heat conduction through a cylindrical wall in the radial direction, the Laplace equation takes the form

$$\frac{d}{dr}\left(r\frac{dT}{dr}\right) = 0 \tag{4-6}$$

Separating variables and integrating, we obtain

$$\frac{dT}{dr} = \frac{c_1}{r} \tag{4-7}$$

and again

$$T = c_1 \ln r + c_2 \tag{4-8}$$

If the system and boundary conditions are as indicated in Figure 4.3, namely

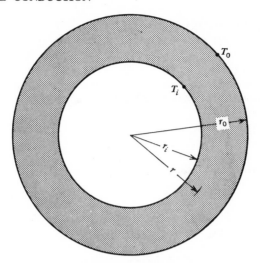

Figure 4.3 Steady-state conduction in a cylindrical wall.

that $T(r_i) = T_i$ and $T(r_0) = T_0$, then the integration constants c_1 and c_2 become

$$c_1 = -\frac{T_i - T_0}{\ln \dfrac{r_0}{r_i}}$$

$$c_2 = T_i - c_1 \ln r_i$$

The expression for $T(r)$ then becomes

$$T_i - T = \frac{\ln \dfrac{r}{r_i}}{\ln \dfrac{r_0}{r_i}} (T_i - T_0) \tag{4-9}$$

In determining the heat flow rate, we apply the Fourier rate equation in cylindrical form

$$q_r = -kA \frac{dT}{dr}$$

The area in this case is $2\pi rL$, and the temperature gradient dT/dr is given by equation (4-7). Substituting these terms into the expression for heat flow rate, we have

$$q_r = \frac{2\pi kL}{\ln \dfrac{r_0}{r_i}} (T_i - T_0) \tag{4-10}$$

which we recognize is equivalent to equation (1-5) obtained in a different way. The application of equation (1-5) was demonstrated in Example 1.1.

4.1–1.2 Hollow Spheres

In the spherical case, the one-dimensional form of Laplace's equation written for radial heat flow is

$$\frac{d}{dr}\left(r^2 \frac{dT}{dr}\right) = 0 \tag{4-11}$$

Separating variables and integrating twice, we obtain

$$\frac{dT}{dr} = \frac{c_1}{r^2} \tag{4-12}$$

and

$$T = -\frac{c_1}{r} + c_2 \tag{4-13}$$

The applicable boundary conditions for this spherical case are $T(r_i) = T_i$ and $T(r_0) = T_0$. Applying these boundary conditions, we obtain, for the constants of integration,

$$c_1 = -\frac{T_i - T_0}{\dfrac{1}{r_i} - \dfrac{1}{r_0}}$$

$$c_2 = \frac{c_1}{r_i} + T_i$$

and, finally, for $T(r)$,

$$T(r) - T_i = \frac{\dfrac{1}{r_i} - \dfrac{1}{r}}{\dfrac{1}{r_i} - \dfrac{1}{r_0}}(T_i - T_0) \tag{4-14}$$

The expression for radial heat flow in a spherical shell is

$$q_r = -kA\frac{dT}{dr}$$

in which $A = 4\pi r^2$ and dT/dr is given by equation (4-12). When these substitutions are made, q_r becomes

$$q_r = \frac{4\pi k}{\dfrac{1}{r} - \dfrac{1}{r_0}}(T_i - T_0) \tag{4-15}$$

or, equivalently,

$$q_r = \frac{4\pi k r_i r_0}{r_0 - r_i}(T_i - T_0) \qquad (4\text{-}16)$$

From equation (4-16) the thermal resistance for a hollow sphere is seen to be

$$R_t = \frac{r_0 - r_i}{4\pi k r_i r_0} \qquad (4\text{-}17)$$

and the shape factor is

$$S = \frac{4\pi r_i r_0}{r_0 - r_i} \qquad (4\text{-}18)$$

For the reader's convenience, the expressions for heat flow rate, thermal resistance, and shape factor for the three one-dimensional steady-state cases considered thus far are summarized in Table 4.1.

Table 4.1 A Summary of Results for One-Dimensional Steady-State Heat Conduction

Configuration	Heat flow rate q_x	Thermal Resistance R_t	Shape Factor S
Plane wall	$q_x = \dfrac{kA}{L}\,\Delta T$	$\dfrac{L}{kA}$	$\dfrac{A}{L}$
Hollow cylinder	$q_r = \dfrac{2\pi k L}{\ln\dfrac{r_0}{r_i}}\,\Delta T$	$\dfrac{\ln\dfrac{r_0}{r_i}}{2\pi k L}$	$\dfrac{2\pi L}{\ln\dfrac{r_0}{r_i}}$
Hollow sphere	$q_r = \dfrac{4\pi k r_0 r_i}{r_0 - r_i}\,\Delta T$	$\dfrac{r_0 - r_i}{4\pi k r_i r_0}$	$\dfrac{4\pi r_0 r_i}{r_0 - r_i}$

4.1-2 One-Dimensional Steady-State Heat Conduction with Internal Generation

In this section two examples involving systems with internal thermal energy generation will be given. The most obvious effects of this added consideration, in an analytical sense, are the different forms of the governing equation and results from those encountered in the previous section.

Energy which is added throughout a medium as internal generation is termed a *homogeneous* source while that added at a boundary or specific location in the medium is a *heterogeneous* source. A homogeneous energy addition will alter the form of the governing equation. The heterogeneous energy addition will not change the governing equation but will, rather, show up as a boundary condition, thus affecting the result in a different manner. In this section we will be concerned with the homogeneous type of energy addition.

Example 4.3

For the plane wall shown in Figure 4.4, compare the expressions for $T(x)$ which apply to the cases of
(a) a uniform volumetric generation rate $\dot{q}$.
(b) a generation rate which varies linearly with temperature in the form

$$\dot{q} = \dot{q}_0[1 + \beta(T - T_0)]$$

In both cases, the surface temperature may be taken to be constant.

The one-dimensional form of the Poisson equation applies. Equation (2-54) written for the x direction is

$$\frac{d^2T}{dx^2} + \frac{\dot{q}}{k} = 0 \tag{4-19}$$

The first separation of variables and integration yields

$$\frac{dT}{dx} + \frac{\dot{q}}{k}x = c_1$$

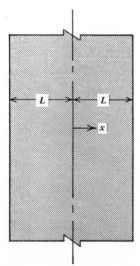

Figure 4.4 Plane wall configuration with internal generation of thermal energy.

The integration constant c_1 is evaluated by applying the symmetry boundary condition $dT/dx = 0$ at $x = 0$. The constant c_1 is thus found to be zero.

A second separation of variables and integration yields the expression

$$T + \frac{\dot{q}x^2}{2k} = c_2$$

The applicable boundary condition in this case is $T(\pm L) = T_L$, from which c_2 is evaluated as

$$c_2 = T_L + \frac{\dot{q}L^2}{2k}$$

The resulting expression for $T(x)$ is

$$T(x) = T_L + \frac{\dot{q}}{2k}(L^2 - x^2) \qquad (4\text{-}20)$$

For case (b), the equation which applies is

$$\frac{d^2T}{dx^2} + \frac{\dot{q}_0}{k}[1 + \beta(T - T_0)] = 0$$

This expression can be put in more convenient form by a slight change in variable if we let $\theta = T - T_0$. The temperature T_0 is a constant reference value. The equation in θ is now

$$\frac{d^2\theta}{dx^2} + \frac{\dot{q}_0}{k}(1 + \beta\theta) = 0$$

The boundary conditions which apply are

$$\frac{d\theta}{dx} = 0 \quad \text{at } x = 0 \text{ (symmetry)}$$

$$\theta = \theta_L = T_L - T_0 \quad \text{at } x = L$$

Introducing the constants A and B, defined as

$$A^2 = \frac{\dot{q}_0}{k}\beta$$

$$B = \frac{\dot{q}_0}{k}$$

we have a yet simpler form for the governing equation

$$\frac{d^2\theta}{dx^2} + A^2\theta + B = 0$$

The general solution to this second-order differential equation is

$$\theta = c_1 \cos Ax + c_2 \sin Ax - B/A^2$$

Applying the boundary conditions, we obtain

$$c_1 = \frac{\theta_L + \dfrac{B}{A^2}}{\cos AL}$$

$$c_2 = 0$$

The final expression for θ may now be written as

$$\theta = \left(\theta_L + \frac{B}{A^2}\right) \frac{\cos Ax}{\cos AL} - \frac{B}{A^2} \tag{4-21}$$

The additional consideration of variable generation with temperature has the obvious result of more complicated mathematics in obtaining the final answer.

Example 4.4

A concrete column used in bridge construction is cylindrical in shape with a diameter of 3 ft. The column is sufficiently long that temperature variation along the column length may be neglected. Treating the column as solid concrete with an average thermal conductivity of 0.54 Btu/hr-ft-°F, determine the temperature at the center of the cylinder at a time when the outside surface temperature is measured to be 180°F. The heat of hydration of concrete may be taken to be 1.1 Btu/lb$_m$hr, and an average density for concrete of 150 lb/ft³ may be assumed.

The governing equation is the one-dimensional Poisson equation in cylindrical coordinates

$$\frac{1}{r}\frac{d}{dr}\left(r\frac{dT}{dr}\right) + \frac{\dot{q}}{k} = 0$$

Boundary conditions are

$$\frac{dT}{dr} = 0 \quad \text{at } r = 0 \text{ (symmetry)}$$

and

$$T = T_L \text{ at } r = R.$$

Separating variables and integrating once, we obtain

$$\frac{dT}{dr} + \frac{\dot{q}}{k}\frac{r}{2} = \frac{c_1}{r}$$

The constant c_1 is found to be 0 from the symmetry condition. A second separation of variables and integration yields, for T,

$$\int_{T_0}^{T_L} dT + \frac{\dot{q}}{k} \int_0^R \frac{r}{2}\, dr = 0$$

$$T_L - T_0 + \frac{\dot{q}}{k}\frac{R^2}{4} = 0$$

or

$$T_0 = T_L + \frac{\dot{q}}{k}\frac{R^2}{4} \tag{4-22}$$

Inserting the appropriate numerical values for T_L, $\dot{q}$, k, and R, we obtain

$$T_0 = 180°F + \frac{(1.1 \text{ Btu/lb}_m\text{hr})(150 \text{ lb}_m/\text{ft}^3)}{0.54 \text{ Btu/hr-ft-}°F} \frac{(1.5 \text{ ft})^2}{4}$$

$$= 180 + 172 = 352°F$$

4.1-3 Heat Transfer from Extended Surfaces

One method of increasing the heat transfer between a surface and an adjacent fluid is by increasing the surface area in contact with the fluid. This increase in area may be in the form of spines, fins, or other types of extended surfaces with various configurations.

Figure 4.5 depicts a general situation which we will use in formulating our analyses of extended surfaces.

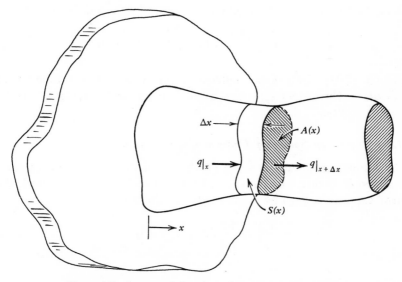

Figure 4.5 An extended surface of general configuration.

Under steady-state conditions a first-law analysis of the control volume of thickness Δx, cross-sectional area $A(x)$, and surface area $S(x)$, reduces to

$$\frac{\delta Q}{dt} = 0$$

Assuming heat conduction to be in the positive x direction (i.e., the fin is at a higher temperature than the adjacent fluid), our first-law relationship may be written as

$$q_x|_x - q_x|_{x+\Delta x} - q_{\text{conv}} = 0$$

or, in words, the energy in by conduction at x is equal to the energy out by conduction at $x + \Delta x$ plus the energy out by convection.

If we substitute the appropriate rate expression for each of the q's, assuming a lumped form in all but the x direction, we have

$$kA \frac{dT}{dx}\bigg|_{x+\Delta x} - kA \frac{dT}{dx}\bigg|_{x} - hS(T - T_\infty) = 0$$

We now express $S(x)$ as the product of the perimeter $P(x)$ and Δx, then divide through by Δx to obtain

$$\frac{kA \dfrac{dT}{dx}\bigg|_{x+\Delta x} - kA \dfrac{dT}{dx}\bigg|_{x}}{\Delta x} - hP(T - T_\infty) = 0$$

Taking the limit as $\Delta x \to 0$, we obtain the general differential equation

$$\frac{d}{dx}\left[kA(x) \frac{dT}{dx} \right] - hP(x)[T - T_\infty] = 0 \qquad (4\text{-}23)$$

Equation (4-23) yields many different forms when applied to different geometries. Three representative configurations are considered in the following sections.

4.1-3.1   Fins or Spines of Uniform Cross Section

Two possible configurations where $P(x) = P$ and $A(x) = A$, both constant, are shown in Figure 4.6. With k and h taken to be constant as well, the general expression given by equation (4-23) reduces to

$$\frac{d^2T}{dx^2} - \frac{hP}{kA}(T - T_\infty) = 0 \qquad (4\text{-}24)$$

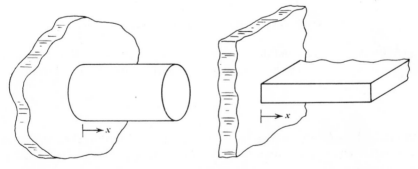

Figure 4.6   Two configurations of extended surfaces with uniform cross sections.

4.1-3.2 Straight Surfaces with Uniformly Varying Cross Section

Figure 4.7 shows a fin of rectangular cross section with linearly varying thickness. For values of A and P designated A_0 and P_0 at the root and A_L and P_L at the end, the variation in $A(x)$ and $P(x)$ may be expressed in a linear fashion as

$$A = A_0 - (A_0 - A_L)\frac{x}{L}$$

$$P = P_0 - (P_0 - P_L)\frac{x}{L}$$

With the semi-thickness of the fin denoted t_0 and t_L at $x = 0$ and $x = L$, respectively, the expressions for A_0, A_L, P_0, and P_L per unit length of fin are

$$A_0 = 2t_0 \qquad P_0 = 2(2t_0 + 1)$$
$$A_L = 2t_L \qquad P_L = 2(2t_L + 1)$$

The governing differential equation which applies in this case, with k and h constant, is

$$\frac{d}{dx}\left\{\left[A_0 - (A_0 - A_L)\frac{x}{L}\right]\frac{dT}{dx}\right\} - \frac{h}{k}\left[P_0 - (P_0 - P_L)\frac{x}{L}\right](T - T_\infty) = 0 \quad (4\text{-}25)$$

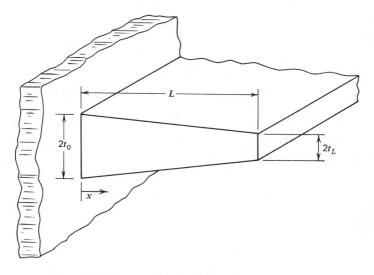

Figure 4.7 A rectangular fin of linearly varying thickness.

4.1-3.3 Curved Surfaces with Uniform Thickness

The circular fin of constant thickness, as shown in Figure 4.8, is a common type of extended surface. For this configuration, $A(r)$ and $P(r)$ may be written

$$A(r) = 4\pi rt$$
$$P(r) = 4\pi r \qquad r_0 < r < r_L$$

Substitution of these quantities into equation (4-23) gives us the applicable expression for this case. With k and h constant, we have

$$\frac{d}{dr}\left(r\frac{dT}{dr}\right) - \frac{hr}{kt}(T - T_\infty) = 0 \qquad (4\text{-}26)$$

In addition to the obvious complexities introduced for more involved geometry, the consideration of variable k and h would make any analysis much more difficult.

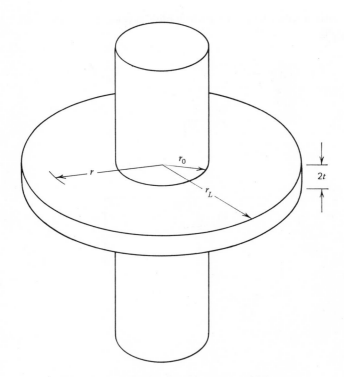

Figure 4.8   A circular fin with constant thickness.

Solutions to each of the equations developed are of interest. We will go into detail regarding only the first of the three configurations considered, the straight fin of constant cross section.

4.1-3.4 Solutions for Straight Fins of Constant Cross Section

The relation to be solved is equation (4-24). Combining terms in the coefficient, we write $hP/kA = m^2$, and transforming the dependent variable as $\theta = T - T_\infty$, we obtain, for the governing relation,

$$\frac{d^2\theta}{dx^2} - m^2\theta = 0 \qquad (4\text{-}27)$$

The general solution to equation (4-27) is of the form

$$\theta = c_1 e^{mx} + c_2 e^{-mx} \qquad (4\text{-}28)$$

or

$$\theta = A \cosh mx + B \sinh mx \qquad (4\text{-}29)$$

Either form for the general solution is correct; one may be preferable for a given case.

Two boundary conditions are necessary to evaluate the constants in equation (4-28) or (4-29). We shall consider four sets:

1. A very long fin

$$\theta = \theta_0 \quad \text{at } x = 0$$
$$\theta = 0 \quad \text{at } x \to \infty$$

2. A known temperature at $x = L$

$$\theta = \theta_0 \quad \text{at } x = 0$$
$$\theta = \theta_L \quad \text{at } x = L$$

3. An insulated end condition

$$\theta = \theta_0 \quad \text{at } x = 0$$
$$\frac{d\theta}{dx} = 0 \quad \text{at } x = L$$

4. Conduction to the end equal to convection from the end

$$\theta = \theta_0 \quad \text{at } x = 0$$
$$k\frac{d\theta}{dx} = h\theta \quad \text{at } x = L$$

The first boundary condition in each of these sets is the same, stating that the temperature at the base of the fin is the same as that of the primary surface to which it is attached. The second condition in each set represents the condition on the other end, at $x = L$.

Using the first set of boundary conditions with equation (4-27), we obtain, for the temperature profile,

$$\frac{\theta}{\theta_0} = \frac{T - T_\infty}{T_0 - T_\infty} = e^{-mx} \tag{4-30}$$

For boundary conditions set (2), the resulting expression for θ is

$$\frac{\theta}{\theta_0} = \frac{T - T_\infty}{T_0 - T_\infty} = \left(\frac{\theta_L}{\theta_0} - e^{-mL}\right)\left(\frac{e^{mx} - e^{-mx}}{e^{mL} - e^{-mL}}\right) + e^{-mx} \tag{4-31}$$

Observe that as $L \to \infty$, equation (4-31) approaches equation (4-30). For set (3), the temperature profile expression which results is

$$\frac{\theta}{\theta_0} = \frac{T - T_\infty}{T_0 - T_\infty} = \frac{e^{mx}}{1 + e^{2mL}} + \frac{e^{-mx}}{1 + e^{-2mL}} \tag{4-32}$$

or

$$\frac{\theta}{\theta_0} = \frac{T - T_\infty}{T_0 - T_\infty} = \frac{\cosh\,[m(L - x)]}{\cosh\,mL} \tag{4-33}$$

Again it may be noticed that both of these expressions reduce to equation (4-30) as $L \to \infty$.

The solution for θ using set (4) of the temperature boundary conditions becomes

$$\frac{\theta}{\theta_0} = \frac{T - T_\infty}{T_0 - T_\infty} = \frac{\cosh\,[m(L - x)] + (h/mk)\,\sinh\,[m(L - x)]}{\cosh\,mL + (h/mk)\,\sinh\,mL} \tag{4-34}$$

Equation (4-34) also reduces to equation (4-30) as $L \to \infty$.

To evaluate the heat transfer from a fin whose temperature profile is given by one of the preceding equations, either of two approaches can be taken. One possible method is to integrate over the fin surface according to

$$q = \int_s h[T(x) - T_\infty]\,dS = \int_s h\theta\,dS \tag{4-35}$$

A second method is to evaluate the rate of heat transfer into the fin by conduction at the root. The equation which applies in this case is

$$q = -kA\,\frac{dT}{dx}\bigg|_{x=0} = -kA\,\frac{d\theta}{dx}\bigg|_{x=0} \tag{4-36}$$

Equation (4-36) is normally the easier of these two expressions to use.

Table 4.2 relates temperature profiles and heat transfer rates from straight fins which apply to each of the four sets of boundary conditions considered earlier.

For more complicated geometries, the expressions for temperature variation and heat flow become more complex than those given in Table 4.2. Certain of these cases will be left as exercises for the reader.

Clearly, a fin would be most effective if the temperature everywhere along it were the same as that of the primary surface to which it is attached. For a 100% effective fin, the heat transfer to the adjacent fluid is given by

$$q = hS(T_0 - T_\infty) \tag{4-37}$$

For a fin with $T(x) < T_0$ for $x > 0$, the heat transfer will be less than that given by equation (4-37). The ratio of actual heat transfer from an extended surface to the maximum possible, as expressed in equation (4-37), is designated *fin effectiveness*, symbolized η_f.

$$\eta_f = \frac{q_{\text{actual}}}{q_{\text{maximum}}} \tag{4-38}$$

A plot of fin effectiveness as a function of significant parameters for a straight fin and for a circular fin of three different lengths is given in Figure 4.9.

For a finned surface, the total heat transfer is given by

$$q_{\text{total}} = q_{\text{primary surface}} + q_{\text{fins}}$$

$$= hA_0(T_0 - T_\infty) + \int_s h(T - T_\infty) \, dS \tag{4-39}$$

Expressing the second term, the heat transfer from the fins, in terms of fin effectiveness, we have

$$q_{\text{total}} = hA_0(T_0 - T_\infty) + hA_f\eta_f(T_0 - T_\infty)$$

or

$$q_{\text{total}} = h(A_0 + A_f\eta_f)(T_0 - T_\infty) \tag{4-40}$$

In equation (4-40) A_0 represents the area of the primary surface which is in contact with the surrounding fluid, and A_f is the total fin surface area. The temperature difference $T_0 - T_\infty$ and the surface coefficient h are assumed constant.

An application of equation (4-40) is illustrated in Example 4.5.

Table 4.2 Temperature Profiles and Heat Transfer for Straight Fins with Constant Area

Boundary Conditions	$\theta(x) = \dfrac{T(x) - T_\infty}{T_0 - T_\infty}$	$q(x)$
1. $\quad \theta(0) = \theta_0$ $\quad \theta(\infty) = 0$	$\dfrac{\theta}{\theta_0} = e^{-mx}$	$q_x = kAm\theta_0$
2. $\quad \theta(0) = \theta_0$ $\quad \theta(L) = \theta_L$	$\dfrac{\theta}{\theta_0} = \left(\dfrac{\theta_L}{\theta_0} - e^{-mL}\right)\left(\dfrac{e^{mx} - e^{-mx}}{e^{mL} - e^{-mL}}\right) + e^{-mx}$	$q_x = kAm\theta_0\left[1 - \dfrac{2(\theta_L - \theta_0 e^{-mL})}{e^{mL} - e^{-mL}}\right]$
3. $\quad \theta(0) = \theta_0$ $\quad \dfrac{d\theta}{dx}(L) = 0$	$\dfrac{\theta}{\theta_0} = \dfrac{\cosh[m(L-x)]}{\cosh\, mL}$	$q_x = kAm\theta_0 \tanh\, mL$
4. $\quad \theta(0) = \theta_0$ $\quad -k\dfrac{d\theta}{dx}(L) = h\theta(L)$	$\dfrac{\theta}{\theta_0} = \dfrac{\cosh[m(L-x)] + (h/mk)\sinh[m(L-x)]}{\cosh\, mL + (h/mk)\sinh\, mL}$	$q_x = kAm\theta_0\,\dfrac{\sinh\, mL + (h/mk)\cosh\, mL}{\cosh\, mL + (h/mk)\sinh\, mL}$

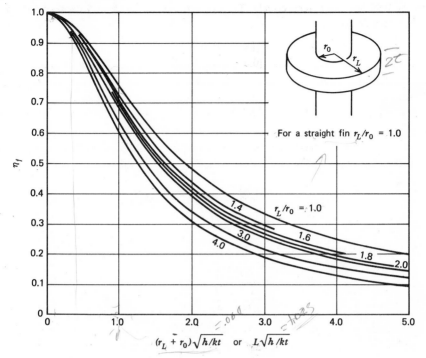

Figure 4.9 Fin effectiveness for straight and circular fins of constant thickness.

Example 4.5

Air and water are separated by a plane wall made of mild steel. It is proposed to increase the heat transfer rate between these two fluids by adding straight rectangular mild steel fins 0.05 in. thick, 1 in. long, and spaced 0.5 in. between centers, to the wall. What percent increase in heat transfer can be realized by adding fins to (a) the air side, (b) the water side, and (c) both sides, of the plane wall? The air- and water-side coefficients may be taken as 2 and 45 Btu/hr-ft²-°F, respectively.

For one square foot of original wall area, the areas of primary and finned surfaces are

$$A_0 = 1 \text{ ft}^2 - (24 \text{ fins})(1 \text{ ft}) \left(\frac{0.05/12 \text{ ft}}{\text{fin}} \right)$$

$$= 0.9 \text{ ft}^2$$

$$A_f = (24 \text{ fins})(1 \text{ ft})(2 \times 1/12 \text{ ft}) + 0.1 \text{ ft}^2$$

$$= 4.1 \text{ ft}^2$$

The values of η_f for the air- and water-side conditions are found, from Figure 4.9, to be 0.92 and 0.39, respectively. The corresponding heat transfer rates with fins on each side are

$$q = h_A \Delta T_A [A_0 + A_f \eta_f]$$
$$= 2\Delta T_A [0.9 + (4.1)(0.92)] = 9.34 \Delta T_A$$

for the air side, and

$$q = 45\Delta T_w [0.9 + (4.1)(0.39)] = 112.5 \Delta T_w$$

for the water side.

The total rate of heat transfer per square foot of unfinned surface, neglecting the conductive resistance of the steel wall, is

$$q_{w/o\,\mathrm{fins}} = \frac{\Delta T_{\mathrm{total}}}{\dfrac{1}{2} + \dfrac{1}{45}} = 1.915 \Delta T_{\mathrm{total}}$$

With fins on the air side only, we have

$$= \frac{\Delta T_{\mathrm{total}}}{\dfrac{1}{9.34} + \dfrac{1}{45}} = 7.74 \Delta T_{\mathrm{total}}$$

an increase of 304%.

With fins on the water side only, the heat transfer rate is

$$q = \frac{\Delta T_{\mathrm{total}}}{\dfrac{1}{2} + \dfrac{1}{112.5}} = 1.965 \Delta T_{\mathrm{total}}$$

an increase of 2.6%.

The heat transfer rate with fins added to both sides is

$$q = \frac{\Delta T_{\mathrm{total}}}{\dfrac{1}{9.34} + \dfrac{1}{112.5}} = 8.63 \Delta T_{\mathrm{total}}$$

an increase of 350%.

The greatest effect of adding fins is seen to occur on the air side, where h is lowest. This is because air-side resistance is controlling in the plane wall case, and any change altering the part of the heat transfer path which controls will have a most significant effect on the total heat transfer capability.

It is true in general that fins should be added to those surfaces where h is low. If air in natural convection is on the outside of a tube or pipe, fins are often added to increase heat transfer because of the low value of h usually associated with this condition.

4.1-4   Steady-State Heat Conduction in Two and Three Dimensions

With more than one significant space variable involved, the solution to Laplace's or Poisson's equation becomes much more involved. Techniques of solving the two- or three-dimensional Laplace equation will be emphasized in this section. These will be considered in the following order: analytical, graphical, integral, and numerical.

4.1-4.1   Two- and Three-Dimensional Steady-State Heat Conduction: Analytical Solutions

Examples of analytical solutions to multidimensional steady-state heat transfer problems will be considered in some detail in this section. We will deal with two-dimensional systems for brevity; the extension from two to three dimensions is quite direct, and no loss of generality is involved in restricting our considerations to the two-dimensional case.

Example 4.6

Find the steady-state temperature distribution in the two-dimensional fin shown in Figure 4.10. At its base the fin temperature is given by $F(y)$. The surface coefficient h may be considered very large. The fin is infinitely long and has a thickness of $2l$.

The differential formulation of this problem requires a statement of the governing differential equation

$$\frac{\partial^2 T}{\partial x^2} + \frac{\partial^2 T}{\partial y^2} = 0 \tag{4-41}$$

and boundary conditions

$$T(0, y) = F(y)$$
$$T(x, \pm l) = T_\infty$$
$$T(\infty, y) = T_\infty$$

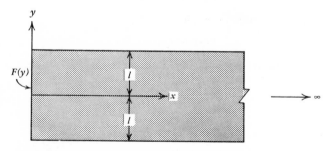

Figure 4.10   Steady-state conduction in a two-dimensional fin.

The first step in the problem solution is to transform the dependent variable, letting $\theta = T - T_\infty$. This transformation does not change the governing equation but reduces three of the boundary conditions to homogeneous form. The new problem in terms of θ is now

$$\frac{\partial^2 \theta}{\partial x^2} + \frac{\partial^2 \theta}{\partial y^2} = 0 \qquad (4\text{-}42)$$

$$\theta(0, y) = f(y) \qquad (4\text{-}43)$$

$$\theta(x, \pm l) = 0 \qquad (4\text{-}44)$$

$$\theta(\infty, y) = 0 \qquad (4\text{-}45)$$

A separation-of-variables technique will be employed to solve equation (4-42). Assuming a product solution of the form

$$\theta(x, y) = X(x) Y(y) \qquad (4\text{-}46)$$

to exist, then a substitution of this product into equation (4-42) and subsequent division of each term by XY yields

$$\frac{1}{X} \frac{d^2 X}{dx^2} = -\frac{1}{Y} \frac{d^2 Y}{dy^2} = \pm \lambda^2 \qquad (4\text{-}47)$$

where λ^2 is a constant whose sign is determined by the physical requirement of symmetry of the function $Y(y)$ about $y = 0$ and of homogeneity of $Y(\pm l)$. The appropriate sign of λ^2 is positive in this case.

The original two-dimensional problem is now reduced to solving two ordinary differential equations. The x- and y-directional problems may now be written as

$$\frac{d^2 X}{dx^2} - \lambda^2 X = 0 \qquad X(\infty) = 0 \qquad (4\text{-}48)$$

$$\frac{d^2 Y}{dy^2} + \lambda^2 Y = 0 \qquad Y(\pm l) = 0, \frac{dY}{dy}(0) = 0 \qquad (4\text{-}49)$$

The nonhomogeneous boundary condition at $X = 0$ is not separable and will be left out until the last stage of the solution.

First solving the y-directional problem, equation (4-49), we have, for the general solution,

$$Y = A \cos \lambda y + B \sin \lambda y$$

and, introducing the boundary conditions, we have

$$Y(y) = A_n \cos \lambda_n y \qquad (4\text{-}50)$$

where A_n is an arbitrary constant associated with the characteristic values λ_n given by

$$\lambda_n = \frac{n\pi}{2l} \qquad \text{for } n = 1, 3, 5, \ldots$$

In the x direction the solution to equation (4-48) is

$$X = Ce^{\lambda_n x} + De^{-\lambda_n x}$$

which becomes, upon applying the boundary condition at $x = \infty$,

$$X(x) = D_n e^{-\lambda_n x} \tag{4-51}$$

The assumed product solution given in equation (3-46) is now

$$\theta(x, y) = X(x) Y(y) = \sum_{n=1,3...}^{\infty} a_n e^{-\lambda_n x} \cos \lambda_n y \tag{4-52}$$

where $a_n = A_n D_n$.

The remaining constant a_n may now be determined from the nonhomogeneous boundary condition at $x = 0$. Doing so, we obtain

$$f(y) = \sum_{n=1,3...}^{\infty} a_n \cos \lambda_n y$$

Multiplying both sides of this equation by $\cos \lambda_m y \, dy$ and observing that

$$\int_0^l \cos \lambda_n y \cos \lambda_m y \, dy = 0 \qquad \text{for} \qquad m \neq n$$

we obtain, for a_n, the expression

$$a_n = \frac{\displaystyle\int_0^l f(y) \cos \lambda_n y \, dy}{\displaystyle\int_0^l \cos^2 \lambda_n y \, dy} \tag{4-53}$$

The denominator in equation (4-53) is equal to $l/2$ for all positive integer values of n. Thus the constant a_n is expressed as

$$a_n = \frac{2}{l} \int_0^l f(y) \cos \lambda_n y \, dy \tag{4-54}$$

and the complete solution for $\theta(x, y)$ becomes

$$\theta(x, y) = \frac{2}{l} \sum_{n=1,3...}^{\infty} \left[\int_0^l f(\xi) \cos \lambda_n \xi \, d\xi \right] e^{-\lambda_n x} \cos \lambda_n y \tag{4-55}$$

Certain forms for the function $f(y)$ are more likely than others. Consider the case of a uniform base temperature

$$f(y) = \theta(0, y) = \theta_0 \qquad \text{(a constant)}$$

The constant a_n becomes

$$a_n = \frac{2}{l} \int_0^l \theta_0 \cos \lambda_n y \, dy = (-1)^{(n-1)/2} \frac{4\theta_0}{n\pi}$$

for $n = 1, 3, 5, \ldots$. The solution for (x, y) is then

$$\theta(x, y) = \frac{4\theta_0}{\pi} \sum_{n=1,3...}^{\infty} \frac{1}{n} (-1)^{(n-1)/2} e^{-\lambda_n n} \cos \lambda_n y \tag{4-56}$$

Example 4.7

Find the steady-state temperature distribution in the two-dimensional fin described in Example 4.6 for the case of heat transfer by convection at the top and bottom surfaces with a finite value for the convective heat transfer coefficient h.

The problem is shown diagrammatically in Figure 4.11. Using the same definition for θ as in the previous example, the problem may be stated as

$$\frac{\partial^2 \theta}{\partial x^2} + \frac{\partial^2 \theta}{\partial y^2} = 0 \tag{4-57}$$

$$\theta(0, y) = F(y) \tag{4-58}$$

$$\theta(\infty, y) = 0 \tag{4-59}$$

$$\frac{\partial \theta}{\partial y}(x, 0) = 0 \quad \text{(symmetry)} \tag{4-60}$$

$$\frac{\partial \theta}{\partial y}(x, \pm l) = h\theta(x, \pm l) \tag{4-61}$$

The separation-of-variables technique, assuming a product solution as given by equation (4-46), allows the original problem to be stated in terms of x- and y-directional components in the form

$$\frac{d^2 X}{dx^2} - \lambda^2 X = 0 \qquad X(\infty) = 0 \tag{4-62}$$

$$\frac{d^2 Y}{dy^2} + \lambda^2 Y = 0 \qquad \frac{dY(o)}{dy} = 0$$

$$-k\frac{dY}{dy}(\pm l) = hY(\pm l) \tag{4-63}$$

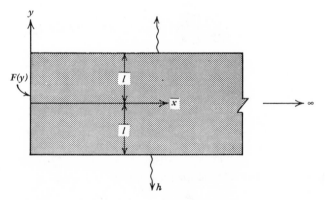

Figure 4.11 Steady-state conduction in a two-dimensional fin with convection to the surrounding fluid.

We again consider the y direction first. The problem given by equation (4-63) with the associated boundary conditions results in the solution

$$Y_n(y) = A_n \cos \lambda_n y \qquad (4\text{-}64)$$

where the characteristic values λ_n are those satisfying the transcendental equation

$$k\lambda_n \sin \lambda_n l = h \cos \lambda_n l \qquad (4\text{-}65)$$

The solution of the x direction is identical to that of Example 4.6, specifically,

$$X(x) = D_n e^{-\lambda_n x} \qquad (4\text{-}66)$$

and $\theta(x, y)$ is given by the product

$$\theta(x, y) = X(x)\,Y(y) = \sum_{n=1}^{\infty} a_n e^{-\lambda_n x} \cos \lambda_n y \qquad (4\text{-}67)$$

The constant a_n, which is the product of $A_n D_n$, is evaluated precisely as given in equation (4-54), and the general expression for $\theta(x, y)$ stated in equation (4-55) applies for the present case also.

For the special case of a constant base temperature, $F(y) = \theta(0, y) = \theta_0$, the constant a_n becomes

$$a_n = \frac{2\theta_0 \sin \lambda_n l}{\lambda_n l + \sin \lambda_n l \cos \lambda_n l} \qquad (4\text{-}68)$$

and the steady-state temperature profile is

$$\theta(x, y) = 2\theta_0 \sum_{n=1}^{\infty} \left(\frac{\sin \lambda_n l}{\lambda_n l + \sin \lambda_n l \cos \lambda_n l} \right) e^{-\lambda_n x} \cos \lambda_n y \qquad (4\text{-}69)$$

4.1-4.2 Two-Dimensional Steady-State Heat Conduction; Graphical Solutions

For steady-state heat conduction in a two-dimensional system with isothermal boundaries, a graphical solution technique known as *flux plotting* is convenient and rapid. This technique is a valid approach to solving the two-dimensional Laplace equation

$$\frac{\partial^2 T}{\partial x^2} + \frac{\partial^2 T}{\partial y^2} = 0$$

where $T(x, y)$ has a constant value on all boundaries. Similar problems where the dependent variable is the stream function ψ, the velocity potential ϕ, the mass concentration of a given constituent C_A, or the electric potential V may be solved by the same approach as we shall now discuss.

In a semi-infinite wall with constant surface temperature, the isotherms and heat flow lines will appear as shown in Figure 4.12. In the wall the isotherms are straight vertical lines, and heat flow lines are horizontal, perpendicular to the isotherms.

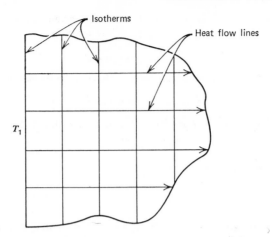

Figure 4.12   Isotherms and heat flow lines in a semi-infinite wall.

If a constant temperature difference exists between any two adjacent isotherms, then the distance between isotherms is indicative of the magnitude of the temperature gradient; close spacing would indicate a relatively large gradient, and a small gradient would be associated with broad spacing.

Since heat transfer occurs by conduction under the influence of a temperature gradient, it is clear that no driving force for heat transfer exists along an isotherm. Thus a line of heat flow may be thought of as an insulated boundary across which heat may not be conducted.

The above concepts are important when considering a more general two-dimensional conduction problem. In Figure 4.13 a set of two isotherms and

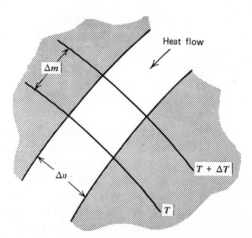

Figure 4.13   A general element for flux plotting analysis.

two heat flow lines is shown. The two isotherms differ from one another by the amount ΔT. The heat flow lines are perpendicular to the isotherms and form a passage, or tube, through which an amount of heat Δq may flow.

If, in a total cross section, there are N total heat flow passages, each with an amount Δq of heat flowing, the total transfer is given by

$$q_{\text{total}} = N\Delta q \qquad (4\text{-}70)$$

For the individual heat flow passage shown, the temperature gradient, in finite difference form, is $[(T + \Delta T) - T]/\Delta m = \Delta T/\Delta m$. The Fourier rate equation written for this passage then gives, for Δq,

$$\Delta q = k\Delta n \frac{\Delta T}{\Delta m} \qquad (4\text{-}71)$$

If the isotherm/heat-flow-line grid is constructed so that $\Delta m = \Delta n$, i.e., a system of curvilinear squares, the expression for Δq becomes simply

$$\Delta q = -k\Delta T \qquad (4\text{-}72)$$

regardless of the size of the squares!

For heat flow between two isothermal boundaries at temperatures T_h and T_c where $T_h > T_c$, with isotherms dividing each flow tube into M divisions, the temperature difference between adjacent isotherms is expressed as

$$\Delta T = \frac{T_h - T_c}{M} \qquad (4\text{-}73)$$

Finally, for a total of N flow passages between boundaries at T_h and T_c, each with the amount Δq of heat flow, the total heat transfer is

$$q_{\text{total}} = N\Delta q = Nk\Delta T$$

$$= \frac{N}{M} k(T_h - T_c) \qquad (4\text{-}74)$$

The ratio N/M, by comparison with equation (1-17), is seen to be the *shape factor S*. The flux plotting technique is then simply a means of determining a value for the shape factor in a two-dimensional steady-state conduction problem in which the boundaries are at constant temperature.

The technique for flux plotting is illustrated in Example 4.8.

Example 4.8

The inside and outside surfaces of the rectangular chimney depicted are determined to be 300°F and 100°F, respectively. How much heat transfers through the brick ($k = 0.40$ Btu/hr-ft-°F) chimney wall per foot of height?

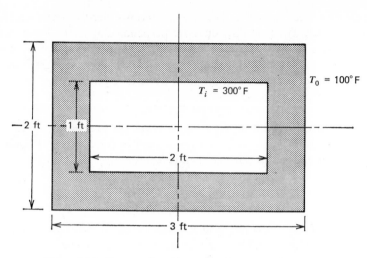

Figure 4.14 Rectangular chimney for flux plotting analysis.

Because the chimney cross section is symmetrical, we need to analyze only a quarter section. An enlarged representation of the lower right quarter is shown in Figure 4.15 to scale.

It is now necessary to construct the grid of orthogonal isotherms and heat flow lines filling the cross section. The general procedure used in drawing the flux plot is as follows:

1. Begin drawing "curvilinear squares" at the location where the isotherms and heat flow lines form patterns that are most nearly true squares. The size of the grid is not critical; it should be large enough to avoid needless tedium but small enough to yield reasonable accuracy.

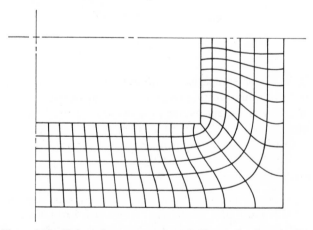

Figure 4.15 Enlarged quarter section of chimney for flux plotting.

2. Once beginning the construction of isotherms and heat flow lines at a particular boundary, the process is continued until the flux plot fills the cross section. Certain rules must be observed in the construction process:
   a. All boundaries are isotherms; heat flow lines intersect the boundaries at right angles.
   b. All boundary corners, representing the intersection of isotherms, are bisected by a heat flow line.
   c. All lines of symmetry are heat flow lines; isothermal lines intercept lines of symmetry at right angles.

By adhering to the above rules of construction and filling the cross section with curvilinear squares, the desired heat transfer rate may be determined. In the present example the number of temperature increments M is 6; the number of heat flow channels for the quarter section is 22, and for the total cross section is 4(22) = 88; thus $N = 88$. The shape factor is thus

$$S = \frac{N}{M} = \frac{88}{6} = 14.7$$

and the heat loss through the chimney becomes

$$q = kS\Delta T = (0.40 \text{ Btu/hr-ft-}°\text{F})(14.7)(200°\text{F})$$

$$= 1147 \text{ Btu/hr per foot of chimney}$$

Shape factors for some common two-dimensional configurations are given in Table 4.3.

4.1-4.3 Two-Dimensional Steady-State Heat Conduction; Integral Solutions

The integral approach to solving multidimensional, steady-state heat conduction problems will be illustrated by means of Example 4.9.

Example 4.9

The two-dimensional fin considered in Example 4.6 is to be solved by the integral technique. All specifications and dimensions are as given previously. The function $f(y)$ is assumed parabolic as given by $\theta(0, y) = \theta_{\max}[1 - (y/l)^2]$. Figure 4.10 is reproduced below for convenience as Figure 4.16. The governing equation and boundary conditions are

$$\frac{\partial^2\theta}{\partial x^2} + \frac{\partial^2\theta}{\partial y^2} = 0$$

$$\theta(0, y) = f(y) = \theta_{\max}\left[1 - \left(\frac{y}{l}\right)^2\right]$$

$$\theta(x, \pm l) = 0$$

$$\theta(\infty, y) = 0$$

Table 4.3 Shape Factors for Conduction

Configuration	Shape factor, S $q = kS(T_i - T_o)$
Concentric circular cylinders	$\dfrac{2\pi L}{\ln (r_o/r_i)}$
Eccentric circular cylinders	$\dfrac{2\pi L}{\cosh^{-1}\left(1 + \dfrac{\rho^2 - \varepsilon^2}{2\rho}\right)}$ $\rho = r_i/r_o$ $\varepsilon = e/r_o$
Circular cylinder in a square cylinder	$\dfrac{2\pi L}{\ln (r_o/r_i) - 0.27079}$
Circular cylinder in a hexagonal cylinder	$\dfrac{2\pi L}{\ln (r_o/r_i) - 0.10669}$
Buried horizontal cylinder of length L	$\dfrac{2\pi L}{\ln (2\rho/r)}$ $\dfrac{2\pi L}{\cosh^{-1}(\rho/r)}$
Buried horizontal thin disc of radius r, thickness $\ll r$	$\dfrac{2.22r}{1 - r/2.83\rho}$
Buried sphere	$\dfrac{4\pi r}{1 - r/20}$
Two cylinders in an infinite, homogeneous medium $L \gg r_1, r_2, \rho$	$\dfrac{2\pi L}{\cosh^{-1}(D^2 - r_1^2 - r_2^2/2r_1 r_2)}$

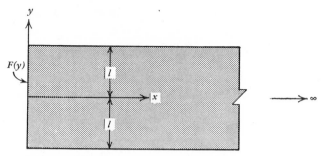

Figure 4.16 Steady-state conduction in a two-dimensional fin.

The integral technique involves solving the integral of the governing differential equation. For the present case, the integral to be solved is

$$\int_0^l \int_0^\infty \left(\frac{\partial^2 \theta}{\partial x^2} + \frac{\partial^2 \theta}{\partial y^2} \right) dx \, dy = 0 \tag{4-75}$$

The task is now to choose an approximate form for the dependent variable, $\theta(x, y)$, which in itself satisfies the boundary conditions. The assumed form of $\theta(x, y)$ will then be made compatible with the governing equation via one of two methods, the *Ritz* or the *Kantorovich* method. These will now be considered in order.

1. Employing the Ritz method, we assume the form of $\theta(x, y)$ in both the x and y dimensions subject to the boundary conditions that must be satisfied. For the present case, a quadratic function will be used in the y direction and an exponential in the x direction. The assumed Ritz profile is

$$\theta(x, y) = A(l^2 - y^2)e^{-Bx}$$

where the parameter A is required by the boundary condition at $x = 0$ to be $\theta_{\max}/l^2$. Thus the function to be used is

$$\theta(x, y) = \frac{\theta_{\max}}{l^2} (l^2 - y^2)e^{-Bx} \tag{4-76}$$

The parameter B will be established as equation (4-76) is substituted into the integral relation, equation (4-75).

The integration will now proceed. The first integration, term by term,[1] of equation (4-75) yields the expression

$$\int_0^l \frac{\partial \theta}{\partial x} \bigg|_0^\infty dy + \int_0^\infty \frac{\partial \theta}{\partial y} \bigg|_0^l dx = 0 \tag{4-77}$$

[1] Integration procedures follow Liebnitz' rule.

Solving for each of these terms separately, using equation (4-76), we have, for the first,

$$\int_0^l \frac{\partial \theta}{\partial x}\bigg|_0^\infty dy = -\int_0^l \frac{B\theta_{\max}}{l^2}(l^2 - y^2)e^{-Bx}\bigg|_0^\infty dy$$

$$= \frac{B\theta_{\max}}{l^2}\int_0^l (l^2 - y^2)\, dy$$

$$= \frac{B\theta_{\max}}{l^2}\left[l^2 y - \frac{y^3}{3}\right]_0^l = \frac{2}{3}Bl\theta_{\max}$$

and, for the second,

$$\int_0^\infty \frac{\partial \theta}{\partial y}\bigg|_0^l dx = -\int_0^\infty \theta_{\max}\frac{2y}{l^2}\bigg|_0^l e^{-Bx}\, dx$$

$$= -\frac{2\theta_{\max}}{l}\int_0^\infty e^{-Bx}\, dx$$

$$= -\frac{2\theta_{\max}}{l}\left[-\frac{e^{-Bx}}{B}\right]_0^\infty = -\frac{2\theta_{\max}}{Bl}$$

With these terms evaluated, equation (4-77) may now be written

$$\frac{2}{3}Bl\theta_{\max} - \frac{2\theta_{\max}}{Bl} = 0$$

and the parameter B is evaluated as

$$B = \frac{\sqrt{3}}{l}$$

The solution, using the Ritz method, is this given as

$$\theta(x, y) = \theta_{\max}\left(\frac{l^2 - y^2}{l^2}\right)e^{-\sqrt{3}x/l} \qquad (4\text{-}78)$$

2. In the Kantorovich method the procedure is much the same except that the functional form for $\theta(x, y)$ is assumed in one direction only. In our present case, let this be the y direction, the form being quadratic as before. The assumed Kantorovich profile is now

$$\theta(x, y) = (l^2 - y^2)X(x) \qquad (4\text{-}79)$$

where the function $X(x)$ will be obtained from the integral formulation and boundary conditions. Substitution of equation (4-79) into equation (4-75) yields

$$\int_0^l \int_0^\infty \left(\frac{\partial^2 T}{\partial x^2} + \frac{\partial^2 T}{\partial y^2}\right) dx\, dy = \int_0^l \int_0^\infty [(l^2 - y^2)X'' - 2X]\, dx\, dy = 0$$

The first integration in the y direction yields

$$\int_0^\infty \left[\left(l^2 y - \frac{y^3}{3} \right) X'' - 2Xy \right]_0^l dx = 0$$

or, evaluating the limits, we have

$$\int_0^\infty \left[\frac{2}{3} l^3 X'' - 2lX \right] dx = 0$$

For this integral, evaluated between the limits 0 and ∞, to be true in general the integrand must identically equal zero. Thus we have the ordinary differential equation

$$X'' - \frac{3}{l^2} X = 0 \tag{4-80}$$

and the applicable boundary conditions

$$X(\infty) = 0, \qquad X(0) = \frac{\theta_{max}}{l^2}$$

The solution to equation (4-80) is

$$X = Me^{-\sqrt{3}x/l} + Ne^{\sqrt{3}x/l}$$

By applying the boundary conditions, we obtain, for the constants M and N,

$$M = \frac{\theta_{max}}{l^2} \qquad N = 0$$

Thus we have, for the function,

$$X(x) = \frac{\theta_{max}}{l^2} e^{-\sqrt{3}x/l}$$

and, for the complete solution,

$$\theta(x, y) = \theta_{max} \left(\frac{l^2 - y^2}{l^2} \right) e^{-\sqrt{3}x/l} \tag{4-81}$$

The Kantorovich method is seen to yield the same result, equation (4-81), as did the Ritz method, equation (4-78), for this example. The two methods will not always yield identical results.

The distinction between the two methods has been illustrated. In the Ritz method a reasonably complete assumption is made regarding the form of the dependent variable in all directions. The assumed profile includes at least one unspecified parameter. Substitution of the assumed Ritz profile into the integral form of the given equation yields an algebraic equation in the unknown parameter. Solving the algebraic equation to determine this parameter completes the solution. In the Kantorovich method the dependent variable is assumed in only one direction with an unspecified function left to describe the variation in the other direction. Substitution of this assumed Kantorovich profile into the integral formulation results in a differential equation which must then be solved to complete the solution.

In general the Ritz method is quicker; the Kantorovich method is the more accurate. It happens that the more general assumed form as in the Kantorovich approach yields a more accurate solution at the expense of greater (sometimes considerably) effort.

A thorough description of the integral technique and many examples are given in Arpaci.[2]

4.1-4.4   Two-Dimensional Steady-State Heat Conduction; Numerical Solution

Example 4.10 gives a numerical solution to a two-dimensional steady-state conduction problem.

Example 4.10

Find the steady-state temperature distribution in a two-dimensional square plate with boundary conditions as shown in Figure 4.17. The applicable differential equation in this case is the two-dimensional form of Laplace's equation

$$\nabla^2 T = \frac{\partial^2 T}{\partial x^2} + \frac{\partial^2 T}{\partial y^2} = 0 \qquad (2\text{-}54)$$

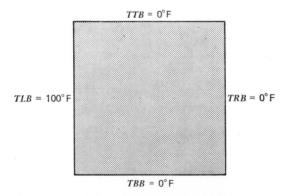

Figure 4.17 Two-dimensional plate for numerical analysis of steady-state conduction.

[2] V. Arpaci, *Conduction Heat Transfer*, (Reading, Mass. Addison-Wesley, 1966).

The finite difference form of this equation is written, using central differences, as

$$\frac{T_{i-1,j} - 2T_{i,j} + T_{i+1,j}}{\Delta x^2} + \frac{T_{i,j-1} - 2T_{i,j} + T_{i,j+1}}{\Delta y^2} = 0$$

For the case of a square grid with $\Delta x = \Delta y$, the solution for $T_{i,j}$ becomes

$$T_{i,j} = \frac{T_{i-1,j} + T_{i+1,j} + T_{i,j-1} + T_{i,j+1}}{4} \tag{4-82}$$

which states that the temperature at a given node is the arithmetic mean of the temperatures at adjacent nodes.

The flow chart (Figure 4.18), program listing, and computer solution follow. Gauss-Seidel iteration is the solution technique employed.

This program is written[3] in sufficiently general form that various node sizes, boundary conditions, and convergence criteria may be used.

```
      PROGRAM GS2D
      DIMENSION T(50,50)
C     THIS PROGRAM COMPUTES THE STEADY STATE TEMP. DISTR.
C     IN A SQUARE FLAT PLATE USING GAUSS SEIDEL ITERATION.
      N =TTYIN(4HN  = )
      M =TTYIN(4HM  = )
      EPS =TTYIN(4HEPS ,2H = )
      K =TTYIN(4HK  = )
      TBB =TTYIN(4HTBB ,2H = )
      TTB =TTYIN(4HTTB ,2H = )
      TLB =TTYIN(4HTLB ,2H = )
      TRB =TTYIN(4HTRB ,2H = )
      TI =TTYIN(4HTI  =,1H )
      N1 =N +1
      M1 =M +1
C     ASSIGN TEMP. ALONG ENTIRE BOUNDARY
      DO 12 I =1,N1
      T(I,1) =TTB
   12 T(I,M1) =TBB
      DO 11 J =1,M1
      T(1,J) =TLB
   11 T(N1,J) =TRB
```

[3] This and subsequent programs were written for the Oregon State University time-shared system. Logical execution and FORTRAN statements should not vary with other systems; however, certain input/output features and other characteristics of the OSU system may require minor modification before the programs listed will compile on other computer systems.

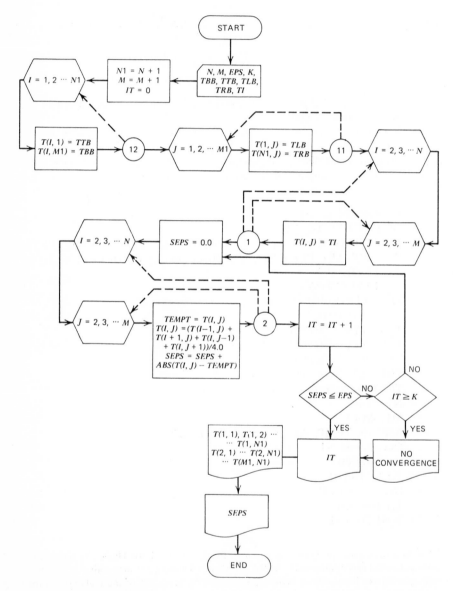

Figure 4.18 Flow diagram for solving a two-dimensional steady-state conduction problem, using Gauss-Seidel iteration.

```
C    ASSIGN INITIAL TEMPERATURES
     DO 1 I=2,N
     DO 1 J=2,M
   1 T(I,J)=TI
C    COMPUTE NEW TEMPERATURES
     IT=0
   3 SEPS=0.0
     DO  2 I=2,N
     DO  2 J=2,M
     TEMPT=T(I,J)
     T(I,J)=(T(I-1,J)+T(I+1,J)+T(I,J-1)+T(I,J+1))/4.
   2 SEPS=SEPS + ABS(T(I,J)-TEMPT)
     IT=IT+1
     IF(SEPS.LE.EPS)GO TO 4
     IF(IT.GE.K)GO TO 5
     GO TO 3
   5 WRITE(61,100)
 100 FORMAT(1H0,14HNO CONVERGENCE)
   4 WRITE(61,99)IT
  99 FORMAT(1H0,16HNO. ITERATIONS  =,I7)
     WRITE(61,101)
 101 FORMAT(1H0,24HTEMPERATURE AT GRID PNTS)
     DO 6 J=1,M1
   6 WRITE(61,102)(T(I,J),I=1,N1)
 102 FORMAT(1H0,9F7.2)
     WRITE(61,103)SEPS
 103 FORMAT(1H0,12HSUM OF EPS  =,E9.2)
     END
```

The stated problem was specified with six increments in both the x and y directions ($N = 5$, $M = 5$). The constant temperature boundaries were established ($TBB = 0.0$, $TTB = 0.0$, $TLB = 100.0$, $TRB = 0.0$); the convergence criterion was set ($EPS = 0.10$); the maximum number of iterations allowed was 50 ($K = 50$); and the temperature at all interior node points was initialized at 50°F ($TI = 50.0$)

The calculation scheme follows the Gauss-Seidel procedure, as discussed in Chapter 3.

Systematically all values of $T_{i,j}$ are recalculated, using equation (4-82). Each newly calculated value $T_{i,j}^{n+1}$ replaces the preceding value $T_{i,j}^{n}$, and this process is repeated until

$$|T_{i,j}^{n+1} - T_{i,j}^{n}| \leq EPS$$

In this example problem, 19 iterations were required before the convergence criterion was satisfied. Had the convergence not been obtained within 50 iterations, the calculation would have been terminated, the statement "No Convergence" written, and the most recent values of $T_{i,j}$ displayed.

A listing of the solution to this example problem follows.

N = 5
M = 5
EPS =   0.10
K = 50
TBB =   0.0
TTB =   0.0
TLB =   100.0
TRB =   0.0
TI =   50.0
NO. ITERATIONS =   19
TEMPERATURE AT GRID PNTS
    100.00   0   0   0   0   0
    100.00   45.47 22.36 11.00   4.55   0
    100.00   59.49 32.98 17.06   7.21   0
    100.00   59.48 32.97 17.06   7.20   0
    100.00   45.46 22.36 10.99   4.55   0
    100.00   0   0   0   0   0
SUM OF EPS = 9.80E−02

END OF FORTRAN EXECUTION

Example 4.10 is representative of the solution when interior temperatures
are desired. The grid spacing was somewhat coarse and the convergence
criterion not extremely rigid; however, the salient points of such a determination are observable. The symmetry of nodal temperatures across the horizontal centerline may be noted.

An additional bit of information may be obtained from the result which
is a further check of solution accuracy.

Breaking the cross section down into nodes in a square array, we are
assuming that heat transfer occurs from node to node *only* and then only
along paths which connect adjacent nodes. Since heat must be supplied to
the left-hand boundary to maintain the constant temperature of 100°F, the
amount of heat supplied is equal to that conducted from the left-hand
boundary to each of the internal nodes adjacent to it. The total heat conducted away from the left-hand boundary is thus seen to be, per foot of
depth (in the z direction),

$$q_{\text{total}} = k \frac{\Delta y}{\Delta x} \sum_{j=2}^{M} (T_{1,j} - T_{2,j})$$

Since this is a square array, $\Delta y / \Delta x = 1$, and the expression for q becomes

$$q_{\text{total}} = k \sum_{j=2}^{M} (T_{1,j} - T_{2,j})$$

$$= k(54.53 + 40.51 + 40.52 + 54.54) = 190.10k$$

The actual value of q_{total} depends on the material involved, i.e., the appropriate value of thermal conductivity.

The preceding discussion considered the heat added. In order that the top, bottom, and right-hand boundaries remain at 0°F, heat must be *removed*, and the amount of heat removed will be the summation of that conducted from all adjacent internal nodes to those nodes at 0°F. The total rate of heat removal is determined by

$$q_{total} = k\left[\frac{\Delta x}{\Delta y}\sum_{i=2}^{N}(T_{i,M} - T_{i,M1}) + \frac{\Delta x}{\Delta y}\sum_{i=2}^{N}(T_{i,2} - T_{i,1}) + \frac{\Delta y}{\Delta x}\sum_{j=2}^{M}(T_{N,j} - T_{N1,j})\right]$$

Again, since the grid array is square, $\Delta x/\Delta y = \Delta y/\Delta x = 1$, and q_{total} per foot of depth is given by

$$q_{total} = k\left[\sum_{i=2}^{N}(T_{i,M} - T_{i,M1}) + \sum_{i=2}^{N}(T_{i,2} - T_{i,1}) + \sum_{j=2}^{M}(T_{N,j} - T_{N1,j})\right]$$

where the summations are for the top, bottom, and right-hand surfaces, respectively. Putting in the appropriate numerical values, we have

$$\begin{aligned}
q_{total} &= k[(45.47 + 22.36 + 11.00 + 4.55) + (4.55 \\
&\quad + 7.21 + 7.20 + 4.55) + (45.46 + 22.36 \\
&\quad + 10.99 + 4.55)] \\
&= 190.25k.
\end{aligned}$$

This result should equal the value obtained earlier for total heat input. The difference is due to the convergence criterion used and to the coarse grid spacing. In light of these specifications, the agreement of 0.15 in 190, or approximately 0.08%, is satisfactory.

An additional output listing is given below, using the same program as before but different boundary conditions, node spacings, and convergence criterion.

```
N   = 8
M   = 8
EPS =      .08
K   = 100
TBB =      50.0
TTB =      180.0
TLB =      150.0
TRB =      0.0
TI  =      100.0

NO. ITERATIONS =    49
```

TEMPERATURE AT GRID PNTS

150.00	180.00	180.00	180.00	180.00	180.00	180.00	170.00	0
150.00	160.13	160.99	158.12	152.56	143.24	126.24	90.35	0
150.00	149.52	145.71	138.91	128.89	114.14	91.39	55.16	0
150.00	142.25	133.40	122.92	109.96	93.02	70.01	38.89	0
150.00	136.08	122.70	109.41	95.02	77.97	56.73	30.41	0
150.00	129.35	111.92	97.01	82.71	67.10	48.54	26.00	0
150.00	119.41	98.63	83.97	71.72	59.20	44.31	25.05	0
150.00	99.65	79.21	68.54	60.98	53.66	44.46	29.88	0
150.00	50.00	50.00	50.00	50.00	50.00	50.00	50.00	0

SUM OF EPS = 6.85E−02

The bottom, top, left-hand, and right-hand boundaries were assigned temperatures of 50°F, 180°F, 150°F, and 0°F, respectively; forty-nine interior nodes were calculated; and a convergence criterion of $\varepsilon \leq 0.08$ was specified. A heat balance check yields

$$\sum q_{in} = 417.77k$$
$$\sum q_{out} = 417.91k$$

which agree within 0.04%.

In Example 4.11 an internal heat source is present. The method of solution will be Gauss elimination rather than Gauss-Seidel iteration, as in the previous example.

Example 4.11

Investigate the effect of the strength of an internal heat source on the steady-state temperature distribution in a plane wall 1 ft thick made of mild steel. One side of the wall is exposed to air at 70°F with the surface conductance equal to 23 Btu/hr-ft²-°F. Furnace gases at 450°F are on the other side with a surface conductance of 125 Btu/hr-ft²-°F.

In this two-dimensional problem the plane steel wall is arbitrarily divided into six subvolumes; thus there are seven locations where the temperatures are desired (two surface nodes and five interior nodes).

A heat balance, written for each node in turn, will yield the system of equations to be solved. The nodal arrangement and quantities of interest are shown in Figure 4.19.

The equations for each node are

node 1: $q_{2-1} + q_c \quad + \dot{u} = 0$

node 2: $q_{1-2} + q_{3-2} + \dot{u} = 0$

node 3: $q_{2-3} + q_{4-3} + \dot{u} = 0$

node 4: $q_{3-4} + q_{5-4} + \dot{u} = 0$

node 5: $q_{4-5} + q_{6-5} + \dot{u} = 0$

node 6: $q_{5-6} + q_{7-6} + \dot{u} = 0$

node 7: $q_{6-7} + q_c \quad + \dot{u} = 0$

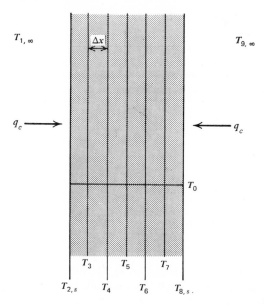

Figure 4.19 Nodal arrangement for plane wall of Example 4.11.

Each of the above seven equations will now be rewritten in terms of temperatures by expressing the q terms appropriately. Considering node 1 in detail, the heat balance expression becomes

$$-\frac{k(T_3 - T_{2,s})(1)}{\Delta x} + h_1(T_{1,\infty} - T_2)(1) + \frac{\dot{u}(\Delta x)(1)}{2} = 0$$

With some rearrangement we obtain the form

$$\frac{h_1 \Delta x}{k} T_{1,\infty} - \left(\frac{1 + h_1 \Delta x}{k}\right) T_{2,s} + T_3 = -\frac{\dot{u}(\Delta x)^2}{2k}$$

In similar fashion the expressions at the remaining six nodes are rewritten. The set of equations so generated follows:

$$c_1 T_{1,\infty} + c_2 T_{2,s} + T_3 \qquad\qquad\qquad\qquad = \frac{c_3}{2}$$

$$T_{2,s} - 2T_3 + T_4 \qquad\qquad\qquad = c_3$$

$$T_3 - 2T_4 + T_5 \qquad\qquad = c_3$$

$$T_4 - 2T_5 + T_6 \qquad\qquad = c_3$$

$$T_5 - 2T_6 + T_7 \qquad\qquad = c_3$$

$$T_6 - 2T_7 + T_{8,s} \qquad = c_3$$

$$T_7 + c_4 T_{8,s} + c_5 T_{9,\infty} = \frac{c_3}{2}$$

where the constants c_n, for $n = 1$ to 5, are

$$c_1 = \frac{h_1 \, \Delta x}{k}$$

$$c_2 = -\left(\frac{1 + h_1 \, \Delta x}{k}\right) = -(1 + c_1)$$

$$c_3 = \frac{-\dot{u} \, \Delta x^2}{k}$$

$$c_4 = -\left(1 + \frac{h_9 \, \Delta x}{k}\right)$$

$$c_5 = \frac{h_9 \, \Delta x}{k} = -(1 + c_4)$$

It is the above set of algebraic equations in T_n which is solved numerically, using Gauss elimination. The flow chart (Figure 4.20), program listing, and computer solution follow.

```
         PROGRAM GELIM
C        THIS PROGRAM DETERMINES THE STEADY STATE TEMP.
C        DIST. IN AN INFINITE PARALLEL SIDED FLAT
C        PLATE WITH INTERNAL ENERGY GENERATION U.  THE
C        SOLUTION IS OBTAINED USING GAUSS ELIMINATION
         DIMENSION A(7,8),T(7)
         N=TTYIN(4HNUMB,4HER O, 4HF NO,4HDES ,3H=  )
         N1=N+1
         H1=TTYIN(4HH1  =,1H )
         T1=TTYIN(4HT1  =,1H )
         H2=TTYIN(4HH2  =,1H )
         T2=TTYIN(4HT2  =,1H )
         U=TTYIN(4H U  =,1H )
         TK=TTYIN(4HTK  =,1H )
         AL=TTYIN(4HTHIC,4HKNES,4HS  = )
         NM1=N-1
         RN=NM1
         DX=AL/RN
         C1=H1*DX/TK
         C2=-(1.+C1)
         C3=-U*DX*DX/TK
         C5=H2*DX/TK
         C4=-(1.+C5)
```

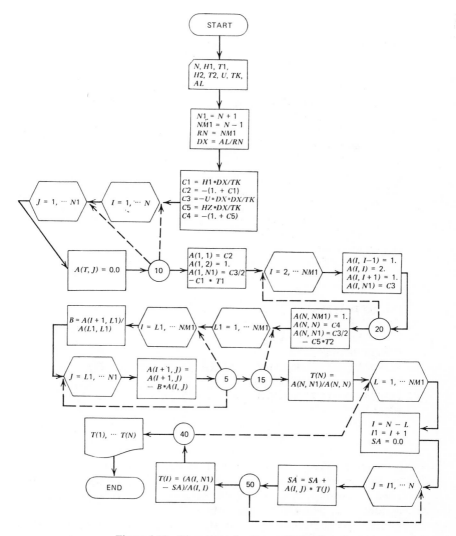

Figure 4.20 Flow chart for Gauss elimination.

```
C         INITIALIZE MATRIX A
          DO 10 I=1,N
          DO 10 J=1,N1
   10 A(I,J)=0.0
C         ASSIGN VALUES TO AUGMENTED MATRIX A
          A(1,1)=C2
          A(1,2)=1.
          A(1,N1)=C3/2.-C1*T1
          DO 20 I=2,NM1
          A(I,I-1)=1.
          A(I,I)=-2.
          A(I,I+1)=1.
   20 A(I,N1)=C3
          A(N,NM1)=1.
          A(N,N)=C4
          A(N,N1)=C3/2.-C5*T2
C         REDUCE MATRIX A
          DO 15 L1=1,NM1
          DO 5 I=L1,NM1
          B=A(I+1,L1)/A(L1,L1)
          DO 5 J=L1,N1
    5 A(I+1,J)=A(I+1,J)-B*A(I,J)
   15 CONTINUE
C         BACK SUBSTITUTION
          T(N)=A(N,N1)/A(N,N)
          DO 40 L=1,NM1
          I=N-L
          I1=I+1
          SA=0.0
          DO 50 J=I1,N
   50 SA=SA+A(I,J)*T(J)
   40 T(I)=(A(I,N1)-SA)/A(I,I)
          WRITE(61,100)
  100 FORMAT(1H ,26HTEMPERATURE AT NODE POINTS)
          WRITE(61,101)(T(I),I=1,N)
  101 FORMAT(1H0,7(3X,F7.1))
          END
```

The program is written so that various node sizes, materials, boundary temperatures, surface conductances, and generation rates may be inserted. In this example the node spacing is 1/6 ft, the thermal conductivity is taken to have the constant value of 21 Btu/hr-ft-°F, the surface conductances and boundary temperatures are those given in the problem statement, and the internal generation rates used are 0; 20,000; 40,000; 60,000; 80,000; and 100,000 Btu/hr-ft^3.

NUMBER OF NODES =   7

H1 =   23.0
T1 =   70.0
H2 =   125.0
T2 =   450.0
 U =   0.0
TK =   21.0
THICKNESS = 1.0

TEMPERATURE AT NODE POINTS

    236.7   267.2   297.6   328.0   358.5   388.9   419.3

NUMBER OF NODES =   7

H1 =   23.0
T1 =   70.0
H2 =   125.0
T2 =   450.0
 U =   20000.0
TK =   21.0
THICKNESS = 1.0

TEMPERATURE AT NODE POINTS

    515.8   584.0   625.7   641.0   629.7   592.1   528.0

NUMBER OF NODES =   7

H1 =   23.0
T1 =   70.0
H2 =   125.0
T2 =   450.0
 U =   40000.0
TK =   21.0
THICKNESS = 1.0

TEMPERATURE AT NODE POINTS

    795.0   900.9   953.8   953.9   901.0   795.3   636.6

NUMBER OF NODES =   7

H1 =   23.0
T1 =   70.0
H2 =   125.0
T2 =   450.0
 U =   60000.0
TK =   21.0
THICKNESS = 1.0

TEMPERATURE AT NODE POINTS

    1074.1   1217.7   1281.9   1266.8   1172.3   998.5   745.2
NUMBER OF NODES =   7
H1 =   23.0
T1 =   70.0
H2 =   125.0
T2 =   450.0
 U =   80000.0
TK =   21.0
THICKNESS = 1.0

TEMPERATURE AT NODE POINTS

    1353.2   1534.5   1610.1   1579.7   1443.6   1201.7   853.9
NUMBER OF NODES =   7
H1 =   23.0
T1 =   70.0
H2 =   125.0
T2 =   450.0
 U =   100000.0
TK =   21.0
THICKNESS = 1.0

TEMPERATURE AT NODE POINTS

    1632.3   1851.4   1938.2   1892.7   1714.9   1404.9   962.5

The results are summarized in Figure 4.21, which shows the temperature profiles generated for each value of $\dot{u}$. One may notice the effect of generation rate on
temperatures at various locations, including each surface; and the manner in which
the maximum interior temperature moves away from the high-temperature gas,
where the surface coefficient is high, toward the side adjacent to the lower-tem-
perature gas, where the surface coefficient is low.

The two numerical examples just discussed were chosen so that the formulation, solution techniques, and types of information given and desired were as
different as possible. The reader should be aware of the variability possible
in solving steady-state heat transfer problems, using the digital computer.

Numerical solutions of some unsteady-state conduction problems will be
considered in Section 4.2-4.

4.2 UNSTEADY-STATE HEAT CONDUCTION

A process designated "unsteady state" or "transient" is one which is time
dependent. Each of the topics and problems considered in Section 4.1 has a
transient counterpart; all steady-state situations have, at one time, undergone
a transient phase. In many cases, the transient phase is a very small portion

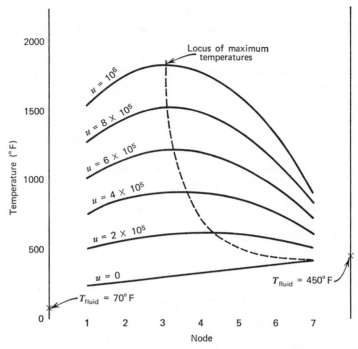

Figure 4.21 Comparison of temperature profiles for different values of internal energy generation.

of the total time a heat transfer process occurs and is thus relatively unimportant. This would be the case in starting up a large plant which, when steady-state conditions are reached, operates with conditions unchanged for very long periods of time, such as weeks or months. In other operations, such as heat treating a metallic casting or vulcanizing a rubber tire, the transient situation is of primary concern and steady-state operation may never occur. Our task in this section is to investigate the time-dependent heat transfer process in a conducting medium and to acquire a facility in analyzing and solving problems where time as well as space coordinates comprise the independent variables.

4.2-1 Transient Heat Conduction in One-Dimensional Systems without Generation

A variant of equation (4-1) applies in the present case, the variation being the addition of a transient term. The governing equation which applies is

$$\frac{1}{x^i}\frac{\partial}{\partial x}\left(x^i\frac{\partial T}{\partial x}\right) = \frac{1}{\alpha}\frac{\partial T}{\partial t} \qquad (4\text{-}83)$$

where i assumes values of 0, 1, and 2 in rectangular, cylindrical, and spherical coordinates, respectively. The parameter α is the *thermal diffusivity* introduced in Chapter 2.

Equation (4-83) is obviously a partial differential equation in contrast to equation (4-1), its steady-state counterpart. Thus, even for this simplest-possible geometry, we are faced with the solution to a partial differential equation. The only exception to this is in those cases where a lumped-parameter analysis is sufficient to describe the physical situation. The lumped-parameter approach will thus be our first consideration.

4.2-1.1 One-Dimensional Transient Conduction; Lumped-Parameter Analysis

Figure 4.22 shows a general "lump" of material which comprises the system of interest. The temperature is taken to be a function of time only, the usual lumped-parameter assumption; thus temperature is uniform throughout the system at any time. Heat is transferred between the system and its surroundings by convection.

By referring to equation (2-14), which is repeated below for reference, we may obtain the appropriate governing equation.

$$\frac{\delta Q}{dt} - \frac{\delta W_s}{dt} - \frac{\delta W_\mu}{dt} = \frac{dE_\sigma}{dt} - \sum_{i=1}^{n}\left(e_i + \frac{p}{\rho}\right)\frac{dM_i}{dt} \qquad (2\text{-}14)$$

For the present situation, each term assumes the following values for the reasons stated:

$\dfrac{\delta W_s}{dt} = 0$ —no shaft work is done

$\dfrac{\delta W_\mu}{dt} = 0$ —no flow occurs within the system; thus no viscous work is done

$\displaystyle\sum_{i=1}^{n}\left(e_i + \frac{P}{\rho}\right)\frac{dM_i}{dt} = 0$ —no mass crosses the control volume boundary

$\dfrac{\delta Q}{dt} = hS(T_\infty - T)$ —heat transfer is by convection from the surroundings to the control volume

$\dfrac{dE_\sigma}{dt} = \rho V c \dfrac{dT}{dt}$

The applicable lumped form of the energy equation is thus

$$\rho V c \frac{dT}{dt} = -hS(T - T_\infty) \qquad (4\text{-}84)$$

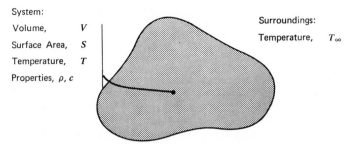

Figure 4.22 A general system for unsteady-state conduction, lumped-parameter analysis.

Letting $\theta(t)$ equal $T(t) - T_\infty$, where T_∞ is constant, we obtain

$$\rho V c \frac{d\theta}{dt} = -hS\theta \qquad (4\text{-}85)$$

A rearrangement of terms and separation of variables yields

$$\frac{d\theta}{\theta} = -\frac{hS}{\rho V c}\,dt$$

which, when integrated from the initial condition $\theta(0) = \theta_0$ to the general condition at time t, gives

$$\int_{\theta_0}^{\theta} \frac{d\theta}{\theta} = -\frac{hS}{\rho V c}\int_0^t dt$$

$$\ln\frac{\theta}{\theta_0} = \ln\frac{T - T_\infty}{T_0 - T_\infty} = -\frac{hS}{\rho V c}\,t \qquad (4\text{-}86)$$

or

$$\frac{\theta}{\theta_0} = \frac{T - T_\infty}{T_0 - T_\infty} = \exp\left(-\frac{hS}{\rho V c}\,t\right) \qquad (4\text{-}87)$$

Equation (4-87) represents a general response of a system with uniform temperature $T(t)$ to a convective heat exchange with its surroundings. The dimensionless argument of the exponential may be arranged in different forms as follows:

$$\frac{hS}{\rho V c}\,t = \left(\frac{hV}{kS}\right)\left(\frac{S^2 k}{\rho V^2 c}\,t\right)$$

$$= \left(\frac{h\,V/S}{k}\right)\left[\frac{\alpha t}{(V/S)^2}\right] \qquad (4\text{-}88)$$

Each of the bracketed terms on the right of equation (4-88) is seen to be nondimensional. Each involves the ratio V/S, which has units of length. The first term is designated the *Biot modulus*, abbreviated Bi.

$$Bi \equiv \frac{h\ V/S}{k} \qquad (4\text{-}89)$$

By analogy with the concepts of thermal resistance discussed both in Chapter 1 and in Section 4.1-1, the Biot modulus is seen to be the ratio of $(V/S)/k$, the conductive (internal) resistance to heat transfer, to $1/h$, the convective (external) resistance to heat transfer. The magnitude of the Biot modulus thus has some physical significance in relating where the greater resistance to heat transfer occurs. A large value for Bi indicates that the conductive resistance controls, i.e., there is more capacity for heat transfer to the surface by convection than for the transfer of heat away from the surface by conduction. A small value for Bi is representative of the case of negligible internal resistance, where greater capacity exists to transfer heat by conduction than by convection. In this latter case, the controlling phenomenon is convection, and temperature gradients within the system are relatively small;

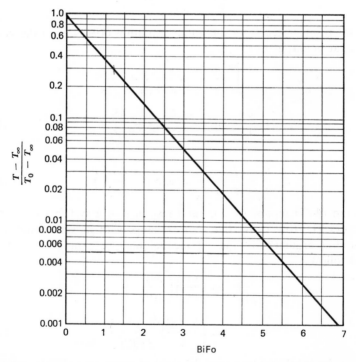

Figure 4.23   Temperature variation in a lumped-parameter system without internal energy generation.

thus this is the case of a uniform temperature, which is the basic assumption in a lumped-parameter analysis.

In light of the foregoing discussion, it is reasonable that the magnitude of the Biot modulus determines whether or not a lumped-parameter analysis of a transient heat conduction problem is sufficiently accurate. A commonly used rule of thumb is that the error inherent in a lumped-parameter analysis will be less than 5% for a value of Bi less than 0.1. The evaluation of the Biot modulus should thus be the initial step in solving any transient heat conduction problem.

The second bracketed term on the right-hand side of equation (4-88) is designated the *Fourier modulus*, and abbreviated Fo.

$$\text{Fo} \equiv \frac{\alpha t}{(V/S)^2} \qquad (4\text{-}90)$$

The Fourier modulus representation given in equation (4-90) is a common way of nondimensionalizing the time variable. A convenient way of stating the lumped-parameter solution for transient conduction is

$$\frac{\theta}{\theta_0} = \frac{T - T_\infty}{T_0 - T_\infty} = e^{-\text{BiFo}} \qquad (4\text{-}91)$$

Figure 4.23 is a graphical portrayal of equation (4-91).

The use of equation (4-91) is illustrated in Example 4.12.

Example 4.12

A cylindrical stainless steel ingot, 4 in. in diameter and 1 ft long, passes through a heat treating furnace which is 20 ft in length. The initial ingot temperature is 200°F, and it must reach 1500°F in preparation for working. The furnace gas is at 2300°F, and the combined radiant and convective surface coefficient is 18 Btu/hr-ft^2-°F. In order that the required conditions be satisfied, what must be the maximum speed with which the ingot moves through the furnace?

Initially, the Biot modulus is calculated to be

$$\text{Bi} = \frac{h\,V/S}{k}$$

$$= \frac{h}{k}\left[\frac{D^2}{4}L \Big/ \left(DL + 2\frac{D^2}{4}\right)\right]$$

$$= \frac{h}{k}\left(\frac{DL/4}{L + D/2}\right)$$

$$= \frac{18\ \text{Btu/hr-ft}^2\text{-}°F}{13\ \text{Btu/hr-ft-}°F}\left[\frac{(4/12\ \text{ft})(1\ \text{ft})/4}{1\ \text{ft} + 1/2(4/12\ \text{ft})}\right]$$

$$= \frac{18}{13}\left(\frac{1}{14}\right) = 0.099$$

With this value for Bi, a lumped-parameter analysis will yield an error of no more than 5%. We will now proceed to use equation (4-91), calculating values for

$$\frac{\theta}{\theta_0} = \frac{T - T_\infty}{T_0 - T_\infty} = \frac{1500 - 2300}{200 - 2300} = 0.381$$

and

$$\text{Fo} = \frac{\alpha t}{(V/S)^2} = \frac{(0.17 \text{ ft}^2/\text{hr})(t, \text{hr})}{(1/14 \text{ ft})^2}$$

$$= 33.3t$$

which, when substituted, gives

$$0.381 = e^{-(0.099)(33.3t)}$$

Solving for t, we obtain $t = 0.2925 \text{ hr} = 17.55 \text{ min}$

The required ingot velocity thus becomes

$$v = \frac{20 \text{ ft}}{17.55 \text{ min}} = 1.14 \text{ ft/min}$$

In Example 4.13 the situation including internal generation of thermal energy is examined.

Example 4.13

The soleplate of a household iron has a surface area of 0.5 ft² and is made of stainless steel with a total weight of 3 lb. With a surface coefficient of 3 Btu/hr-ft²-°F between the iron and its surroundings at 80°F, how long after being turned on will the iron reach 240°F? The iron is rated at 500 watts and is originally at the temperature of the surroundings.

A first step will be the evaluation of the Biot modulus:

$$\text{Bi} = \frac{h\,V/S}{k}$$

$$= \frac{(3 \text{ Btu/hr-ft}^2\text{-}°F)}{(13 \text{ Btu/hr-ft-}°F)} \left[\frac{3 \text{ lb}/488 \text{ lb/ft}^3}{0.5 \text{ ft}^2} \right]$$

$$= 0.00284$$

The consideration of the soleplate as a lumped system will introduce very little error for a value of Bi this small.

A term-by-term evaluation of equation (2-14) gives the following:

$$\frac{\delta W_s}{dt} = \frac{\delta W_\mu}{dt} = \sum_{i=1}^n \left(e_i + \frac{P_i}{\rho} \right) \frac{dM_i}{dt} = 0$$

$$\frac{\delta Q}{dt} = hS(T_\infty - T) + \dot{q}V$$

$$\frac{dE_\sigma}{dt} = \rho Vc \frac{dT}{dt}$$

Letting $\theta = T - T_\infty$, we have

$$\rho c V \frac{d\theta}{dt} = \dot{q} V - h S \theta$$

A rearrangement yields

$$\frac{d\theta}{dt} = a - b\theta$$

where $a = \dot{q}/\rho c$ and $b = hS/\rho c V$
Separating variables and integrating, we obtain

$$\int_0^\theta \frac{-b\,d\theta}{a - b\theta} = -b \int_0^t dt$$

$$\ln \frac{a - b\theta}{a} = -bt$$

or

$$t = \frac{1}{b} \ln \frac{1}{1 - \frac{b}{a}\theta}$$

The constants are evaluated as follows:

$$a = \frac{(500 \text{ w})(3.413 \text{ Btu/w-hr})}{(3 \text{ lb})(0.11 \text{ Btu/lb-°F})} = 5170°\text{F/hr}$$

$$b = \frac{(3 \text{ Btu/hr-ft}^2\text{-°F})(0.5 \text{ ft}^2)}{(3 \text{ lb})(0.11 \text{ Btu/lb-°F})} = 4.54 \text{ hr}^{-1}$$

Substituting these values into the expression for t and solving yields

$$t = \frac{\text{hr}}{4.54} \ln \frac{1}{1 - \frac{4.54 \text{ hr}^{-1}}{5170°\text{F/hr}} (160°\text{F})}$$

$$= \frac{1}{4.54} \ln \frac{1}{0.859} = 0.0247 \text{ hr} = 1.48 \text{ min}$$

4.2-1.2 One-Dimensional Transient Conduction; Systems with Bi > 0.1

When the applicable value of the Biot modulus is greater than 0.1, a lumped-parameter analysis will introduce more than 5% error, which is generally unacceptable. In cases such as these, equation (4-83) applies and one must solve the partial differential equation. The one-dimensional configurations of interest remain the plane wall, the cylinder, and the sphere. Each of these shapes will be considered in the following sections.

Transient Conduction in an Infinite Plane Wall. The case to be considered is shown in Figure 4.24. The wall is of thickness $2L$ and extends to infinity

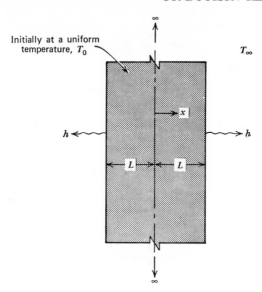

Figure 4.24 Analysis of transient heat conduction in an infinite plane wall.

in the y and z directions. The space variable x is measured from the center-line. Initially, the wall is at uniform temperature T_0; then the adjacent medium is raised to and maintained at temperature T_∞. The convective surface coefficient is h. A complete statement of this transient problem, including the governing differential equation and the initial and boundary conditions, follows:

governing differential equation

$$\frac{1}{\alpha}\frac{\partial T}{\partial t} = \frac{\partial^2 T}{\partial x^2} \tag{4-92}$$

initial condition

$$T(x, 0) = T_0 \tag{4-93}$$

boundary conditions

$$\frac{\partial T}{\partial x}(0, t) = 0 \tag{4-94}$$

$$-k\frac{\partial T}{\partial x}(L, t) = h[T(L, t) - T_\infty] \tag{4-95}$$

Equations (4-94) and (4-95) express, respectively, the symmetrical nature of the temperature profile within the plane wall at any time t, and the fact that conductive heat transfer is equal to the convective heat transfer at the wall surface.

The problem will lend itself to solution more simply if the dependent variable is transformed to $\theta(x, t)$, where

$$\theta(x, t) \equiv T(x, t) - T_\infty \tag{4-96}$$

The complete problem, in terms of $\theta(x, t)$, may now be expressed as

$$\frac{1}{\alpha} \frac{\partial \theta}{\partial t} = \frac{\partial^2 \theta}{\partial x^2} \tag{4-97}$$

$$\theta(x, 0) = \theta_0 \tag{4-98}$$

$$\frac{\partial \theta}{\partial x}(0, t) = 0 \tag{4-99}$$

$$-k \frac{\partial \theta}{\partial x}(L, t) = h\theta(L, t) \tag{4-100}$$

The problem will be solved by our usual separation-of-variables technique letting

$$\theta(x, t) = X(x)\tau(t) \tag{4-101}$$

the variables $X(x)$ and $\tau(t)$ being functions of the single parameters x and t, respectively.

Substitution of equation (4-101) into equation (4-96) yields

$$\frac{1}{\alpha} X \frac{d\tau}{dt} = \tau \frac{d^2X}{dx^2}$$

Dividing both sides of this expression by $X\tau$, we obtain

$$\frac{1}{\alpha} \frac{1}{\tau} \frac{d\tau}{dt} = \frac{1}{X} \frac{d^2X}{dx^2} \tag{4-102}$$

Each side of equation (4-102) is a function of one independent variable only, thus both must equal some constant which we shall call λ^2. Letting each side in turn equal λ^2, we have the two ordinary differential equations

$$\frac{d\tau}{dt} = \alpha\lambda^2\tau \tag{4-103}$$

and

$$\frac{d^2X}{dx^2} = \lambda^2 x \tag{4-104}$$

The general solution to equation (4-103) is

$$\tau = Ae^{\alpha\lambda^2 t}$$

which indicates τ to be a rapidly increasing function of t, approaching infinity for large values of t. Physically, of course, this is an impossible situation; the constant λ^2 thus must have a negative sign, and the ordinary differential equations to be solved are

$$\frac{d\tau}{dt} = -\alpha\lambda^2\tau \qquad (4\text{-}105)$$

and

$$\frac{d^2X}{dx^2} = -\lambda^2 X \qquad (4\text{-}106)$$

The general solutions to equations (4-105) and (4-106) are

$$\tau = Ae^{-\alpha\lambda^2 t} \qquad (4\text{-}107)$$

and

$$X = B \cos \lambda x + C \sin \lambda x \qquad (4\text{-}108)$$

The complete expression for θ now becomes

$$\theta(x, t) = X(x)\tau(t) = e^{-\alpha\lambda^2 t}[M \cos \lambda x + N \sin \lambda x] \qquad (4\text{-}109)$$

Note that there are two integration constants, $M = AB$ and $N = AC$.

Applying the first boundary condition, that of temperature symmetry about the centerline, expressed by equation (4-99), we have

$$\frac{\partial\theta}{\partial x}(0, t) = (e^{-\alpha\lambda^2 t})(\lambda[-M \sin 0 + N \cos 0]) = 0$$

For this expression to be true, the constant N must be zero. The remaining expression for $\theta(x, t)$ is

$$\theta(x, t) = Me^{-\alpha\lambda^2 t} \cos \lambda x \qquad (4\text{-}110)$$

Substituting our θ expression, equation (4-110), into the wall surface boundary condition, equation (4-100), we have

$$+k\lambda Me^{-\alpha\lambda^2 t} \sin \lambda L = hMe^{-\alpha\lambda^2 t} \cos \lambda L$$

which, upon canceling like terms, becomes

$$\tan \lambda L = \frac{h}{k\lambda} \qquad (4\text{-}111)$$

If the right side of this equation is multiplied and divided by L, the ratio $L/k\lambda$ becomes

$$\frac{hL}{k}\frac{1}{\lambda L} \equiv \frac{\text{Bi}}{\lambda L}$$

where Bi is the Biot modulus introduced in Section 4.2-1.1 and defined by equation (4-89). The transcendental equation may now be written as

$$\tan \lambda L = \frac{\text{Bi}}{\lambda L} \qquad (4\text{-}112)$$

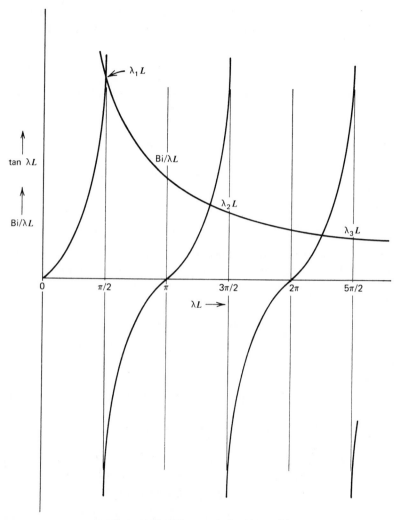

Figure 4.25 Characteristic values of λ.

This equation determines the appropriate values of λ which constitute a solution. At each intersection in Figure 4.25, a value of λ is obtained which satisfies equation (4-111) and is thus a value which provides a solution to our equation.

Each value of λ satisfying equation (4-111) will be denoted $\lambda_1, \lambda_2, \ldots, \lambda_n$, where $1 \leq n < \infty$.

Our solution is now of the form

$$\theta_n(x, t) = M_n e^{-\alpha \lambda_n^2 t} \cos \lambda_n x \qquad (4\text{-}113)$$

where an infinite number of particular solutions exist—one for each value of n from 1 to ∞. The complete solution for θ is given by the sum

$$\theta(x, t) = \sum_{n=1}^{\infty} M_n e^{-\alpha \lambda_n^2 t} \cos \lambda_n x \tag{4-114}$$

To complete the solution, M_n must be determined; this is done by applying the initial condition, equation (4-98).

$$\theta(x, 0) = \theta_0 = \sum_{n=1}^{\infty} M_n \cos \lambda_n x \tag{4-115}$$

Multiplying both sides of equation (4-115) by $\cos \lambda_m x \, dx$ and integrating over the interval $0 \leq x \leq L$, we have

$$\theta_0 \int_0^L \cos \lambda_m x \, dx = \int_0^L \sum_{n=1}^{\infty} M_n \cos \lambda_n x \cos \lambda_m x \, dx$$

We may now dispose of the summation sign since the right-hand side of this expression vanishes for values of $m \neq n$. Utilizing this condition and solving for the constant M_n, we have

$$M_n = \theta_0 \frac{\displaystyle\int_0^L \cos \lambda_n x \, dx}{\displaystyle\int_0^L \cos \lambda_n^2 x \, dx}$$

$$= \theta_0 \left(\frac{2 \sin \lambda_n L}{\lambda_n L + \sin \lambda_n L \cos \lambda_n L} \right) \tag{4-116}$$

The final solution for $\theta(x, t)$ may now be written as

$$\frac{\theta(x, t)}{\theta_0} = 2 \sum_{n=1}^{\infty} e^{-\alpha \lambda_n^2 t} \frac{\sin \lambda_n L \cos \lambda_n x}{\lambda_n L + \sin \lambda_n L \cos \lambda_n L} \tag{4-117}$$

or, letting $\delta_n = \lambda_n L$, we may write equation (4-117) as

$$\frac{\theta(x, t)}{\theta_0} = 2 \sum_{n=1}^{\infty} e^{-\delta_n^2 (\alpha t / L^2)} \frac{\sin \delta_n \cos \delta_n x / L}{\delta_n + \sin \delta_n \cos \delta_n} \tag{4-118}$$

Notice that, according to equation (4-112), δ_n is a function of Bi, the Biot modulus

$$\delta_n \tan \delta_n = \text{Bi} \tag{4-119}$$

and the bracketed term in the exponential, $\alpha t / L^2$, equals Fo, the Fourier modulus. Thus, according to equation (4-118), the temperature at a given time and position in a plane wall is a function of Bi, Fo, and a nondimensional position parameter x/L.

The solution given by equation (4-118) has widespread application; it has been calculated accurately and presented in graphical form (see Appendix B). Figures B.1 and B.4 show values of dimensionless temperature as a function of Bi and Fo for values of $x/L = 0$, 0.2, 0.4, 0.6, 0.8, and 1.0. Figure B.7 portrays the dimensionless temperature at the center of a plane wall as a function of the parameters $\alpha t/L^2$ and k/hL.

One-Dimensional Transient Conduction in Infinite Cylinders. For the case of a cylinder of radius R, extending to infinity in the $\pm z$ directions, with uniform initial temperature T_0, whose surroundings are suddenly raised to and maintained at temperature T_∞, the solution obtained through an analytical procedure exactly as in the previous section is

$$\frac{\theta(r, t)}{\theta_0} = \frac{T(r, t) - T_\infty}{T_0 - T_\infty} = 2 \sum_{n=1}^{\infty} \frac{e^{-\delta_n^2(\alpha t/R^2)}}{\delta_n} \frac{J_1(\delta_n)J_0(\delta_n r/R)}{J_0^2(\delta_n) + J_1^2(\delta_n)} \quad (4\text{-}120)$$

where δ_n is a function of the dimensionless ratio hR/k, according to

$$\delta_n \frac{J_1(\delta_n)}{J_0(\delta_n)} = \frac{hR}{k} \quad (4\text{-}121)$$

and $J_0(\delta_n)$ and $J_1(\delta_n)$ are Bessel functions of the first kind (zero and first order, respectively) of the argument δ_n.

Note that the ratio hR/k is of the form of a Biot modulus, but is not equal to Bi, since the radius R of a cylinder is not the ratio of volume to surface area.

Equation (4-120) relates temperature to the nondimensional parameters hR/k, $\alpha t/R^2$, and r/R. Figures B.2 and B.2 in Appendix B are graphical representations of this equation. The central temperature is shown in dimensionless fashion as a function of the parameters $\alpha t/R^2$ and k/hR in Figure B.8.

One-Dimensional Transient Conduction in a Sphere. The solution for spherical geometry to the same problem considered in plane and cylindrical cases is

$$\frac{\theta(r, t)}{\theta_0} = \frac{T(r, t) - T_\infty}{T_0 - T_\infty} = 4 \frac{R}{r} \sum_{n=1}^{\infty} e^{-\delta_n^2(\alpha t/R^2)} \sin \delta_n \frac{r}{R} \frac{\sin \delta_n - \delta_n \cos \delta_n}{2\delta_n - \sin 2\delta_n}$$

$$(4\text{-}122)$$

where the δ_n are roots of the characteristic equation

$$1 - \delta_n \cot \delta_n = \frac{hR}{k} \quad (4\text{-}123)$$

Solutions of equation (4-122) are portrayed in Figures B.3 and B.6 in Appendix B, with the central temperature history shown in dimensionless form in Figure B.9.

4.2-2 Two- and Three-Dimensional Transient Conduction

Figure 4.26 shows the cross section of a rectangular bar of width $2L$ and height $2l$. The bar is at a uniform initial temperature T_0, and loses heat by convection on all sides to surroundings at T_∞, with a uniform surface conductance h prevailing at all surfaces.

The problem of expressing the temperature in the rectangular bar as a function of x, y, and t parallels the one-dimensional transient conduction analysis described in the previous three sections. The dependent variable $T(x, y, t)$ is first transformed to $\theta(x, y, t)$, according to

$$\theta(x, y, t) \equiv \frac{T(x, y, t) - T_\infty}{T_0 - T_\infty} \tag{4-124}$$

The applicable differential equation is

$$\frac{\partial \theta}{\partial t} = \alpha \left(\frac{\partial^2 \theta}{\partial x^2} + \frac{\partial^2 \theta}{\partial y^2} \right) \tag{4-125}$$

and the initial and boundary conditions are, respectively,

$$\theta(x, y, 0) = 1 \tag{4-126}$$

$$\frac{\partial \theta}{\partial x}(0, y, t) = 0 \tag{4-127}$$

$$\frac{\partial \theta}{\partial y}(x, 0, t) = 0 \tag{4-128}$$

$$\theta(L, y, t) + \frac{k}{h} \frac{\partial \theta}{\partial x}(L, y, t) = 0 \tag{4-129}$$

$$\theta(x, l, t) + \frac{k}{h} \frac{\partial \theta}{\partial y}(x, l, t) = 0 \tag{4-130}$$

We now perform a partial separation of variables on the variable $\theta(x, y, t)$, forming the product solution

$$\theta(x, y, t) = X(x, t) Y(y, t) \tag{4-131}$$

where the variables $X(x, t)$ and $Y(y, t)$ are functions of one space variable and time.

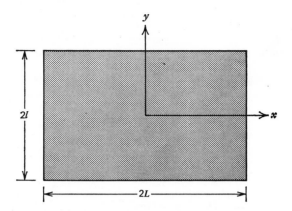

Figure 4.26 Rectangular cross section for two-dimensional transient conduction analysis.

Substituting this product into equation (4-125), we obtain

$$X \frac{\partial Y}{\partial t} + Y \frac{\partial X}{\partial t} = \alpha \left[Y \frac{\partial^2 X}{\partial x^2} + X \frac{\partial^2 Y}{\partial y^2} \right]$$

We now divide through by the product XY and separate terms in X and Y to achieve the expression

$$\frac{1}{X} \left[\frac{\partial X}{\partial t} - \alpha \frac{\partial^2 X}{\partial x^2} \right] = -\frac{1}{Y} \left[\frac{\partial Y}{\partial t} - \alpha \frac{\partial^2 Y}{\partial y^2} \right] \qquad (4\text{-}132)$$

Each side of equation (4-132) is a function of only one space variable, thus both must equal a constant. If this constant is either positive or negative, the solutions for $X(x, t)$ and $Y(y, t)$ will have different functional forms. This is, of course, not consistent physically, so the constant must be zero.

Setting both sides of equation (4-132) equal to zero, according to the argument just presented, our two-dimensional transient conduction problem reduces to two one-dimensional problems whose formulations are
in the x direction

$$\frac{\partial X}{\partial t} = \alpha \frac{\partial^2 X}{\partial x^2}$$

$$X(x, 0) = 1$$

$$\frac{\partial X}{\partial x}(0, t) = 0$$

$$X(L, t) + \frac{k}{h} \frac{\partial X}{\partial x}(L, t) = 0$$

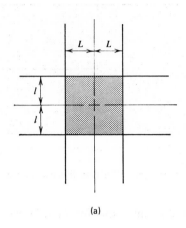

(a)

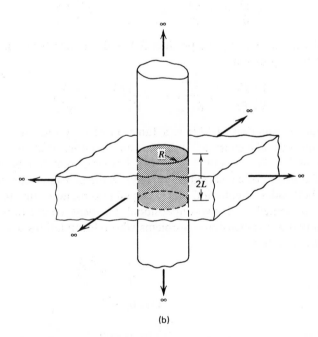

(b)

Figure 4.27 Two-dimensional combinations of one-dimensional forms for transient conduction analysis. (a) A two-dimensional rectangular element. (b) A cylinder of finite length.

in the y direction

$$\frac{\partial Y}{\partial t} = \alpha \frac{\partial^2 Y}{\partial y^2}$$

$$Y(y, 0) = 1$$

$$\frac{\partial Y}{\partial y}(0, t) = 0$$

$$Y(l, t) + \frac{k}{h}\frac{\partial Y}{\partial y}(l, t) = 0$$

Each of these one-dimensional transient problems is seen to be identical with the case presented at length in Section 4.2-1.2 for the infinite plane wall, and whose solution is presented graphically in Figures B.1, B.4, and B.7. The variables $X(x, t)$ and $Y(y, t)$ are dimensionless temperatures given by

$$X = \frac{T(x, t) - T_\infty}{T_0 - T_\infty} \tag{4-133}$$

and

$$Y = \frac{T(y, t) - T_\infty}{T_0 - T_\infty} \tag{4-134}$$

thus the temperature at location (x, y) at time t is given, according to equation (4-131), by

$$\theta(x, y, t) = \frac{T(x, y, t) - T_\infty}{T_0 - T_\infty} = X(x, t)Y(y, t)$$

with values for both $X(x, t)$ and $Y(y, t)$ available in the figures cited above.

The development and discussion presented thus far for a two-dimensional rectangular geometry may be extended to include a finite cylinder or a three-dimensional rectangular solid. Figure 4.27 illustrates the combinations of one-dimensional forms which yield two-dimensional objects of interest.

The use of the charts for solving problems in one or more dimensions will be illustrated in the following example problems.

Example 4.14

Fire clay brick is initially at a uniform temperature of 100°F, and is then exposed to hot gas at 1200°F with a convective coefficient of 4 Btu/hr-ft²-°F applying at all surfaces. After 20 hours' exposure to the high-temperature gas under these conditions, find the center temperature of (a) a 2-foot-thick infinite plane wall, (b) a very long square column measuring 2 ft by 2 ft in cross section, and (c) a 2 ft by 2 ft by 2 ft cubical block with one face lying on an insulated surface.

It is first necessary to calculate the Biot modulus. The values are

(a) $\mathrm{Bi} = \dfrac{h(V/S)}{k} = \dfrac{(4\ \mathrm{Btu/hr\text{-}ft^2\text{-}{}^\circ F})(1\ \mathrm{ft})}{0.65\ \mathrm{Btu/hr\text{-}ft\text{-}{}^\circ F}} = 6.15$

(b) $\mathrm{Bi} = \dfrac{(4)(2 \times 2 \times H)}{(0.65)(4 \times 2 \times H)} = 3.075$

(c) $\mathrm{Bi} = \dfrac{(4)(2 \times 2 \times 2)}{(0.65)(5 \times 2 \times 2)} = 2.46$

Each value for Bi is sufficiently large that a lumped-parameter solution is not valid. The charts in Appendix B apply and will be used.

For (a), the following parametric values apply:

$$\frac{k}{hV/S} = \frac{1}{6.15} = 0.163$$

$$\frac{\alpha t}{(V/S)^2} = \frac{(0.02\ \mathrm{ft^2/hr})(20\ \mathrm{hr})}{1\ \mathrm{ft^2}} = 0.4$$

$$\frac{x}{L} = 0$$

Reading from the charts for an infinite plane, we obtain

$$\frac{T - T_\infty}{T_0 - T_\infty} = 0.60$$

and $T = 1200 - 0.60(1100) = 540^\circ\mathrm{F}$.

For (b), the same values of k/hL, $\alpha t/L^2$ apply as in (a). For a plane which is 2 ft thick in the x-direction, a value of 0.60 is read for $(T - T_\infty)/(T_0 - T_\infty)$. A plane which is 2 ft thick in the y direction will also have a value of $(T - T_\infty)/(T_0 - T_\infty) = 0.60$. For a rectangular cross section measuring 2 ft by 2 ft, the value of $(T - T_\infty)/(T_0 - T_\infty)$ which applies is the product of the two values stated above. Thus, for the two-dimensional case,

$$\frac{T - T_\infty}{T_0 - T_\infty} = \left(\frac{T - T_\infty}{T_0 - T_\infty}\right)_x \left(\frac{T - T_\infty}{T_0 - T_\infty}\right)_y = (0.60)(0.60) = 0.36$$

from which $T = 1200 - 0.36(1100) = 804^\circ\mathrm{F}$.

For (c), we extend the results given in (a) and (b). In (a) convection occurred on two surfaces, in (b) four surfaces were involved, and in (c) five surfaces are to be considered. Values of $(T - T_\infty)/(T_0 - T_\infty)$ obtained for the x and y directions in (b) still apply. In the z direction convection occurs from one surface only, the other being insulated. This is equivalent to convection from both surfaces of a slab 4 ft thick. The parameters in the z direction thus have the following values:

$$\frac{k}{hL} = \frac{(0.65 \text{ Btu/hr-ft-}^\circ\text{F})}{(4 \text{ Btu/hr-ft}^2\text{-}^\circ\text{F})(2 \text{ ft})} = 0.0813$$

$$\frac{\alpha t}{L^2} = \frac{(0.02 \text{ ft}^2/\text{hr})(20 \text{ hr})}{(2 \text{ ft})^2} = 0.10$$

$$\frac{x}{L} = 0$$

From the chart, for these conditions,

$$\frac{T - T_\infty}{T_0 - T_\infty} = 0.98$$

Now, for the cubical solid,

$$\frac{T - T_\infty}{T_0 - T_\infty} = \left(\frac{T - T_\infty}{T_0 - T_\infty}\right)_x \left(\frac{T - T_\infty}{T_0 - T_\infty}\right)_y \left(\frac{T - T_\infty}{T_0 - T_\infty}\right)_z$$

$$= (0.60)(0.60)(0.98) = 0.353$$

yielding, for T,

$$T = 1200 - 0.353(1100) = 814^\circ\text{F}$$

Example 4.15

An asbestos cylinder 5 inches in diameter initially at the uniform temperature of 100°F is placed in a medium at 1200°F with a convective coefficient of 4 Btu/hr-ft^2-$^\circ$F applying at all surfaces. Find the time required for the center to reach 500°F if the cylinder is (a) very long (no end effects) and (b) 2 in. tall with one base sitting on an insulating surface.

The Biot modulus calculation yields, for (a),

$$\text{Bi} = \frac{hV/S}{k} = \frac{(4 \text{ Btu/hr-ft}^2\text{-}^\circ\text{F})\left(\dfrac{\pi D^2}{4} L\right)}{(0.125 \text{ Btu/hr-ft-}^\circ\text{F})(\pi D L)} = 3.33$$

and, for (b),

$$\text{Bi} = \frac{4\left(\dfrac{\pi D^2}{4} L\right)}{0.125\left(\pi D L + \dfrac{\pi D^2}{4}\right)} = 2.67$$

For both parts of this example, a lumped-parameter solution is invalid and the charts must be used.

For (a),

$$\frac{k}{hR} = \frac{0.125}{(4)(2.5/12)} = 0.15$$

$$\frac{\alpha t}{R^2} = \frac{(0.125 \text{ Btu/hr-ft-}^\circ\text{F})t}{\left(\dfrac{36 \text{ lb}_m}{\text{ft}^3}\right)\left(0.25 \dfrac{\text{Btu}}{\text{lb}_m\text{-}^\circ\text{F}}\right)\left(\dfrac{2.5}{12} \text{ ft}\right)^2}$$

$$= 0.32t$$

$$\frac{x}{L} = 0$$

$$\frac{T - T_\infty}{T_0 - T_\infty} = \frac{500 - 1200}{100 - 1200} = 0.637$$

From the chart for an infinite cylinder we read $\alpha t / R^2 = 0.20$, and the required time is

$$t = \frac{0.20}{0.32} = 0.625 \text{ hr} = 37.5 \text{ min}$$

For (b), the finite cylinder is a combination of an infinite cylinder and an infinite plane. Since convection occurs from only one of the bases of the cylinder and the other base is insulated, the equivalent infinite plane is 4 in. thick.

For the cylinder, the following parametric values apply:

$$\frac{k}{hR} = 0.15$$

$$\frac{\alpha t}{R^2} = 0.32t$$

$$\frac{x}{R} = 0$$

$$\frac{T - T_\infty}{T_0 - T_\infty} = \left(\frac{T - T_\infty}{T_0 - T_\infty}\right)_{\text{cyl}}$$

For the equivalent plane,

$$\frac{k}{hL} = \frac{0.125}{(4)(2/12)} = 0.1875$$

$$\frac{\alpha t}{L^2} = \frac{0.125t}{(36)(0.25)(2/12)^2} = 0.5t$$

$$\frac{x}{L} = 0$$

$$\frac{T - T_\infty}{T_0 - T_\infty} = \left(\frac{T - T_\infty}{T_0 - T_\infty}\right)_{\text{plane}}$$

For the finite cylinder, $(T - T_\infty)/(T_0 - T_\infty) = 0.637$; thus

$$\left(\frac{T - T_\infty}{T_0 - T_\infty}\right)_{\text{cyl}} \left(\frac{T - T_\infty}{T_0 - T_\infty}\right)_{\text{plane}} = 0.637$$

The problem may now be solved by a trial-and-error process involving the charts for infinite planes and cylinders, using the following procedure:

1. Assume t

2. Calculate $\dfrac{\alpha t}{R^2}$ and $\dfrac{\alpha t}{L^2}$

3. Read $\left(\dfrac{T - T_\infty}{T_0 - T_\infty}\right)_{\text{cyl}}$ and $\left(\dfrac{T - T_\infty}{T_0 - T_\infty}\right)_{\text{plane}}$

4. Calculate $\left(\dfrac{T - T_\infty}{T_0 - T_\infty}\right)_{\text{cyl}} \left(\dfrac{T - T_\infty}{T_0 - T_\infty}\right)_{\text{plane}}$

5. Continue until $\left(\dfrac{T - T_\infty}{T_0 - T_\infty}\right)_{\text{cyl}} \left(\dfrac{T - T_\infty}{T_0 - T_\infty}\right)_{\text{plane}} = 0.637$

For this problem, the time which satisfies the above algorithm is

$$t = 0.45 \text{ hr} = 27 \text{ min}$$

4.2-3 One-Dimensional Transient Conduction in a Semi-Infinite Wall

In many situations of engineering interest the conducting medium is sufficiently thick that the change in conditions at one boundary defines the transient problem; no other boundary conditions affect $T(x, t)$. In such a case, the medium thickness is treated as infinite.

The situation of interest is illustrated in Figure 4.28. A large plane wall initially at the uniform temperature T_0 has its surface suddenly raised to and maintained at a new temperature T_s. The differential equation to be solved is

$$\frac{1}{\alpha} \frac{\partial T}{\partial t} = \frac{\partial^2 T}{\partial x^2} \qquad (4\text{-}135)$$

subject to the initial and boundary conditions

$$T(x, 0) = T_0 \qquad (4\text{-}136)$$

$$T(0, t) = T_s \qquad (4\text{-}137)$$

$$T(\infty, t) = T_0 \qquad (4\text{-}138)$$

The problem is classic in the heat transfer literature. Techniques of solving it vary considerably; among the more usual approaches are the Laplace and Fourier transformations. In the sections to follow we shall use an analytical technique and an integral approach to this problem.

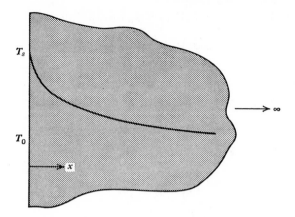

Figure 4.28 Temperature distribution in a semi-infinite wall at time t.

4.2-3.1   Semi-Infinite Wall Analysis; Analytical Solution

From the forms of solutions in Section 4.2-2 we may express the variable s involved in the semi-infinite wall problem in the nondimensional form

$$\frac{T - T_0}{T_s - T_0} = f\left(\frac{x}{L}, \frac{\alpha t}{L^2}\right) \tag{4-139}$$

In the previous section there was a characteristic dimension L for the geometry being considered. In the present case, no such dimension exists; thus we eliminate L between the two parameters in the function on the right-hand side of equation (4-139) to express the temperature as a function of one independent parameter. Our choices for this parameter are

$$\frac{T - T_0}{T_s - T_0} = f\left(\frac{\alpha t}{x^2}\right)$$

or

$$\frac{T - T_0}{T_s - T_0} = f\left(\frac{x}{\sqrt{\alpha t}}\right)$$

We shall express the variables in the form

$$Y = f(\eta)$$

where

$$Y = \frac{T - T_0}{T_s - T_0} \tag{4-140}$$

and

$$\eta = \frac{x}{2\sqrt{\alpha t}} \qquad (4\text{-}141)$$

Substitution of these parameters into equation (4-135) yields the ordinary differential equation

$$\frac{d^2Y}{d\eta^2} + 2\eta \frac{dY}{d\eta} = 0 \qquad (4\text{-}142)$$

and the boundary and initial conditions

$$Y(0) = 1 \qquad (4\text{-}143)$$

$$Y(\infty) = 0 \qquad (4\text{-}144)$$

Equation (4-142) may be integrated once to yield

$$\frac{dY}{d\eta} = c_1 e^{-\eta^2} \qquad (4\text{-}145)$$

and a second time to give

$$Y = c_1 \int e^{-\eta^2} \, d\eta + c_2 \qquad (4\text{-}146)$$

where c_1 and c_2 are integration constants.

The integral in equation (4-146) is related to the *error function*, which is defined as

$$\operatorname{erf} \phi \equiv \frac{2}{\sqrt{\pi}} \int_0^\phi e^{-\eta^2} \, d\eta \qquad (4\text{-}147)$$

The error function is encountered frequently in mathematical physics. Tables of erf ϕ are found in numerous handbooks; a brief tabulation of erf ϕ is included in Appendix C. Particular properties of erf ϕ of note are

$$\operatorname{erf}(0) = 0 \qquad \text{and} \qquad \operatorname{erf}(\infty) = 1$$

Applying the definition of the error function to our solution given by equation (4-147), we have

$$Y = 1 - \operatorname{erf} \eta$$

or

$$\frac{T - T_0}{T_s - T_0} = 1 - \operatorname{erf}\left(\frac{x}{2\sqrt{\alpha t}}\right) \qquad (4\text{-}148)$$

Alternatively, we may write

$$\frac{T_s - T}{T_s - T_0} = \text{erf}\left(\frac{x}{2\sqrt{\alpha t}}\right)$$
(4-149)

If, instead of a fixed wall surface temperature, a fluid at temperature T_∞ is adjacent to the wall at $x = 0$, the boundary condition becomes

$$h[T_\infty - T(0, t)] = -k\frac{\partial T}{\partial x}(0, t)$$
(4-150)

With this modification, the solution to equation (4-135) is

$$\frac{T_\infty - T}{T_\infty - T_0} = \text{erf}\frac{x}{2\sqrt{\alpha t}} + \exp\left(\frac{hx}{k} + \frac{h^2\alpha t}{k^2}\right)\left[1 - \text{erf}\left(\frac{h\sqrt{\alpha t}}{k} + \frac{x}{2\sqrt{\alpha t}}\right)\right]$$
(4-151)

Equation (4-151) is seen to reduce to equation (4-149) for $x = 0$, $h = \infty$.

4.3-3.2   Semi-Infinite Wall Analysis; Integral Solution

The semi-infinite wall may be analyzed with an integral technique by referring to equation (2-26), which is repeated below for reference.

$$\frac{\delta Q}{dt} - \frac{\delta W_s}{dt} - \frac{\delta W_\mu}{dt} = \frac{\partial}{\partial t}\int_{\text{cv}} e\rho \, dV + \int_{\text{cs}}\left(e + \frac{P}{\rho}\right)\rho(\mathbf{v} \cdot \mathbf{n}) \, dA$$
(2-26)

The integral expression will be applied to a control volume defined as shown in Figure 4.29. The control volume shown extends from the wall ($x = 0$) to some depth L in the medium. The distance from the wall where the temperature is still affected by the boundary condition is designated δ the "penetration distance." A requirement on L is that $L > \delta$.

For a control volume as shown within a conducting medium, the terms in equation (2-26) have the following values:

$$\frac{\delta W_s}{dt} = \frac{\delta W_\mu}{dt} = \int_{\text{cs}}\left(e + \frac{P}{\rho}\right)\rho(\mathbf{v} \cdot \mathbf{n}) \, dA = 0$$

$$\frac{\delta Q}{dt} = q_x$$

$$\frac{\partial}{\partial t}\int_{\text{cv}} e\rho \, dV = \frac{d}{dt}\int_0^L u\rho A \, dx = \frac{d}{dt}\int_0^L \rho c T A \, dx$$

The integral expression which applies is thus

$$\frac{q_x}{A} = \frac{d}{dt}\int_0^L \rho c T \, dx$$
(4-152)

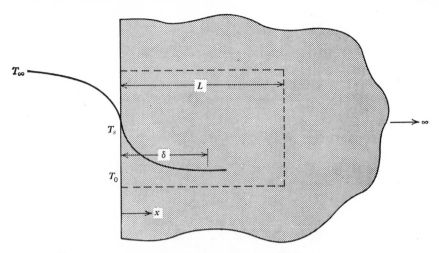

Figure 4.29 Control volume for integral analysis of the semi-infinite wall.

The interval $0 \leq x \leq L$ will now be divided into two increments, giving

$$\frac{q_x}{A} = \frac{d}{dt} \left[\int_0^\delta \rho c T \, dx + \int_\delta^L \rho c T_0 \, dx \right]$$

and since $T_0 \neq T_0(x)$, this becomes

$$\frac{q_x}{A} = \frac{d}{dt} \left[\int_0^\delta \rho c T \, dx + \rho c T_0 (L - \delta) \right]$$

yielding, for the integral equation to be solved,

$$\frac{q_x}{A} = \frac{d}{dt} \int_0^\delta \rho c T \, dx - \rho c T_0 \frac{d\delta}{dt} \qquad (4\text{-}153)$$

To obtain a solution to this problem, one must assume a temperature profile of the form $T = T(x, \delta)$ and substitute the expression into equation (4-153). A differential equation in $\delta(t)$ will be achieved which may then be solved and used to express the temperature profile as $T = T(x, t)$.

Two boundary conditions at the wall will now be considered.

Case 1. Constant Wall Temperature

The wall, initially at uniform temperature T_0 has its surface raised to and maintained at temperature T_s for $t > 0$. At two different times the temperature profiles and penetration depths will appear as shown in Figure 4.30.

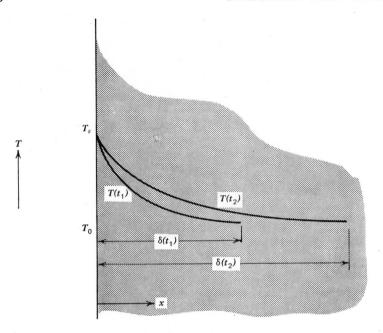

Figure 4.30 Temperature profiles and penetration depths in a semi-infinite wall with constant surface temperature T_s.

If we assume the temperature profile to be parabolic of the form

$$T = a + bx + cx^2$$

the application of the following boundary conditions

$$T(0) = T_s$$

$$T(\delta) = T_0$$

$$\frac{\partial T}{\partial x}(\delta) = 0$$

gives, for $T(x)$, the expression

$$\frac{T - T_0}{T_s - T_0} = \left(1 - \frac{x}{\delta}\right)^2 \tag{4-154}$$

Having assumed a form for the temperature profile resulting in equation (4-154), we are also able to evaluate the heat flux at the wall according to

$$\frac{q_x}{A} = -k\frac{\partial T}{\partial x}(0, t) = \frac{2k}{\delta}(T_s - T_0) \tag{4-155}$$

Equations (4-154) and (4-155) may now be substituted into equation (4-153) to yield

$$\frac{2k}{\delta}(T_s - T_0) = \frac{d}{dt}\int_0^\delta \rho c\left[T_0 + (T_s - T_0)\left(1 - \frac{x}{\delta}\right)^2\right]dx - \rho c T_0 \frac{d\delta}{dt}$$

Dividing each term by the product ρc, which is assumed constant, we obtain

$$\frac{2\alpha}{\delta}(T_s - T_0) = \frac{d}{dt}\int_0^\delta \left[T_0 + (T_s - T_0)\left(1 - \frac{x}{\delta}\right)^2\right]dx - T_0 \frac{d\delta}{dt}$$

which, after performing the integration, becomes

$$\frac{2\alpha}{\delta}(T_s - T_0) = \frac{d}{dt}\left[(T_s - T_0)\frac{\delta}{3}\right]$$

We may cancel the temperature differences $T_s - T_0$ to obtain

$$6\alpha = \delta\frac{d\delta}{dt}$$

and solve for the penetration depth δ.

$$\delta = \sqrt{12\alpha t} \tag{4-156}$$

The corresponding temperature profile may be obtained from equations (4-154) and (4-156) as

$$\frac{T - T_0}{T_s - T_0} = \left[1 - \frac{x}{\sqrt{12\alpha t}}\right]^2 \tag{4-157}$$

Equation (4-157) is the approximate solution to the same case for which equation (4-148) is an exact solution. The two results are compared in Figure 4.31.

Case 2. A Specified Heat Flux at the Wall

The boundary conditions which apply in this case are as follows:

$$T(\delta) = T_0$$

$$-k\frac{\partial T}{\partial x}(0) = \frac{q_{x0}}{A}(t) = F(t)$$

$$\frac{\partial T}{\partial x}(\delta) = 0$$

Assuming a parabolic form, the temperature profile consistent with the above boundary conditions is

$$T - T_0 = \frac{[F(t)](\delta - x)^2}{2k\delta} \tag{4-158}$$

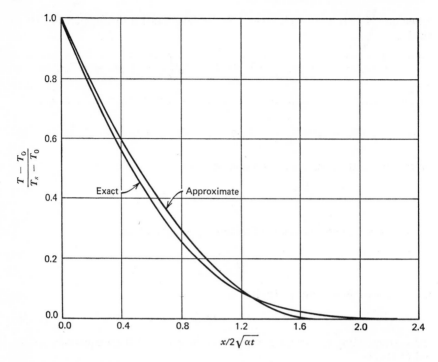

Figure 4.31 A comparison of approximate and exact temperature profiles in a semi-infinite wall with constant surface temperature.

Substituting equation (4-155) into equation (4-150) and solving, we obtain

$$\frac{d}{dt}\left[\frac{\delta^2 F(t)}{6k}\right] = \frac{\alpha F(t)}{k} \tag{4-159}$$

and

$$\delta(t) = \sqrt{6\alpha}\left[\frac{1}{F(t)}\int_0^t F(t)\,dt\right]^{1/2} \tag{4-160}$$

Equation (4-160) applies to any time variation in wall heat flux. For the case where $F(t) = q_0/A$ (a constant), the surface temperature is given by

$$T_s - T_0 = \frac{q_0}{Ak}\sqrt{\frac{3}{2}\alpha t} \tag{4-161}$$

which differs by approximately 8% from the exact value

$$T_s - T_0 = \frac{1.13 q_0}{Ak}\sqrt{\alpha t} \tag{4-162}$$

4.2-4   Transient Heat Conduction; Numerical and Graphical Analysis

The differencing schemes for solving transient heat conduction problems via numerical means were introduced in Chapter 3. We will now use some of the differencing and solution techniques seen earlier to solve transient conduction problems, but consideration will first be given to some graphical methods which are based on results of differencing operations.

4.2-4.1   One-Dimensional Transient Conduction; Explicit Form— the Schmidt Plot

The explicit form of the one-dimensional transient conduction analysis was developed earlier in the form

$$T_i^{n+1} = \frac{\alpha \, \Delta t}{\Delta x^2}\left[T_{i-1}^n + T_{i+1}^n + \dot{q}\,\frac{\Delta x^2}{k} + \left(\frac{\Delta x^2}{\alpha \, \Delta t} - 2\right)T_i^n \right] \qquad (3\text{-}19)$$

Reference is also made to the stability criterion given earlier for the one-dimensional explicit case, which is

$$\frac{\alpha \, \Delta t}{\Delta x^2} \le \frac{1}{2} \qquad (3\text{-}31)$$

The equality is a desirable case to use in equation (3-31) since the T_i^n term on the right-hand side will drop out. With this simplification, namely, $\alpha \Delta t / \Delta x^2 = 1/2$, and with no internal generation ($\dot{q} = 0$), equation (3-19) becomes

$$T_i^{n+1} = \tfrac{1}{2}(T_{i-1}^n + T_{i+1}^n) \qquad (4\text{-}163)$$

which says that the temperature at node i, after some time interval Δt has elapsed, is equal to the arithmetic mean of the temperatures at adjacent node points, Δx units away, at the *start* of the time interval. The time increment and space interval are related by the stability criterion

$$\frac{\alpha \, \Delta t}{\Delta x^2} = \frac{1}{2} \qquad (4\text{-}164)$$

Equations (4-163) and (4-164) are the basis of a graphical technique known as the *Schmidt plotting technique*. The idea is quite simple; consider the region shown in Figure 4.32. The curved line represents the temperature distribution across the portion of the wall depicted at time zero, designated $T^0(x)$. The temperatures at each of five node points are designated T_{i-2}^0, T_{i-1}^0, T_i^0, T_{i+1}^0, and T_{i+2}^0.

Now, according to equation (4-163), the temperature at node i, after a time increment Δt has elapsed, is given by

$$T_i^1 = \tfrac{1}{2}(T_{i-1}^0 + T_{i+1}^0)$$

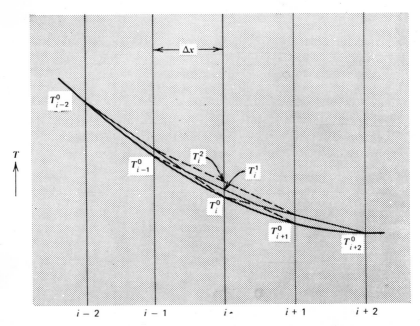

$$\longleftarrow \Delta x \longrightarrow$$

T

T^0_{i-2}

T^2_i

T^0_{i-1}

T^1_i

T^0_i

T^0_{i+1}

T^0_{i+2}

$i-2$ $i-1$ i $i+1$ $i+2$

Figure 4.32 Illustration of the Schmidt plotting technique.

or, in the figure, a straight line connecting T^0_{i-1} and T^0_{i+1} will intersect the line for node i at a point on the temperature scale which represents T^1_i. This same procedure applied to all nodes in the array over the first time interval will generate new values of T^1, and the lines so drawn represent a finite-difference form of the temperature profile after an increment Δt of elapsed time. The process is then continued as many times as one may wish. Example 4.16 illustrates the use of the Schmidt plotting technique.

Example 4.16

A brick wall ($\alpha = 0.018$ ft²/hr), 2 ft thick is initially at a uniform temperature of 70°F. How long will it be until the center of the wall reaches 300°F if the two surfaces are simultaneously raised to 700°F and 300°F, respectively, and maintained at these levels?

This problem is ideally suited to the Schmidt plotting technique. Figure 4.33 shows the wall in cross section.

The graphical procedure is indicated by the lines in the figure. Solid and dashed lines are used alternately so that the reader may follow the steps more clearly.

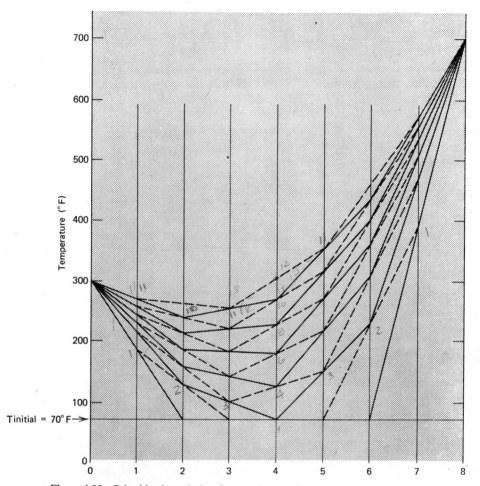

Figure 4.33 Schmidt plot solution for transient conduction across a plane wall.

Spatial increments are chosen for convenience such that $\Delta x = 1/4$ ft; thus there are 7 nodes, including the boundaries.

The procedure is continued until one of the construction lines intersects node 4 (the center) at a temperature value of 300°F or higher. In the present example, the temperature at node 4 is 300°F after 12 time increments.

The length of one time increment is determined from the stability criterion, equation (4-164), and the space increment chosen. In this example, Δt is evaluated as

$$\Delta t = \frac{\Delta x^2}{2\alpha} = \frac{(1/4 \text{ ft})^2}{2(0.018 \text{ ft}^2/\text{hr})} = 1.74 \text{ hr}$$

The desired answer is thus determined as

$$\text{elapsed time} = 12(1.74) = 20.9 \text{ hr}$$

In the preceding example the boundary conditions were the simplest possible, those of prescribed temperature. Other boundary conditions could be accommodated in this technique, such as insulated boundaries where the temperature gradient is zero, prescribed heat flux where the temperature gradient is known, and convection. The matter of convection at a surface will now be considered.

With convection at a boundary, the surface heat flux may be expressed as

$$\frac{q_x}{A}\bigg|_{\text{surface}} = -k \frac{\partial T}{\partial x}\bigg|_{\text{surface}} = h(T_{\text{surf}} - T_\infty) \qquad (4\text{-}165)$$

Referring to Figure 4.34, we may apply equation (4-165) to reference plane (0) to obtain

$$-\frac{\partial T}{\partial x}\bigg|_0 = \frac{T_0 - T_\infty}{k/h} = \frac{T_0 - T_\infty}{\Delta x^*} \qquad (4\text{-}166)$$

where Δx^* is the ratio k/h, having units of length, and is a fictitious or pseudo thickness representing the effect of the convective film. The pseudo thickness k/h is the amount by which the solid boundary is extended so that the effect of convection at a boundary may be included in a graphical solution. This distance k/h is representative of the fluid-surface thermal resistance

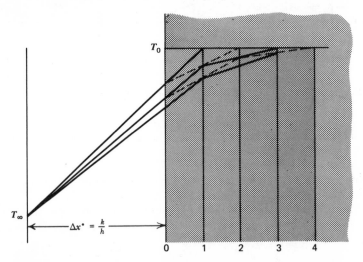

Figure 4.34   Schmidt plotting procedure with convection at a boundary.

and has a single value *independent of the space increment used for the conducting medium*. Example 4.17 illustrates the way convection is treated in the graphical solution of a transient heat conduction problem.

Example 4.17

A stainless steel casting ($k = 13$ Btu/hr-ft-°F, $\alpha = 0.17$ ft²/hr) is heated to 1400°F in a furnace and then allowed to cool in air at 80°F. The casting is 10 in. thick and may be considered flat with sufficient length that end effects need not be accounted for. For a surface conductance of 22 Btu/hr-ft²-°F, determine the length of time until the center reaches 1100°F. What will be the surface temperature at this time?

Since the casting is symmetrical, we need consider only one-half of the cross section, remembering that the temperature gradient is zero at the center (as for an insulated wall at this location). The Schmidt plot construction is shown in Figure 4.35.

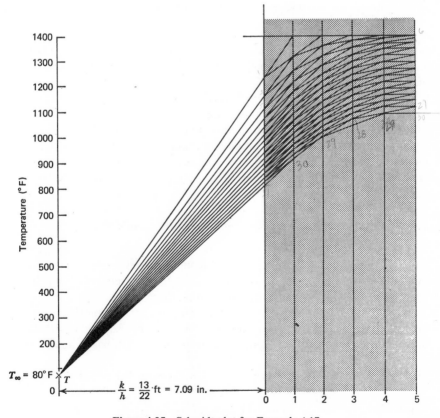

Figure 4.35 Schmidt plot for Example 4.17.

The space increment in the solid is chosen conveniently as 1 in. or 1/12 ft. The time increment consistent with this value of Δx is, from equation (4-164),

$$\Delta t = \frac{\Delta x^2}{2\alpha} = \frac{(1/12 \text{ ft})^2}{2(0.17 \text{ ft}^2/\text{hr})} = 0.020 \text{ hr}$$

The pseudo thickness of the convective film is evaluated as

$$\Delta x^* = \frac{k}{h} = \frac{13 \text{ Btu/hr-ft-}^\circ\text{F}}{22 \text{ Btu/hr-ft}^2\text{-}^\circ\text{F}} = 0.591 \text{ ft} = 7.09 \text{ in.}$$

With these values for Δx, Δt, and Δx^*, the temperature at the center (node 5) reaches 1100°F after approximately 29 time increments; thus the total elapsed time for this condition to be reached is

$$t = (29 \text{ increments})\left(\frac{0.020 \text{ hr}}{\text{increment}}\right) = 0.58 \text{ hr} = 34.8 \text{ min}$$

At this time the surface temperature is observed to be 835°F.

In this example the Schmidt plotting technique was not the only way the problem could be solved; the chart solutions apply in this case, and the answers obtained using the unsteady-state charts are consistent with those obtained by graphical means.

The Schmidt plot technique was developed from the explicit form of the difference equations for one-dimensional transient conduction mentioned earlier. A logical consideration at this point is whether other differencing schemes might lead to other graphical techniques and, if so, how the techniques compare regarding difficulty and accuracy. One other differencing approach, leading to an alternate graphical procedure, will now be considered.

4.2-4.2   Alternating Direction Procedures; the Saul'ev Graphical Solution

No attempt will be made here to present a complete development or explanation of *alternating direction procedures* (ADP). The references which are cited are much more exhaustive in describing these methods.

Alternating direction procedures fall into two primary categories: (1) implicit methods (ADIP) based on ideas proposed by Peaceman and Rachford[4] and by Douglas[5] and (2) explicit methods (ADEP) based on the work

---

[4] D. W. Peaceman and H. H. Rachford, Jr., "The Numerical Solution of Parabolic and Elliptic Differential Equations," *J. Soc. Appl. Math* 3 (1955).

[5] Jim Douglas, Jr., "On the Numerical Integration of $\frac{\partial u}{\partial t} = \frac{\partial^2 u}{\partial x^2}\frac{\partial^2 u}{\partial y^2}$ Σy implicit Methods," *J. Soc. Ind. and Appl. Math.* 3 (1955).

of Saul'ev.[6] These methods afford considerable savings in computer time and require less computer memory than the more familiar techniques presented in Chapter 3. These benefits are particularly great in two- and three-dimensional problems.

ADIP. The prototype of implicit alternating direction methods is illustrated by the two-dimensional transient heat conduction case, the governing differential equation being

$$\frac{1}{\alpha}\frac{\partial T}{\partial t} = \frac{\partial^2 T}{\partial x^2} + \frac{\partial^2 T}{\partial y^2} \qquad (4\text{-}167)$$

This method differs from others in the treatment of the $\nabla^2 T$ terms.

Given a solution at the nth time step, $T_{i,j}^{n+1/2}$ is found by treating one space derivative explicitly (say, $\partial^2 T/\partial x^2$) and the other implicitly. In the next interval of time, from $n + 1/2$ to $n + 1$, the methods are "switched" with $\partial^2 T/\partial x^2$ treated implicitly and $\partial^2 T/\partial y^2$ treated explicitly. The result is a reduction in number of calculations, the number required being about one-seventh as many as required in the Crank-Nicholson method. In difference form the scheme appears, from n to $n + 1/2$, as

$$\frac{T_{i-1,j}^{n+1/2} - 2T_{i,j}^{n+1/2} + T_{i+1,j}^{n+1/2}}{\Delta x^2} + \frac{T_{i,j-1}^{n} - 2T_{i,j}^{n} + T_{i,j+1}^{n}}{\Delta y^2}$$

$$= \frac{2}{\alpha \Delta t}(T_{i,j}^{n+1/2} - T_{i,j}^{n}) \quad (4\text{-}168)$$

and, from $n + 1/2$ to $n + 1$, as

$$\frac{T_{i-1,j}^{n+1/2} - 2T_{i,j}^{n+1/2} + T_{i+1,j}^{n+1/2}}{\Delta x^2} + \frac{T_{i,j-1}^{n+1} - 2T_{i,j}^{n+1} + T_{i,j+1}^{n+1}}{\Delta y^2}$$

$$= \frac{2}{\alpha \Delta t}(T_{i,j}^{n+1} - T_{i,j}^{n+1/2}) \quad (4\text{-}169)$$

A natural extension of these ideas to three dimensions, in which the time step is broken down into the intervals from n to $n + 1/3$, $n + 1/3$ to $n + 2/3$, and $n + 2/3$ to $n + 1$ and two space derivatives are treated explicitly during each subinterval, was suggested by Douglas and Gunn.[7] Some stability and accuracy difficulties are encountered in this extension; however, Douglas

[6] K. Saul'ev, "On Methods of Numerical Integration of Equations of Diffusion," Doklady Akad, Nauk, USSR **115** (1957); 1077.

[7] Jim Douglas, Jr., and J. E. Gunn, "A General Formulation of Alternating Direction Methods," *Numerische Mathematik* **6** (1964): 428.

and Gunn did achieve a satisfactory solution for the three-dimensional case. Their referenced work should be consulted for details.

ADEP.   The prototype method of the explicit alternating direction type is a one-dimensional solution due to Saul'ev. Some differences exist between ADEP and ADIP, these being

1. The most recently calculated values of temperature (in the time variable t) are used as soon as possible.
2. The calculations at each time step are always started from alternate (and opposite) boundaries.

In one dimension the procedure is similar to the simple explicit formulation. In this case, the difference equation is

$$T_i^{n+1} - T_i^n = \frac{\alpha \, \Delta t}{\Delta x^2} \, [T_{i-1}^{n+1} - T_i^{n+1} - T_i^n + T_{i+1}^n] \qquad (4\text{-}170)$$

Proceeding from the left-hand boundary to the right, T_{i-1}^{n+1} will be a known quantity. For the time step from $n + 1$ to $n + 2$, the calculation proceeds from the right-hand boundary to the left, according to

$$T_i^{n+2} - T_i^{n+1} = \frac{\alpha \, \Delta t}{\Delta x^2} \, [T_{i-1}^{n+1} - T_i^{n+1} - T_i^{n+2} + T_{i+1}^{n+2}] \qquad (4\text{-}171)$$

where T_{i+1}^{n+2} is a known quantity.

The procedure described is unconditionally stable.

. Equations (4-170) and (4-171) are in a form which is useful for computer solution. A graphical solution is possible when we set

$$\frac{\alpha \, \Delta t}{\Delta x^2} = 1 \qquad (4\text{-}172)$$

in which case the preceding equations become

$$T_i^{n+1} = \tfrac{1}{2}(T_{i-1}^{n+1} + T_{i+1}^n) \qquad (4\text{-}173)$$

and

$$T_i^{n+2} = \tfrac{1}{2}(T_{i-1}^{n+1} + T_{i+1}^{n+2}) \qquad (4\text{-}174)$$

These equations will be used to solve the same problem considered in Example 4.16.

Example 4.18

Given the brick wall with dimensions, initial conditions, and boundary conditions as specified in Example 4.16, determine the time required for the center of the wall to reach 300°F, using the Saul'ev graphical technique.

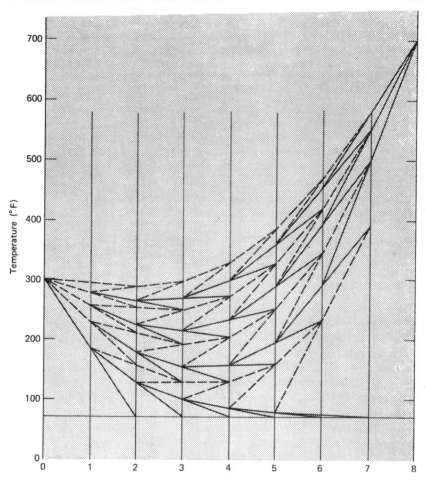

Figure 4.36 Saul'ev graphical solution for transient conduction across a plane wall.

The graphical construction for this solution is shown in Figure 4.36. For the desired condition to be reached, approximately 7.3 time steps are required. The time increment per step is calculated from equation (4-172) as

$$\Delta t = \frac{\Delta x^2}{\alpha} = \frac{(1/4 \text{ ft})^2}{0.018 \text{ ft}^2/\text{hr}} = 3.47 \text{ hr}$$

The total time elapsed is thus evaluated as

$$7.3(3.47) = 25.3 \text{ hr}$$

which is approximately 20% greater than the solution obtained in Example 4.16. For smaller space and time increments the two solution techniques will yield more compatible results.

The time and effort saved in the Saul'ev technique, compared with the Schmidt plot technique, is obvious. The same comparison can be made relative to the computer time needed to solve problems by ADEP, compared with simple explicit difference equations. The benefits from using ADEP in two and three dimensions are even more pronounced.

The same technique for treating convective films, using the pseudo thickness $\Delta x^* = k/h$, may be employed with the Saul'ev procedure as was demonstrated earlier with the Schmidt plot.

Extensions of the Saul'ev ADEP approach to two dimensions have been presented by Larkin[8] and by Barakat and Clark.[9] Alloda and Quon[10] have carried out numerical experiments for three-dimensional applications of ADEP.

4.2-4.3 One- and Two-Dimensional Transient Conduction; Numerical Solutions

An example problem is considered, using the digital computer to solve a transient heat conduction problem formulated numerically.

Example 4.19

A flat magnesium plate is 1 ft thick; other dimensions are sufficiently large that end effects may be considered negligible. The plate is initially at a uniform temperature of 100°F. The temperature of the top surface of the magnesium is suddenly lowered to and maintained at 0°F. The bottom surface of the plate may be considered to be insulated. Describe the temperature distribution in the plate with time for a period of 12 min (0.2 hr) after the surface temperature is lowered to 0°F.

The properties of magnesium will be assumed not to vary significantly over the temperature range from 0°F to 100°F.

The differential equation which applies to the one-dimensional transient case is

$$\frac{\partial T}{\partial t} = \alpha \frac{\partial^2 T}{\partial x^2} \tag{4-92}$$

and the difference equation will be written in explicit form for purposes of achieving a computer solution, where

$$T_i^{n+1} = \frac{\alpha \, \Delta t}{\Delta x^2} (T_{i-1}^n + T_{i+1}^n) + \left(1 - \frac{2\alpha \, \Delta t}{\Delta x^2}\right) T_i^n \tag{3-19}$$

[8] B. K. Larkin, "Some Stable Explicit Difference Approximations to the Diffusion Equation," *Mathematics of Computation* 18 No. 86 (1964): 196.
[9] H. Z. Barakat and J. A. Clark, "On the Solution of the Diffusion Equation by Numerical Methods," *J. Ht. Transfer, Trans. ASME* (Nov. 1966): 421–427.
[10] S. R. Alloda and D. Quon, "A Stable, Explicit Numerical Solution of the Conduction Equation for Multidimensional Nonhomogeneous Media," *Ht. Trans. Los Angeles, Chem. Engr. Prog. Symposium* 62, No. 64 (1966): 151–156.

is used to calculate values of T_i at the end of a time interval Δt with all temperatures known at the beginning of the time interval.

The program listing, flow chart (Figure 4.37), and computer solution follow. The program is written in sufficiently general form that the time-temperature history of the plate may be determined for a different number of locations within the plate, for any material of interest, for any initial and boundary temperature specifications, for any time step, and for any length of time desired.

For this example problem, the following specifications were made:

thermal conductivity	$TK =$	99.0
density	$DEN =$	109.0
heat capacity	$HC =$	0.232
number of internal nodes evaluated	$N =$	5
width of conducting medium	$AL =$	1.0
time increment (for computational purposes)	$DT =$	0.005
initial temperature	$TI =$	100.0
number of elapsed time intervals at which printout is desired	$L =$	4
length of time to be considered	$TMAX =$	0.21

```
        PROGRAM DITRC
C       THIS PROGRAM SOLVES FOR THE TEMP. DISTR. AS A
C       FUNCTION OF TIME FOR A 1 DIMENSIONAL ROD INSULATED
C       ON 1 END, INITIALLY AT A CONSTANT TEMP=T1, AND THEN
C       SUBJECTED TO TEMP.=TBC AT TIME Ø + DELTA TIME ON THE
C       OTHER SIDE.
        DIMENSION T(5ØØ)
        TK=TTYIN(4HTK  =, 1H )
        DEN=TTYIN(4HDEN ,2H= )
        HC=TTYIN(4HHC  =,1H )
        N=TTYIN(4HN  = )
        AL=TTYIN(4HAL =,1H )
        DX=AL/N
        DX2=DX*DX
        DT=TTYIN(4HDT  =,1H )
        TI=TTYIN(4HTI  =,1H )
        L=TTYIN(4HL  = )
        TMAX=TTYIN(4HTMAX,3H  = )
        N1=N+1
        CC=(TK*DT)/(DEN*HC*DX2)
        IF(CC.LE.Ø.5Ø)GO TO 5
        WRITE(61,1Ø3)
        GO TO 2ØØ
```

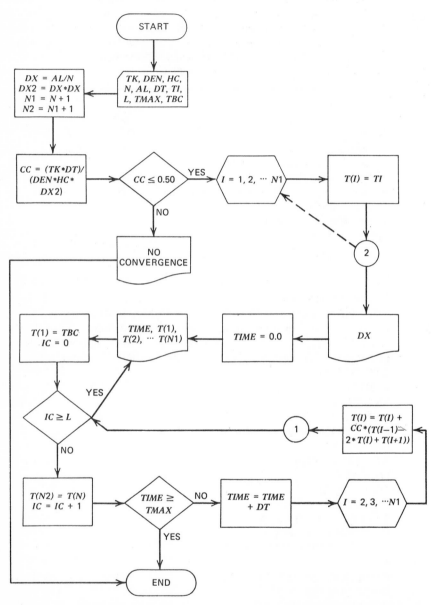

Figure 4.37 Flow diagram for solving a one-dimensional transient heat conduction problem, using an explicit approach.

```
103 FORMAT(1H0,14HNO CONVERGENCE)
  5 CONTINUE
    N2=N1+1
    DO 2 I=1,N1
  2 T(I)=TI
    TBC=TTYIN(4HTBC ,2H= )
    TIME=0.0
    WRITE(61,102)DX
102 FORMAT(1H0,16HINTERVAL WIDTH =,F8.4)
    WRITE(61,100)
  7 WRITE(61,101)TIME,(T(I),I=1,N1)
    T(1)=TBC
    IC=0
  3 IF(IC.GE.L)GO TO 7
    T(N2)=T(N)
    IC=IC+1
    IF(TIME.GE.TMAX)GO TO 200
    TIME=TIME + DT
    DO 1 I=2,N1
  1 T(I)=T(I)+CC*(T(I-1)-2.*T(I)+T(I+1))
    GO TO 3
100 FORMAT(1H0,4HTIME,15X,26HTEMPERATURE AT GRID POINTS)
101 FORMAT(1H0,F8.3,4X,12(2X,F6.2))
200 STOP
    END
```

The printout of this case is given below. One consideration that must be made is the relationship between the time increment and space interval as specified by equation (3-31) for stability of the solution. According to equation (3-31),

$$\frac{\alpha \, \Delta t}{\Delta x^2} \leq \frac{1}{2} \qquad (3\text{-}31)$$

In this case,

$$\frac{k}{\rho c_p} \frac{\Delta t}{\Delta x^2} = \frac{(99 \text{ Btu/hr-ft-}^{\circ}\text{F})(0.005 \text{ hr})}{(109 \text{ lb}_m/\text{ft}^3)(0.232 \text{ Btu/lb}_m\text{-}^{\circ}\text{F})(0.2 \text{ ft})^2}$$
$$= 0.489$$

which is satisfactory. Any time increment less than 0.005 hr will be acceptable for insuring numerical stability and will also likely provide improved accuracy. There is no criterion for convergence in this explicit technique. One must decide how small a time increment to choose balancing computer time versus accuracy.

Two other solutions were obtained for this problem; each is identical to the first except for the time increment, which is 0.0025 in the second case and 0.0005 in the third case. Printouts are made for the same increments in real time so that the results may be compared.

TK   =   99.0
DEN =   109.0
HC   =   0.232
N = 5
AL   =   1.0
DT   =   .005
TI   =   100.0
L = 4
TMAX =   .21
TBC =   0.0
INTERVAL WIDTH =   .2000

TIME   TEMPERATURE AT GRID POINTS

0	100.00	100.00	100.00	100.00	100.00	100.00
.020	0	27.93	50.15	66.16	75.83	79.00
.040	0	19.37	35.64	47.79	55.17	57.57
.060	0	14.03	25.85	34.69	40.06	41.81
.080	0	10.18	18.77	25.18	29.09	30.35
.100	0	7.39	13.63	18.28	21.12	22.04
.120	0	5.37	9.89	13.27	15.33	16.00
.140	0	3.90	7.18	9.64	11.13	11.62
.160	0	2.83	5.21	7.00	8.08	8.43
.180	0	2.05	3.79	5.08	5.87	6.12
.200	0	1.49	2.75	3.69	4.26	4.45

END OF FORTRAN EXECUTION

TK   =   99.0
DEN =   109.0
HC   =   0.232
N = 5
AL   =   1.0
DT   =   .0025
TI   =   100.0
L = 8
TMAX =   .21
TBC =   0.0
INTERVAL WIDTH =   .2000

TIME   TEMPERATURE AT GRID POINTS

0	100.00	100.00	100.00	100.00	100.00	100.00
.020	0	34.08	60.58	78.16	87.90	91.07
.040	0	24.09	44.87	60.63	70.37	73.64
.060	0	18.64	34.98	47.61	55.54	58.21
.080	0	14.63	27.49	37.47	43.74	45.86
.100	0	11.51	21.64	29.49	34.44	36.10

TIME   TEMPERATURE AT GRID POINTS   (*contd.*)

.120	0	9.06	17.03	23.22	27.11	28.42
.140	0	7.13	13.41	18.28	21.34	22.37
.160	0	5.62	10.55	14.39	16.80	17.61
.180	0	4.42	8.31	11.33	13.23	13.87
.200	0	3.48	6.54	8.92	10.41	10.92

END OF FORTRAN EXECUTION

TK =   99.0
DEN =   109.0
HC =   0.232
N = 5
AL =   1.0
DT =   .0005
TI =   100.0
L = 40
TMAX =   0.21
TBC =   0.0
INTERVAL WIDTH =   .2000

TIME   TEMPERATURE AT GRID POINTS

0	100.00	100.00	100.00	100.00	100.00	100.00
.020	0	38.06	67.22	84.96	93.54	96.02
.040	0	27.17	50.75	68.42	79.13	82.69
.060	0	21.56	40.77	55.79	65.28	68.52
.080	0	17.53	33.25	45.65	53.56	56.27
.100	0	14.34	27.21	37.38	43.89	46.12
.120	0	11.74	22.28	30.62	35.95	37.78
.140	0	9.61	18.25	25.08	29.45	30.95
.160	0	7.88	14.95	20.54	24.12	25.35
.180	0	6.45	12.25	16.83	19.76	20.77
.200	0	5.28	10.03	13.78	16.19	17.01

END OF FORTRAN EXECUTION

Figure 4.38 shows the temperature-time history within the magnesium slab as achieved numerically for the three time increments used; the position corresponding to the values plotted is at the insulated wall. The temperature profile across the slab after 6 min (0.10 hr) for each of the time increments is shown in Figure 4.39.

The error encountered in such a solution for relatively large time increments is apparent. The correct temperatures (for $\Delta t \rightarrow 0$) do not differ greatly from the values determined at $\Delta t = 0.0005$ hr. These results illustrate the necessity for one to use caution in overreliance upon numerical results. The fact that a numerical solution is obtained, i.e., that stability criteria are met, does not insure accuracy.

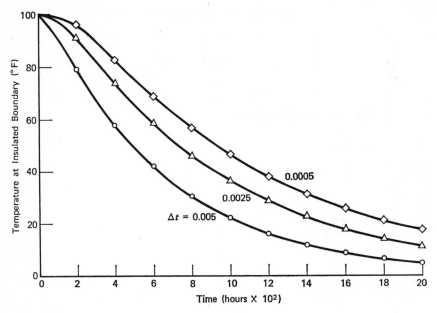

Figure 4.38 Comparison of explicit solutions for various time increments.

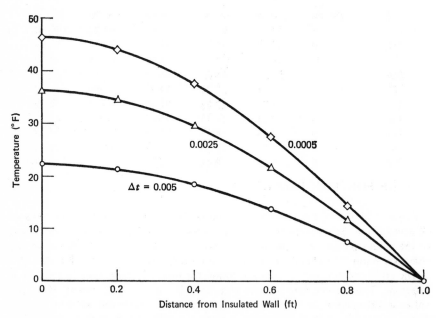

Figure 4.39 Temperature profiles across slab after 0.10 hour elapsed time for three time increments.

In Example 4.20 the same problem is solved as in Example 4.19; however, an *implicit* solution scheme is employed.

Example 4.20

Determine the same information as in the previous example for one-dimensional conduction in a 1-foot magnesium plate. Solve the one-dimensional transient conduction problem, using an implicit formulation, and compare the results with those just obtained for the explicit solution.

The differential equation to be solved is again equation (4-92)

$$\frac{\partial T}{\partial t} = \alpha \frac{\partial^2 T}{\partial x^2} \tag{4-92}$$

This expression is now written in implicit form as

$$\frac{T_i^{n+1} - T_i^n}{\Delta t} = \alpha \frac{T_{i-1}^{n+1} - 2T_i^{n+1} + T_{i+1}^{n+1}}{\Delta x^2}$$

The desired unknown quantity T_i^{n+1} is obtained from this expression in the form

$$T_i^{n+1} = D_i + \frac{A}{B}(T_{i-1}^{n+1} + T_{i+1}^{n+1})$$

with the terms A, B, and D_i defined as

$$A = -\alpha \, \Delta t / \Delta x^2$$
$$B = 1 + 2\alpha \, \Delta t / \Delta x^2$$
$$D_i = T_i^n / B$$

The solution scheme will involve iteration in which the most recently calculated value of T_i^{n+1} is used.

The flow diagram (Figure 4.40), FORTRAN program listing, and output data follow.

Specifications for this problem are similar to those in the previous example. The physical properties of magnesium are the same; other pertinent input data are

number of nodes (not including the left boundary)	$N = 6$
width of conducting medium	$AL = 1.0$
time increment	$DT = 0.01$
	0.005
	0.001
initial temperature	$TI = 100.0$
boundary condition (at left boundary)	$BC = 0.0$
convergence criterion	$EPS = 0.05$

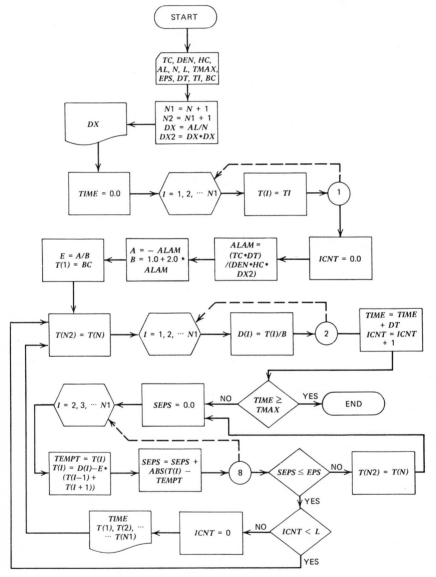

Figure 4.40 Flow diagram for Example 4.20.

```
      PROGRAM DIITC
C     THIS PROGRAM COMPUTES THE TRANSIENT TEMP. DISTRI-
C     BUTION IN A ONE DIMENSIONAL WALL USING ITERATION.
C     THE WALL IS INSULATED ON ONE SURFACE.
      DIMENSION T(50),D(50)
      TC=TTYIN(4HTC =,1H )
      DEN=TTYIN(4HDEN ,2H= )
      HC =TTYIN(4HHC =,1H )
      AL=TTYIN(4HAL =,1H )
      N=TTYIN(4HN = )
      L=TTYIN(4HL = )
      TMAX=TTYIN(4HTMAX,3H = )
      N1=N+1
      N2=N1+1
      EPS=TTYIN(4HEPS ,2H= )
      DX=AL/N
      DX2=DX*DX
      WRITE(61,102)DX
C     READ IN TIME INCREMENT
      DT=TTYIN(4HDT =,1H )
C     INITIALIZE TIME AND TEMPERATURE
      TIME=0.0
      TI=TTYIN(4HTI =,1H )
C     INITIAL TEMPERATURE DISTRIBUTION IS FLAT
      DO 1 I=1,N1
    1 T(I)=TI
      ICNT=0
C     B.C. AT TIME = 0+DELTA TIME
      BC=TTYIN(4HBC =,1H )
      WRITE(61,100)
C     ASSIGN VALUES TO COEFFICIENTS
      ALAM=(TC*DT)/(DEN*HC*DX2)
      A=-ALAM
      B=1.0+2.0*ALAM
      E=A/B
      WRITE(61,101)TIME,(T(I),I=1,N1)
      T(1)=BC
C     COMPUTE D
    5 T(N2)=T(N)
      DO 2 I=1,N1
    2 D(I)=T(I)/B
C     ADVANCE TIME
      TIME=TIME+DT
      ICNT=ICNT+1
      IF(TIME.GE.TMAX)GO TO 6
```

```
C      COMPUTE NEW TEMPERATURE DISTRIBUTION
    3 SEPS=0.0
      DO 8 I=2,N1
      TEMPT=T(I)
      T(I)=D(I)-E*(T(I-1)+T(I+1))
    8 SEPS=SEPS+ABS(T(I)-TEMPT)
      IF(SEPS.LE.EPS)GO TO 4
      T(N2)=T(N)
      GO TO 3
    4 IF(ICNT.LT.L)GO TO 5
      ICNT=0
      WRITE(61,101)TIME,(T(I),I=1,N1)
      GO TO 5
  100 FORMAT(1H0,3X,4HTIME,18X,11HTEMPERATURE)
  101 FORMAT(1H ,F8.3,4X,12(2X,F6.2))
  102 FORMAT(1H ,16HINTERVAL WIDTH  =,F8.4)
    6 END
```

Three different output listings are given for the three time increments specified: 0.01 hr, 0.005 hr, and 0.001 hr, respectively. Comparisons can be made as values for temperatures at equal time intervals for each node, including the insulated boundary and the 0°F boundary. It is apparent for this case, in direct contrast to that in the preceding example, that the solution is not extremely sensitive to the chosen time step. Values of temperature determined for 0.01 hr and for 0.001 hr differ by no more than 3%. It would appear that even for the largest time step used in this case, the results are better than any of those obtained in the previous example by explicit means.

```
TC  =        99.0
DEN =       109.0
HC  =        0.232
AL  =        1.0
N = 6
L = 2
TMAX =       0.30
EPS  =       0.05
INTERVAL WIDTH =      .1667
DT =         0.01
TI =        100.0
BC =         0.0
```

TIME				TEMPERATURE			
0	100.00	100.00	100.00	100.00	100.00	100.00	100.00
.020	0	38.77	65.38	81.21	89.80	93.93	95.14
.040	0	25.54	47.60	64.43	75.82	82.27	84.35
.060	0	19.71	37.64	52.42	63.26	69.82	72.01

TIME   TEMPERATURE   (*contd.*)

.080	0	15.99	30.76	43.26	52.70	58.54	60.52
.100	0	13.19	25.44	35.91	43.89	48.88	50.58
.120	0	10.95	21.14	29.87	36.55	40.75	42.17
.140	0	9.11	17.59	24.87	30.45	33.95	35.15
.160	0	7.59	14.65	20.72	25.37	28.29	29.29
.180	0	6.32	12.21	17.26	21.13	23.57	24.40
.200	0	5.27	10.17	14.38	17.61	19.64	20.34
.220	0	4.39	8.48	11.99	14.68	16.38	16.95
.240	0	3.66	7.07	10.00	12.24	13.65	14.13
.260	0	3.06	5.90	8.34	10.21	11.39	11.79
.280	0	2.55	4.93	6.96	8.52	9.50	9.84
.300	0	2.13	4.11	5.81	7.11	7.93	8.21

TC = 99.0
DEN = 109.0
HC = 0.232
AL = 1.0
N = 6
L = 4
TMAX = 0.30
EPS = 0.05
INTERVAL WIDTH = .1667
DT = 0.005
TI = 100.0
BC = 0.0

TIME   TEMPERATURE

0	100.00	100.00	100.00	100.00	100.00	100.00	100.00
.020	0	35.76	63.04	80.37	89.99	94.59	95.92
.040	0	24.39	46.08	63.27	75.30	82.28	84.55
.060	0	19.13	36.72	51.47	62.48	69.24	71.51
.080	0	15.60	30.08	42.42	51.82	57.68	59.67
.100	0	12.87	24.85	35.12	42.97	47.90	49.57
.120	0	10.66	20.59	29.11	35.63	39.74	41.13
.140	0	8.84	17.07	24.13	29.55	32.96	34.12
.160	0	7.33	14.16	20.02	24.51	27.34	28.30
.180	0	6.08	11.75	16.61	20.34	22.68	23.48
.200	0	5.05	9.75	13.78	16.88	18.82	19.49
.220	0	4.19	8.09	11.44	14.01	15.62	16.17
.240	0	3.48	6.71	9.49	11.62	12.96	13.42
.260	0	2.89	5.58	7.88	9.65	10.76	11.14
.280	0	2.40	4.64	6.55	8.02	8.94	9.26
.300	0	2.00	3.85	5.45	6.67	7.43	7.70

TC =   99.0
DEN =   109.0
HC =   0.232
AL =   1.0
N = 6
L = 20
TMAX =   0.30
EPS =   0.05
INTERVAL WIDTH =   .1667
DT =   0.001
TI =   100.0
BC =   0.0

TIME				TEMPERATURE			
0	100.00	100.00	100.00	100.00	100.00	100.00	100.00
.020	0	33.39	60.75	79.40	90.17	95.32	96.80
.040	0	23.56	44.88	62.26	74.80	82.25	84.71
.060	0	18.72	36.03	50.71	61.80	68.69	71.02
.080	0	15.32	29.56	41.75	51.07	56.91	58.90
.100	0	12.63	24.38	34.47	42.20	47.06	48.71
.120	0	10.43	20.14	28.48	34.87	38.89	40.26
.140	0	8.62	16.64	23.53	28.82	32.14	33.27
.160	0	7.12	13.75	19.44	23.81	26.56	27.49
.180	0	5.88	11.36	16.07	19.68	21.95	22.72
.200	0	4.86	9.39	13.28	16.26	18.14	18.77
.220	0	4.02	7.76	10.97	13.44	14.99	15.51
.240	0	3.32	6.41	9.07	11.10	12.38	12.82
.260	0	2.74	5.30	7.49	9.18	10.23	10.59
.280	0	2.27	4.39	6.21	7.60	8.47	8.77
.300	0	1.89	3.64	5.14	6.30	7.02	7.27

The conclusion just reached regarding a comparison between results of the same problem achieved by explicit and by implicit means is not to be interpreted to imply that implicit techniques are always clearly superior to explicit ones. Considerations of required computer storage, running time numerical stability, length of problem, and other special items may lead to different conclusions in various problems.

We now direct our attention to the case of two-dimensional transient conduction. A final example follows, involving the solution of a transient heat conduction problem via numerical means.

Example 4.21

A long rod, square in cross section, is initially at a uniform temperature of 50°F. Determine the temperature distribution as a function of time for the rod after one side is raised and maintained at a temperature of 200°F and the other three sides are lowered and held at 0°F.

The equation which applies in this case is the two-dimensional form of the heat equation, which is

$$\frac{\partial T}{\partial t} = \frac{k}{\rho c_p}\left(\frac{\partial^2 T}{\partial x^2} + \frac{\partial^2 T}{\partial y^2}\right)$$

This differential equation is next written in *implicit* finite-difference form as

$$\frac{T_{i,j}^{n+1} - T_{i,j}^n}{\Delta t} = \frac{k}{\rho c_p}\left[\frac{T_{i-1,j}^{n+1} - 2T_{i,j}^{n+1} + T_{i+1,j}^{n+1}}{\Delta x^2} + \frac{T_{i,j-1}^{n+1} - 2T_{i,j}^{n+1} + T_{i,j+1}^{n+1}}{\Delta y^2}\right]$$

The above form of the difference equation is next modified if we let $\Delta x = \Delta y$ and group the constant terms into a single constant A, defined as $A = k\,\Delta t/\rho c_p\,\Delta x^2$.

$$[1 + 4A]T_{i,j}^{n+1} = T_{i,j}^n + A[T_{i-1,j}^{n+1} + T_{i+1,j}^{n+1} + T_{i,j-1}^{n+1} + T_{1,j+1}^{n+1}]$$

Finally, solving for $T_{i,j}^{n+1}$, we write

$$T_{i,j}^{n+1} = D_{i,j} + E[T_{i-1,j}^{n+1} + T_{i+1,j}^{n+1} + T_{i,j-1}^{n+1} + T_{i,j+1}^{n+1}]$$

where

$$D_{i,j} = \frac{T_{i,j}^n}{1 + 4A}$$

$$E = \frac{A}{1 + 4A}$$

The last of the difference equations written is the one which is solved to obtain the temperature distribution $T_{i,j}$ at node (i,j) at a succession of time periods. The flow diagram for the computer solution follows as Figure 4.41; a program listing and computer output for the problem solution are included in turn. Input conditions for which the output values apply are the following:

thermal conductivity	$TK = 99.0$
density	$DEN = 109.0$
heat capacity	$HC = 0.232$
number of increments in both the horizontal and vertical directions	$N = 6$
time increment (for computational purposes)	$DT = 0.002$
	$= 0.004$
initial temperature	$TIN = 50.0$
number of elapsed time intervals at which printout is desired	$L = 2$
	10
length of real time to be considered	$TMAX = 0.018$

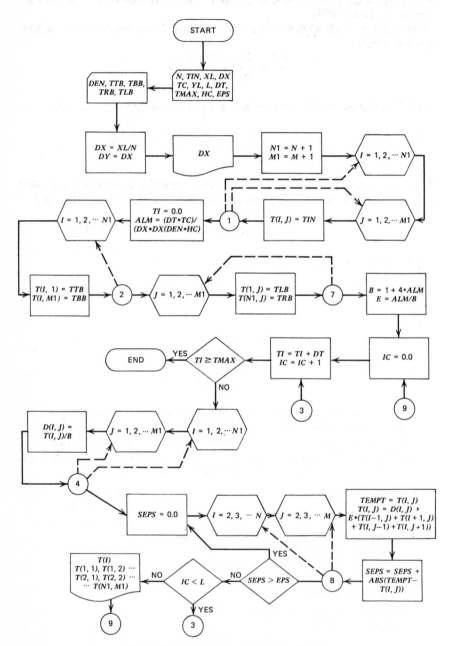

Figure 4.41 Flow diagram for Example 4.21.

```
      PROGRAM D2TC
      DIMENSION T(50,50),D(50,50)
C     THIS PROGRAM SOLVES FOR THE TRANSIENT TEMP. DISTRI-
C     BUTION RESULTING FROM CONDUCTION IN A FLAT PLATE
C     WITH CONSTANT TEMPERATURE BOUNDARIES. THE SOLU-
C     TION IS OBTAINED BY ITERATION AT EACH TIME STEP.
C     DATA INPUT
      N=TTYIN(4HN  = )
      TIN=TTYIN(4HTIN ,2H= )
      XL=TTYIN(4HXL =,1H )
      DX=XL/N
      TC=TTYIN(4HTC =,1H )
C     FOR THIS PROGRAM DELTA X = DELTA Y
      DY=DX
      YL=TTYIN(4HYL =,1H )
      WRITE(61,99)DX
   99 FORMAT(1H ,16HINTERVAL WIDTH =,F7.4)
      L=TTYIN(4HL  = )
      M=YL/DY
      DT=TTYIN(4HDT =,1H )
      TMAX=TTYIN(4HTMAX,3H  = )
      HC=TTYIN(4HHC =,1H )
      EPS=TTYIN(4HEPS ,2H= )
      DEN=TTYIN(4HDEN ,2H= )
      TTB=TTYIN(4HTTB ,2H= )
      TBB=TTYIN(4HTBB ,2H= )
      TRB=TTYIN(4HTRB ,2H= )
      TLB=TTYIN(4HTLB ,2H= )
      N1=N+1
      M1=M+1
C     ASSIGN INITIAL TEMPERATURES
      DO 1 I=1,N1
      DO 1 J=1,M1
      T(I,J)=TIN
    1 CONTINUE
      ALM=(DT*TC)/(DX*DX*HC*DEN)
      TI=0.0
      WRITE(61,101)TI
      WRITE(61,100)((T(I,J),I=1,N1),J=1,M1)
C     SET BOUNDARY VALUES
      DO 2 I=1,N1
      T(I,1)=TTB
      T(I,M1)=TBB
    2 CONTINUE
      DO 7 J=1,M1
```

```
        T(1,J)=TLB
        T(N1,J)=TRB
      7 CONTINUE
C       SET  COEFFICIENTS  B,E
        B=1.0+4.0*ALM
        E=ALM/B
      9 IC=0.0
      3 TI=TI+DT
        IC=IC+1
        IF(TI.GE.TMAX)GO TO 6
        DO 4 I=1,N1
        DO 4 J=1,M1
        D(I,J)=T(I,J)/B
      4 CONTINUE
C       BEGIN  ITERATIONS
      5 SEPS=0.0
        DO 8 I=2,N
        DO 8 J=2,M
        TEMPT=T(I,J)
        T(I,J)=D(I,J)+E*(T(I-1,J)+T(I+1,J)+T(I,J-1)+T(I,J+1))
      8 SEPS=SEPS+ABS(TEMPT-T(I,J))
        IF(SEPS.GT.EPS)GO TO 5
        IF(IC.LT.L)GO TO 3
        WRITE(61,101)TI
        WRITE(61,100)((T(I,J),I=1,N1),J=1,M1)
    100 FORMAT(1H ,7(2X,F6.2))
    101 FORMAT(1H0,6HTIME =,F9.6)
        GO TO 9
      6 END
```

Output temperatures are shown for boundary nodes and twenty-five internal nodes at time increments of 0.004 hr for each of the two cases computed. No significant difference exists between the two cases after 0.012 hr of real time; i.e., the courseness of the time increment has no appreciable effect on the output past this time.

```
TIN   =       50.0
XL    =        0.40
TC    =       99.0
YL    =        0.40
INTERVAL WIDTH   =        .0667
L  =  2
DT    =        0.002
TMAX  =        0.018
HC    =        0.232
EPS   =        0.08
DEN   =       109.0
```

TTB =   0.0
TBB =   0.0
TRB =   0.0
TLB =   200.0
TIME =   0

50.00	50.00	50.00	50.00	50.00	50.00	50.00
50.00	50.00	50.00	50.00	50.00	50.00	50.00
50.00	50.00	50.00	50.00	50.00	50.00	50.00
50.00	50.00	50.00	50.00	50.00	50.00	50.00
50.00	50.00	50.00	50.00	50.00	50.00	50.00
50.00	50.00	50.00	50.00	50.00	50.00	50.00
50.00	50.00	50.00	50.00	50.00	50.00	50.00

TIME =   .004000

200.00	0	0	0	0	0	0
200.00	91.01	47.48	27.88	17.23	9.00	0
200.00	120.83	72.07	44.67	27.95	14.49	0
200.00	128.00	79.26	50.01	31.45	16.27	0
200.00	120.82	72.07	44.66	27.94	14.49	0
200.00	91.00	47.48	27.87	17.22	9.00	0
200.00	0	0	0	0	0	0

TIME =   .008000

200.00	0	0	0	0	0	0
200.00	93.51	48.91	26.97	14.61	6.50	0
200.00	125.43	75.39	44.26	24.63	11.07	0
200.00	133.40	83.40	50.01	28.11	12.69	0
200.00	125.43	75.38	44.26	24.63	11.07	0
200.00	93.51	48.91	26.96	14.61	6.50	0
200.00	0	0	0	0	0	0

TIME =   .012000

200.00	0	0	0	0	0	0
200.00	93.72	49.09	26.93	14.39	6.29	0
200.00	125.81	75.73	44.24	24.29	10.73	0
200.00	133.85	83.82	50.01	27.74	12.32	0
200.00	125.81	75.73	44.24	24.29	10.73	0
200.00	93.72	49.09	26.93	14.38	6.29	0
200.00	0	0	0	0	0	0

TIME =   .016000

200.00	0	0	0	0	0	0
200.00	93.73	49.10	26.93	14.36	6.27	0
200.00	125.84	75.76	44.24	24.25	10.70	0
200.00	133.88	83.85	50.01	27.70	12.28	0
200.00	125.84	75.76	44.24	24.25	10.70	0
200.00	93.74	49.10	26.93	14.36	6.27	0
200.00	0	0	0	0	0	0

```
TIN  =        50.0
XL   =        0.40
TC   =        99.0
YL   =        0.40
INTERVAL WIDTH  =      .0667
L  =  10
DT   =        0.0004
TMAX =        0.018
HC   =        0.232
EPS  =        0.05
DEN  =        109.0
TTB  =        0.0
TBB  =        0.0
TRB  =        0.0
TLB  =        200.0

TIME  =          0
     50.00     50.00     50.00     50.00     50.00     50.00     50.00
     50.00     50.00     50.00     50.00     50.00     50.00     50.00
     50.00     50.00     50.00     50.00     50.00     50.00     50.00
     50.00     50.00     50.00     50.00     50.00     50.00     50.00
     50.00     50.00     50.00     50.00     50.00     50.00     50.00
     50.00     50.00     50.00     50.00     50.00     50.00     50.00
     50.00     50.00     50.00     50.00     50.00     50.00     50.00

TIME  =       .004000
    200.00         0         0         0         0         0         0
    200.00     93.12     48.54     26.99     15.01      6.88         0
    200.00    124.74     74.71     44.27     25.30     11.72         0
    200.00    132.60     82.61     50.00     28.86     13.43         0
    200.00    124.74     74.71     44.27     25.29     11.72         0
    200.00     93.12     48.54     26.99     15.01      6.88         0
    200.00         0         0         0         0         0         0

TIME  =       .008000
    200.00         0         0         0         0         0         0
    200.00     93.72     49.09     26.93     14.38      6.28         0
    200.00    125.82     75.73     44.24     24.27     10.72         0
    200.00    133.85     83.82     50.00     27.72     12.30         0
    200.00    125.82     75.73     44.23     24.27     10.72         0
    200.00     93.72     49.09     26.93     14.37      6.28         0
    200.00         0         0         0         0         0

TIME  =       .012000
    200.00         0         0         0         0         0         0
    200.00     93.74     49.10     26.92     14.36      6.26         0
    200.00    125.84     75.76     44.23     24.25     10.70         0
```

TIME =   .Ø12ØØØ   (*contd.*)

2ØØ.ØØ	133.88	83.85	5Ø.ØØ	27.69	12.27	Ø
2ØØ.ØØ	125.84	75.76	44.23	24.25	1Ø.7Ø	Ø
2ØØ.ØØ	93.74	49.1Ø	26.92	14.36	6.26	Ø
2ØØ.ØØ	Ø	Ø	Ø	Ø	Ø	Ø

TIME =   .Ø16ØØØ

2ØØ.ØØ	Ø	Ø	Ø	Ø	Ø	Ø
2ØØ.ØØ	93.74	49.1Ø	26.92	14.36	6.26	Ø
2ØØ.ØØ	125.85	75.76	44.23	24.24	1Ø.69	Ø
2ØØ.ØØ	133.89	83.85	5Ø.ØØ	27.69	12.27	Ø
2ØØ.ØØ	125.85	75.76	44.23	24.24	1Ø.69	Ø
2ØØ.ØØ	93.74	49.1Ø	26.92	14.36	6.26	Ø
2ØØ.ØØ	Ø	Ø	Ø	Ø	Ø	Ø

A heat balance check at each time is verification of solution accuracy. A check of the temperature values at 0.016 hr for a time increment of 0.0004 hr yields the following values:

$$\sum_{j=1}^{5} (T_{0,j} - T_{1,j}) = 426.93$$

$$2 \sum_{i=1}^{5} (T_{i,5} - T_{i,6}) + \sum_{j=1}^{5} (T_{5,j} - T_{6,j}) = 426.93$$

The heat balance check is exact insofar as the number of significant figures carried in this computation is concerned.

The last example was a two-dimensional case where those considered previously were for one space dimension. The extension of the methods presented to three dimensions is straightforward.

With the inclusion of the numerical examples in this and previous sections of the current chapter, it is hoped that the reader has gained an appreciation of the power and versatility of numerical techniques and computer solutions. The examples shown have not been, nor were they intended to be, exhaustive.

Many newer and more sophisticated techniques are being developed at a very rapid pace. With the background acquired in this and subsequent chapters, the reader is better prepared to perceive these more advanced schemes and techniques.

4.3 CLOSURE

In this chapter the conduction mechanism of heat transfer has been considered in considerable depth. Numerous example problems have been included to illustrate the manner of application of the basic concepts being considered.

Steady-state conduction was considered for one- and multi-dimensional systems both without and with an internal heat source. Some considerable detail was gone into with regard to extended surfaces. In the case of conduction in more than one dimension, solutions to typical problems were obtained by analytical, graphical, integral, and numerical methods.

Transient, or unsteady-state, conduction was considered in approximately the same order as for steady-state conduction. In one dimension a lumped-parameter analysis was determined to be sufficiently accurate for values of the Biot modulus less than 0.1. In cases where a lumped-parameter analysis is unsatisfactory, a class of one-dimensional solutions was seen to be particularly versatile. Combinations of the solutions to the one-dimensional heat equation with special initial and boundary conditions, presented in easily read graphical form, allow two- and three-dimensional composite systems to be analyzed also. Special consideration was given to solving the case of a semi-infinite wall by analytical and by integral techniques.

Several example problems were included in which both steady-state and transient conditions were solved by numerical means. In each such case, the problem formulation was given and discussed, the flow diagram for the computer solution was illustrated, a listing of the FORTRAN program was supplied, and the computer output was presented along with a discussion of the limitations and special considerations necessary for the techniques employed. A direct consequence of numerical formulation of transient conduction led to two graphical solution techniques; these are designated the Schmidt plotting technique and the Saul'ev alternating direction technique. Some discussion of numerical stability and solution accuracy was also included.

CHAPTER 5 _____

CONVECTION HEAT TRANSFER

Some of the fundamental ideas involved in heat transfer by convection were introduced in Chapter 1. Many of the basic terms and concepts were discussed. In this chapter the convective mode of heat transfer will be considered in depth, including techniques of solving associated problems.

In all convective heat transfer situations the energy exchange occurs between a surface and an adjacent fluid. The nature of the fluid flow must therefore be understood and adequately described before the associated heat transfer problem can be solved.

5.1 FUNDAMENTAL CONCEPTS IN CONVECTIVE HEAT TRANSFER

5.1-1 Fluid Flow Considerations

When a fluid flows past a solid surface, certain things occur. If the solid body represents an obstruction, then the fluid must change its flow path to flow around the solid. If the flow is parallel to a plane solid surface, then the effect of the surface will be transmitted for some distance into the fluid.

An initial distinction to be made regarding surface-fluid interaction is whether the flow is internal or external. For *internal flow* the fluid is confined to flow in a passage such as a circular pipe or rectangular duct. The walls

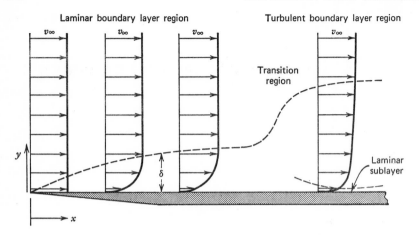

Figure 5.1 Flow considerations parallel to a plane surface.

of the flow passage comprise the surface of interest in this case. In *external flow* the fluid is not confined but flows over the outside surface of a solid body. Flow may be parallel to a plane surface, normal to a cylinder, or with any orientation relative to a solid of any geometric configuration.

5.1-1.1 Boundary Layer Flow Parallel to a Flat Rate

Consider first the situation shown in Figure 5.1, that of steady uniform flow in a direction parallel to a plane solid surface.

One may observe the velocity profiles in the fluid at various positions along the surface. In the figure the x direction is along the surface measured from the *leading edge*, i.e., the point of first contact; and the y direction is normal to the surface. The *free stream velocity*, designated v_∞, remains unchanged at relatively large values of y. For smaller values of y, however, the velocity is seen to vary from a value of zero at the surface, $y = 0$, to the free stream velocity v_∞ at some value of y which increases for larger values of x.

The region close to the surface, where the velocity gradient dv_x/dy is non-zero, is the *boundary layer*. The interaction between the flowing fluid and the boundary is a function of the velocity gradient at the solid-fluid interface, $(dv_x/dy)|_{y=0}$. The force per unit area of contact, or the shear stress, is proportional to the velocity gradient at the surface, the proportionality constant being the *viscosity* of the fluid

$$\tau_0 = \mu \frac{dv_x}{dy}\bigg|_{y=0}. \tag{5-1}$$

where

τ_0 is the shear stress at the wall in lb_F/ft^2

$\dfrac{dv_x}{dy}\bigg|_{y=0}$ is the velocity gradient at the wall in sec^{-1}

and

μ is the viscosity in $lb_F sec/ft^2$ or $lb_m/sec\text{-}ft$

The similarity between equation (5-1) for fluid flow and equation (1-1) for heat transfer is obvious. Analogous quantities between the two expressions are shear stress and heat flux; velocity gradient and temperature gradient; and viscosity and thermal conductivity. Both of these equations represent a quantity of transfer (τ_0 or q/A) in terms of a driving force (dv_x/dy or dT/dy) and a transport property (μ or k) for the situation where transfer is by molecular means. The nature of conduction heat transfer through molecular interaction was discussed in Chapter 1.

In a fluid at relatively low velocities there is a distinct layer-like effect where layers (lamina) of fluid particles move past one another with exchange between adjacent layers at a molecular level only. This very orderly, predictable flow is designated *laminar* and is described by equation (5-1). Since there is no bulk movement of particles between fluid layers, the transfer of energy must also be by molecular means; thus, in laminar flow, heat transfer normal to the flow direction is by conduction and is described by the Fourier rate equation, equation (1-1).

As the fluid velocity increases, the tendency for the bulk transport of fluid particles between fluid layers increases also. For all flow situations, the velocity exists above which there is significant movement of fluid particles in the direction normal to that of bulk flow. Such motion is characterized by very irregular, chaotic fluid behavior and is termed *turbulent*. In *turbulent* flow the movement of fluid in a direction normal to a solid boundary greatly enhances the rate of heat transfer when the surface and the fluid are at different temperatures.

We now observe again Figure 5.1. The dashed line divides the region close to the surface where velocity gradients are nonzero (the boundary layer) from the region outside the boundary layer where the velocity profile is unaffected by the presence of the solid boundary (the free stream). The boundary layer thickness is greatly exaggerated in the figure for clarity. Observe that the boundary layer thickness is zero at the leading edge and grows regularly as x increases. Within the boundary layer, near the leading edge, the velocity profile is a smooth function of y with $0 < v_x < v_\infty$ for $0 < y < \delta$, where δ represents the boundary layer thickness. Flow in the boundary layer is always laminar near the leading edge, regardless of the nature of flow in the free stream.

At some value of x along the surface, the boundary layer thickness is observed to increase quite markedly and then to continue to increase regularly for larger values of x. This is due to a change in the nature of flow within the boundary layer from laminar to turbulent. In a turbulent boundary layer the velocity profile is relatively flat for all but very small values of y. In the near vicinity of the surface, flow is still hypothesized to be laminar. Within this very thin layer, designated the *laminar sublayer*, equations (5-1) and (1-1) apply; even when the boundary layer flow is primarily turbulent, heat transfer at the surface is by conduction with equation (1-1) still applying. Conduction heat transfer, therefore, still represents a significant (in fact, controlling) consideration even though the heat transfer mode is designated convection.

For the situation depicted in Figure 5.1, there are three distinct regions of boundary layer flow. The quantity which is significant in this regard is the Reynolds number, symbolized Re. The Reynolds number is defined, for the case shown, as

$$\text{Re}_x \equiv \frac{x v_\infty \rho}{\mu} = \frac{x v_\infty}{\nu} \tag{5-2}$$

where x and v_∞ have the meanings already discussed and ρ and μ are the fluid density and viscosity, respectively. The symbol ν is seen to equal the ratio μ/ρ and is designated the *kinematic viscosity*, or the *molecular diffusivity of momentum*. The Reynolds number is seen to be dimensionless. The regions of boundary layer flow and the range of Re_x for each are as follows:

(a) $0 < \text{Re}_x < 2 \times 10^5$ boundary layer flow is laminar

(b) $2 \times 10^5 < \text{Re}_x < 3 \times 10^6$ boundary layer flow may be either laminar or turbulent

(c) $3 \times 10^6 < \text{Re}_x$ boundary layer flow is turbulent

The Reynolds number, physically, can be thought of as the ratio of inertial forces to viscous forces. At low values of Re, the viscous forces predominate, and flow is regular or laminar. At high values of Re, the inertial forces are controlling, and turbulent flow prevails. At intermediate values of Re, flow may fluctuate, being sometimes laminar and other times turbulent, as the viscous and inertial forces fluctuate in their dominance.

5.1-1.2 External Flow Past Bluff Bodies

In the preceding discussion of external flow, the free stream velocity was parallel to the plane surface considered. Quite obviously, the flow conditions would change if the free stream were directed normal to, or at some oblique angle to, the surface. Our considerations now will be for those cases when

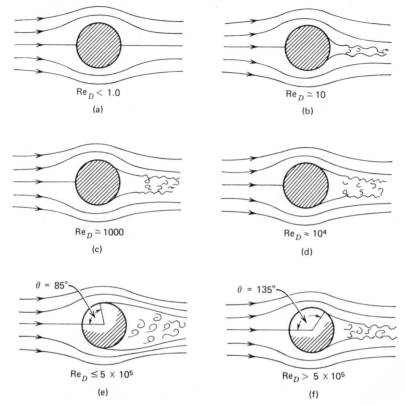

Figure 5.2 Streamline flow past a cylinder oriented normal to v_∞.

the body past which flow occurs is bluff and the associated external flow is very irregular.

As an initial consideration we shall consider a cylinder whose axis is oriented normal to the free stream flow, as depicted in Figure 5.2. The streamline pattern is shown for various values of Reynolds number, which is defined as

$$\text{Re}_D \equiv \frac{D v_\infty \rho}{\mu} = \frac{D v_\infty}{\nu} \qquad (5\text{-}3)$$

in similar fashion to the Reynolds number for flat plate flow, except that the significant length in this case is the diameter of the cylinder.

The forwardmost point on the cylinder is designated the *forward stagnation point;* and 180° away, at the back of the cylinder, is the *aft,* or *rear, stagnation point.* These are the locations on the surface where the adjacent free stream flow tangent to the surface is theoretically zero. At the forward stagnation point, for instance, all of the kinetic energy of the fluid particles on the *stagnation streamline* has been converted to pressure. This *stagnation pressure* includes both static pressure and velocity pressure components.

For streamlines which conform to the shape of the cylinder, the pressure and free stream velocity are smoothly varying functions of the angle θ. For $0 < \theta < 90°$, the pressure decreases and v_∞ increases to minimum and maximum values, respectively, at $\theta = 90°$. For $90° < \theta < 180°$, the pressure increases and free stream velocity decreases, and stagnation conditions are reached once again at $\theta = 180°$. The case of fully attached streamlines is shown in Figure 5.2(a). This situation will occur only at very small values of Re_D, aptly termed "creeping flow."

As Re_D increases above that for creeping flow, the phenomenon of *boundary layer separation* occurs.

The streamlines near the surface of a bluff body begin to "separate" from the surface near the aft stagnation point, as shown in Figure 5.2(b). As the Reynolds number increases, the separation point moves farther forward, as shown in (c) and (d). In each of these instances the flow in the boundary layer, i.e., that part which remains attached, is laminar. Flow behind the separation point is very chaotic and turbulent. This region is called the *turbulent wake*.

For values of Re_D up to approximately 5×10^5, the separation point moves ever forward with increasing Re_D, and boundary layer flow is laminar. At $Re_D \cong 5 \times 10^5$ the separation point is at its most forward position, $\theta \cong 85°$. For larger values of Re_D, flow in the attached portion of the boundary layer is partially turbulent, and the separation point moves to the rear. At transition the value of θ changes relatively abruptly from 85° to 135°. Streamline patterns for values of Re_D below and above the transition point are shown in (e) and (f) of Figure 5.2. At flow conditions with Re_D above the transition value, the separation point again moves forward quite slowly.

A part of the total consideration of flow past bluff bodies and boundary layer separation is the force required to hold a bluff body in place under various conditions of flow. Drag forces due to fluid effects are grouped into two general categories: *viscous drag* and *pressure*, or *form*, *drag*.

Viscous drag is that due solely to the viscous nature of the fluid flowing past the body. This is the effect present in an attached boundary layer.

Pressure, or form, drag is present whenever boundary layer separation occurs. Pressure drag is the result of unequal pressures on the forward and rearward portions of a bluff body. Usually, viscous effects are negligible in comparison to pressure effects when both are present.

These effects are expressed quantitatively in terms of two coefficients, which are defined as follows:

$$C_f \equiv \frac{F_d/A_{\text{contact}}}{\rho v_\infty^2/2} \qquad (5\text{-}4)$$

$$C_D \equiv \frac{F_d/A_{\text{proj}}}{\rho v_\infty^2/2} \qquad (5\text{-}5)$$

where C_f is the *frictional drag coefficient* or the *coefficient of skin friction* (dimensionless)

C_D is the *drag coefficient* (dimensionless)

F_d is the drag force

A_{contact} is the area of contact between the surface and fluid

A_{proj} is the projected area of the bluff body normal to the flow direction

$\dfrac{\rho v_\infty^2}{2}$ is the kinetic energy of the fluid at free stream velocity v_∞

There are obvious differences in C_f and C_D, both in the condition they describe and in the equations which define them. The coefficient of skin friction may be determined analytically if the velocity profile is known, according to

$$C_f = \frac{F_d/A_{\text{contact}}}{\rho v_\infty^2/2} = \frac{1}{\rho v_\infty^2/2}\left(\mu \left.\frac{dv_\infty}{dy}\right|_{y=0}\right) \qquad (5\text{-}6)$$

where $(dv_\infty/dy)|_{y=0}$ is the velocity gradient at the fluid-solid interface.

The drag coefficient, on the other hand, is associated with a flow condition which is not at all amenable to analysis. Empirical evaluation of C_D is necessary. Figures 5.3 and 5.4 are representative of large amounts of experimental data for flow past cylinders, flat plates and disks, and spheres. Both figures show the drag coefficient as a function of Re_D.

For both cylinders and spheres, there is a noticeable abrupt decrease in C_D near a value for Re_D of 5×10^5. This is the case of boundary layer flow transition from laminar to turbulent and the associated rearward movement of the separation point.

5.1-1.3 Internal Flow

Thus far the cases of external flow have been considered. In Figure 5.5 the velocity profiles and boundary layer thickness are shown for an *internal flow* situation, namely, flow in a circular conduit, i.e., a tube or a pipe. In both (a) and (b) of this figure the boundary layer flow is laminar and the growth of the boundary layer is regular near the entrance. For increasing

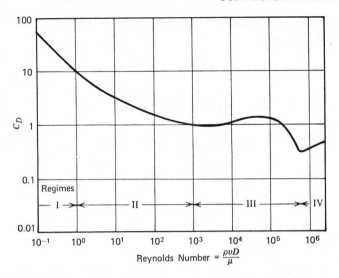

Figure 5.3 Drag coefficient versus Re_D for flow past cylinders.

values of x, the boundary layer grows to include an ever-increasing portion of the flow cross section. In every case, the situation is reached where the boundary layer "fills" the entire pipe cross section, or—thinking of it in a slightly different manner—the effects of the wall, manifested by the fluid viscosity, are felt throughout the cross section.

For values of x greater than that at which the boundary layers meet, the velocity profiles no longer change along the flow path. This is the region of

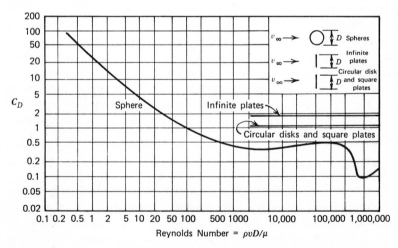

Figure 5.4 Drag coefficient versus Re_D for flow past spheres, flat plates, and disks.

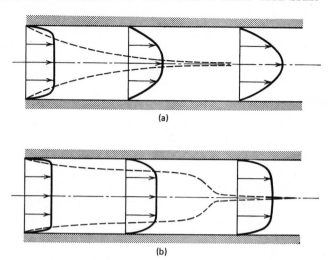

(a)

(b)

Figure 5.5 Velocity profiles and boundary layer growth in a circular conduit.

fully developed flow; and we may write, in this case, $dv_x/dx = 0$. For smaller values of x, the velocity profile is still developing; this region is designated the *entrance region*. The value of x at which flow becomes fully developed is designated the *entrance length*. Figure 5.5(a) and (b) show velocity profiles and boundary layer growth for the cases where the fully developed flow condition is laminar and turbulent, respectively.

The criterion for whether fully developed conduit flow is laminar or turbulent is, once again, the Reynolds number. For conduit flow, however, the Reynolds number form which pertains is

$$\mathrm{Re}_D \equiv \frac{Dv_{\mathrm{avg}}\rho}{\mu} = \frac{Dv_{\mathrm{avg}}}{\nu} \tag{5-7}$$

This is very similar to the Reynolds number considered for flow parallel to a flat surface, given by equation (5-2), except that the length parameter is the conduit diameter in the present case.

For conduit flow, the flow will be laminar for Re_D less than 2300. Above this value, flow may be either laminar or turbulent, with the probability of turbulent flow becoming greater as Re_D increases. At values of Re_D of 10,000 and above, flow may be safely assumed as turbulent. Below a value of Re_D of 2300, viscous forces are dominant, and any disturbances in the flow—such as those caused by a protrusion from the pipe wall—will be dampened out by the viscous forces. At large values of Re_D inertial forces are predominant over the viscous forces, and random, chaotic, turbulent flow is to be expected.

Quantitatively, the force of interaction between the walls of a conduit and a flowing fluid is expressed by

$$\frac{\Delta P}{\rho} = 2f_f \frac{L}{D} v_{avg}^2 \tag{5-8}$$

where

ΔP is the change in pressure resulting from viscous effects in a flow passage of length L and diameter D

ρ is fluid density

v_{avg} is the average fluid velocity

f_f is the *Fanning friction factor* (dimensionless)

Equation (5-8) is, specifically, a definition of the Fanning friction factor. This term is equivalent to the coefficient of skin friction discussed in the previous section. The Fanning friction factor is shown as a function of Reynolds number in Figure 5.6.

In this figure the regions of laminar, transition, and fully turbulent flow are labeled. In laminar flow a single line is sufficient to relate f_f and Re_D. For the transition and fully turbulent flow regimes, however, f_f is a function of both Re_D and the *relative roughness* e/D. The relative roughness is a ratio of the mean height e, of protrusions above the conduit wall, to the conduit diameter D. Values of e/D are shown in Figure 5.7 for various common construction materials.

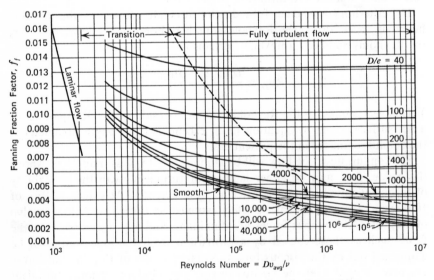

Figure 5.6 Fanning friction factor versus Re_D for internal flow.

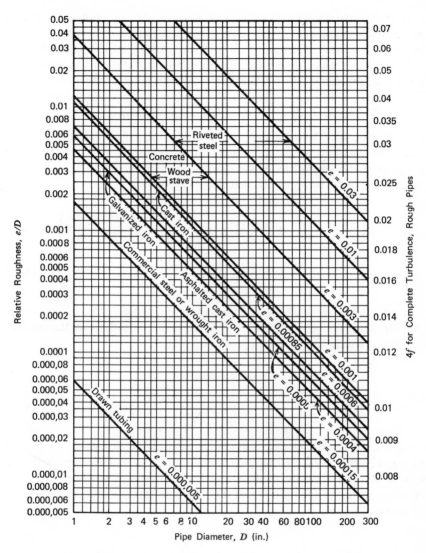

Figure 5.7 Relative roughness of conduits for various materials.

In the transition region f_f varies with Re_D and e/D, the minimum value at any Reynolds number corresponding to hydraulically smooth pipe or tubing. In the fully turbulent range, at large values of Re_D, pipe roughness is the only parameter which determines f_f. The minimum value of f_f at a Reynolds number in this range is for hydraulically smooth pipe or tubing.

5.1-1.4 The Hydrodynamic Boundary Layer Equations

For the case of external, incompressible flow of a fluid parallel to a flat plate, the equations which apply in a laminar boundary layer are the x- and y- component forms of the Navier-Stokes equation, given in vector form by equation (2-66). The appropriate scalar equations are

$$\rho\left(\frac{\partial v_x}{\partial t} + v_x\frac{\partial v_x}{\partial x} + v_y\frac{\partial v_x}{\partial y}\right) = \rho g_x - \frac{\partial P}{\partial x} + \mu\left(\frac{\partial^2 v_x}{\partial x^2} + \frac{\partial^2 v_x}{\partial y^2}\right) \qquad (5\text{-}9)$$

$$\rho\left(\frac{\partial v_y}{\partial t} + v_x\frac{\partial v_y}{\partial x} + v_y\frac{\partial v_y}{\partial y}\right) = \rho g_y - \frac{\partial P}{\partial y} + \mu\left(\frac{\partial^2 v_y}{\partial x^2} + \frac{\partial^2 v_y}{\partial y^2}\right) \qquad (5\text{-}10)$$

In a boundary layer which is tissue-paper thin, certain order-of-magnitude simplifications can be made. The basis for these simplifications are the assumptions that, within the boundary layer,

$$v_x \gg v_y \qquad (5\text{-}11)$$

$$\frac{\partial v_x}{\partial y} \gg \frac{\partial v_x}{\partial x} \qquad (5\text{-}12)$$

On the basis of the relative magnitudes of v_x and v_y expressed in equation (5-11), we observe that all terms involving velocities in equation (5-10) are very small in comparison to the corresponding terms in equation (5-9). On an order-of-magnitude basis, then, the velocity terms in equation (5-10) may be deleted, leaving

$$0 = \rho g_y - \frac{\partial P}{\partial y} \qquad (5\text{-}13)$$

and, with negligible gravitational effects, we may write

$$\frac{\partial P}{\partial y} = 0 \qquad (5\text{-}14)$$

This is a significant result, in that the pressure is now completely determined by effects in the free stream.

Equation (5-9) may now be written in simplified form, incorporating equations (5-12) and (5-14) and neglecting gravity, to yield

$$\rho\left(\frac{\partial v_x}{\partial t} + v_x \frac{\partial v_x}{\partial x} + v_y \frac{\partial v_x}{\partial y}\right) = -\frac{dP}{dx} + \mu \frac{\partial^2 v_x}{\partial y^2} \tag{5-15}$$

In the free stream, under conditions of negligible gravitational effects, the pressure and velocity are related by Bernoulli's equation, which enables us to write

$$-\frac{dP}{dx} = -\frac{d}{dx}\left(\frac{v_\infty^{\,2}}{2}\right) = -v_\infty \frac{dv_\infty}{dx} \tag{5-16}$$

Making this substitution, we may rewrite equation (5-15) in the form

$$\rho\left(\frac{\partial v_x}{\partial t} + v_x \frac{\partial v_x}{\partial x} + v_y \frac{\partial v_x}{\partial y}\right) = -v_\infty \frac{dv_\infty}{dx} + \mu \frac{\partial^2 v_x}{\partial y^2} \tag{5-17}$$

Equation (5-17), along with the two-dimensional, incompressible form of the equation of continuity

$$\frac{\partial v_x}{\partial x} + \frac{\partial v_y}{\partial y} = 0 \tag{5-18}$$

are known as the *boundary layer equations.*

5.1-1.5 The Laminar, Incompressible Boundary Layer on a Flat Plate; the Blasius Solution

For steady, constant, free stream flow parallel to a flat surface, the boundary layer equations reduce to

$$v_x \frac{\partial v_x}{\partial x} + v_y \frac{\partial v_x}{\partial y} = \nu \frac{\partial^2 v_x}{\partial y^2} \tag{5-19}$$

and

$$\frac{\partial v_x}{\partial x} + \frac{\partial v_y}{\partial y} = 0 \tag{5-20}$$

the applicable boundary conditions being

$$v_x = v_y = 0 \quad \text{at } y = 0 \tag{5-21}$$

$$v_x = v_\infty \quad \text{at } y = \infty \tag{5-22}$$

and

$$v_x = v_\infty \quad \text{at } x = 0 \tag{5-23}$$

The solution to equations (5-19) and (5-20) which satisfies these boundary conditions has been obtained in numerous ways by several investigators. The original solution was obtained by Blasius,[1] and the general result is termed the Blasius solution.

It is first necessary to use the stream function ψ, defined according to

$$v_x \equiv \frac{\partial \psi}{\partial y} \qquad v_y \equiv -\frac{\partial \psi}{\partial x} \tag{5-24}$$

which automatically satisfies continuity, equation (5-20). Thus our set of differential equations may be reduced to a single equation which, written in terms of ψ, is

$$\frac{\partial \psi}{\partial y}\frac{\partial^2 \psi}{\partial x \, \partial y} - \frac{\partial \psi}{\partial x}\frac{\partial^2 \psi}{\partial y^2} = \nu \frac{\partial^3 \psi}{\partial y^3} \tag{5-25}$$

The next step is a transformation of the variables x, y, and ψ in such a way that this partial differential equation becomes an ordinary differential equation. The parameters involved in this transformation are

$$\eta \equiv y\left(\frac{v_\infty}{\nu x}\right)^{1/2} \tag{5-26}$$

and

$$f(\eta) \equiv \psi/(\nu x v_\infty)^{1/2} \tag{5-27}$$

The introduction of f and η into equation (5-25) does, in fact, yield the ordinary differential equation

$$f''' + \frac{1}{2}ff'' = 0 \tag{5-28}$$

with boundary conditions

$$f(0) = f'(0) = 0$$

$$f'(\infty) = 1$$

Several approaches to solving equation (5-28) may be found in the literature. One may consult Schlichting,[2] Howarth,[3] Goldstein,[4] and Kays[5]

[1] H. Blasius, "Grenzschichten in Fliisigkeiten mit Kleiner Reibung, 2," *Math. u. Phys.* **56** (1908): 1. NACA Tech. Memo. No. 1256 gives the English translation.

[2] H. Schlichting, *Boundary Layer Theory*, 4th ed. (New York: McGraw-Hill, 1960), p. 117.

[3] L. Howarth, "On the Solution of the laminar boundary layer equations," *Proc. Roy. Soc. London,* **A164** (1938): 547.

[4] S. Goldstein, *Modern Developments in Fluid Dynamics*, (London: Oxford Univ. Press, 1938), p. 135.

[5] W. M. Kays, *Convective Heat and Mass Transfer*, (New York: McGraw-Hill, 1966), p. 83.

Table 5.1   Values of $f, f', $ and f'' for the Boundary
Layer along a Flat Plate

η	f	f'	f''
0	0	0	0.3321
0.4	0.0266	0.133	0.3315
0.8	0.1061	0.265	
1.2	0.238	0.394	
1.6	0.420	0.517	
2.0	0.650	0.630	
2.4	0.922	0.729	
2.8	1.23	0.812	
3.2	1.57	0.876	
3.6	1.93	0.923	
4.0	2.31	0.956	
5.0	3.28	0.992	
6.0	4.28	0.999	
7.0	5.28	0.9999	
8.0	6.28	1.0000	

for examples of different approaches. We shall dwell only on the results which are tabulated in Table 5.1. These particular numerical values are those of Howarth.

Of particular significance are the following results:

(a) The boundary layer thickness δ is normally taken to be where $v_x/v_\infty = 0.99$. An accepted value of η at this condition is 5.0; thus in equation form

$$\delta = y|_{\eta=5.0} = 5.0 \sqrt{\frac{vx}{v_\infty}} \tag{5-29}$$

or

$$\frac{\delta}{x} = \frac{5.0}{\sqrt{\dfrac{xv_\infty}{v}}} = 5.0 \, \mathrm{Re}_x^{-1/2} \tag{5-30}$$

According to equation (5-29) or (5-30), the thickness of the boundary layer is proportional to $x^{1/2}$.

(b) The local coefficient of skin friction is related to the velocity gradient at the surface according to

$$C_{fx} \equiv \frac{\tau}{\dfrac{\rho v_\infty^2}{2}} = \frac{\mu \dfrac{\partial v_x}{\partial y}\bigg|_{y=0}}{\dfrac{\rho v_\infty^2}{2}} \tag{5-31}$$

and $\partial v_x / \partial y |_{y=0}$ can be found from the Blasius solution as follows:

$$\frac{\partial v_x}{\partial y}\bigg|_{y=0} = \frac{\partial}{\partial y} v_\infty f'(0)$$

$$= v_\infty \frac{d}{d\eta} f'(0) \frac{\partial \eta}{\partial y}$$

$$= v_\infty f''(0) \left(\frac{v_\infty}{vx}\right)^{1/2}$$

$$= 0.332 v_\infty \left(\frac{v_\infty}{vx}\right)^{1/2} \tag{5-32}$$

Making the appropriate substitution into equation (5-31), we obtain

$$C_{fx} = \frac{0.332 \mu v_\infty \left(\frac{v_\infty}{vx}\right)^{1/2}}{\dfrac{\rho v_\infty^{\,2}}{2}}$$

$$= 0.664 \left(\frac{v}{xv_\infty}\right)^{1/2} = 0.664\, \mathrm{Re}_x^{-1/2} \tag{5-33}$$

The local coefficient of skin friction is thus seen to vary inversely as the square root of the local Reynolds number.

(c) The mean coefficient of skin friction is defined as

$$C_{fL} = \frac{\dfrac{F_d}{A}}{\dfrac{\rho v_\infty^{\,2}}{2}}$$

where F_d is the total viscous drag force imposed on a flat surface of area A. The mean and local coefficients are related by expressing the drag force in terms of both, according to

$$F_d = C_{fL} A \frac{\rho v_\infty^{\,2}}{2} = \int_A C_{fx} \frac{\rho v_\infty^{\,2}}{2} \, dA \tag{5-34}$$

Making the appropriate cancellation, we may solve for C_{fL} to obtain

$$C_{fL} = \frac{1}{A} \int_A C_{fx} \, dA \tag{5-35}$$

which, for a plate of constant width and length L, becomes

$$C_{fL} = \frac{1}{L} \int_0^L C_{fx} \, dx \tag{5-36}$$

Upon substituting the expression for C_{fx} given by equation (5-33), we have, for the mean coefficient for the laminar boundary layer along a flat plate with constant v_∞,

$$C_{fL} = \frac{0.664}{L} \sqrt{\frac{\nu}{v_\infty}} \int_0^L x^{-1/2} \, dx$$

$$= 1.328 \sqrt{\frac{\nu}{Lv_\infty}} = 1.328 \, \mathrm{Re}_L^{-1/2} \tag{5-37}$$

5.1-1.6 Integral Analysis of the Hydrodynamic Boundary Layer

In the preceding section an "exact" solution was obtained for the case of a laminar boundary layer along a flat plate with constant free stream velocity. In a physical situation with flow or geometry of a more complex nature, an analytical solution is either extremely difficult or impossible. One must, in such cases, obtain useful information via approximate methods. The integral approach presented in Section 2.1-2 is useful in this regard.

The governing integral relations are equations (2-21) and (2-28), which are repeated below for reference.

Conservation of mass

$$\frac{\partial}{\partial t} \int_{\mathrm{cv}} \rho \, dV + \int_{\mathrm{cs}} \rho (\mathbf{v} \cdot \mathbf{n}) \, dA = 0 \tag{2-21}$$

Newtons second law

$$\sum \mathbf{F} = \frac{\partial}{\partial t} \int_{\mathrm{cv}} \mathbf{v} \rho \, dV + \int_{\mathrm{cs}} \mathbf{v} \rho (\mathbf{v} \cdot \mathbf{n}) \, dA \tag{2-28}$$

These equations will be applied to the control volume shown in Figure 5.8. In the figure the control volume width is Δx and its height is Y, where the restriction on Y is that it must be greater than the boundary layer thickness δ. The x axis is tangent to the solid surface at the lower left corner of the control volume; the y axis is normal to the solid surface at the point of tangency.

The integral expression for conservation of mass, applied to this control volume, yields the following:

$$\frac{\partial}{\partial t} \int_{\mathrm{cv}} \rho \, dV = 0 \qquad \text{(steady flow)}$$

$$\int_{\mathrm{cs}} \rho (\mathbf{v} \cdot \mathbf{n}) \, dA = \int_0^Y \rho v_x \, dy \big|_{x+\Delta x} - \int_0^Y \rho v_x \, dy \big|_x - \dot{m} \big|_{\mathrm{top}} \Delta x$$

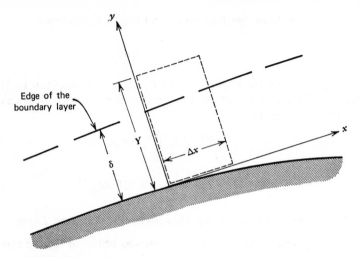

Figure 5.8   Control volume for integral analysis of the hydrodynamic boundary layer.

Setting this sum equal to zero, as dictated by equation (2-21), and dividing by Δx, we have

$$\frac{\int_0^Y \rho v_x \, dy \big|_{x+\Delta x} - \int_0^Y \rho v_x \, dy \big|_x}{\Delta x} - \mathring{m}\big|_{\text{top}} = 0$$

In the limit as $\Delta x \to 0$, the mass flow rate across the top portion of the control surface may be written as

$$\mathring{m}\big|_{\text{top}} = \frac{d}{dx} \int_0^Y \rho v_x \, dy \tag{5-38}$$

We next evaluate the momentum fluxes for this control volume. Equation (2-28) is our governing relationship. Evaluating each term in equation (2-28), we have, in the x direction,

$$\sum F_x = -\tau_{yx} \Delta x + \int_0^Y P \, dy \big|_x - \int_0^Y P \, dy \big|_{x+\Delta x}$$

$$\frac{\partial}{\partial t} \int_{\text{cv}} v_x \rho \, dv = 0 \quad \text{(steady flow)}$$

$$\int_{\text{cs}} v_x \rho (\mathbf{v} \cdot \mathbf{n}) \, dA = \int_0^Y \rho v_x^2 \, dy \big|_{x+\Delta x} - \int_0^Y \rho v_x^2 \, dy \big|_x - v_\infty \Delta x \mathring{m}\big|_{\text{top}}$$

The applicable momentum expression for steady flow is achieved when these terms are related to one another as indicated by equation (2-28). Making this substitution, as well as introducing equation (5-38) for the mass flow rate across the top of the control volume, we have

$$-\tau_{yx}\Delta x + \int_0^Y P\,dy\Big|_x - \int_0^Y P\,dy\Big|_{x+\Delta x}$$

$$= \int_0^Y \rho v_x^2\,dy\Big|_{x+\Delta x} - \int_0^Y \rho v_x^2\,dy\Big|_x - v_\infty\,\Delta x\,\frac{d}{dx}\int_0^Y \rho v_x\,dy$$

We next divide through by Δx and evaluate in the limit as $\Delta x \to 0$ to obtain

$$-\tau_{yx} - \frac{d}{dx}\int_0^Y P\,dy = \frac{d}{dx}\int_0^Y \rho v_x^2\,dy - v_\infty\frac{d}{dx}\int_0^Y \rho v_x\,dy \qquad (5\text{-}39)$$

Allowing the previously made boundary layer assumption to stand, that $P \neq P(y)$, we may rewrite the pressure term in equation (5-39), obtaining

$$-\tau_{yx} - Y\frac{dP}{dx} = \frac{d}{dx}\int_0^Y \rho v_x^2\,dy - v_\infty\frac{d}{dx}\int_0^Y \rho v_x\,dy$$

The pressure gradient dP/dx is related to the free stream velocity according to Bernoulli's equation. Replacing dP/dx by $-\rho v_\infty(dv_\infty/dx)$ and combining the two integral terms, we have, finally, for the momentum integral expression,

$$\tau_{yx} - Y v_\infty \rho\frac{dv_\infty}{dx} = \frac{d}{dx}\int_0^Y \rho v_x(v_\infty - v_x)\,dy \qquad (5\text{-}40)$$

Equation (5-40) is generally referred to as the von Kàrmàn momentum integral expression. An analysis employing this equation requires that the velocity profile $v_x(y)$ be known before the integral term can be evaluated. This velocity profile must be assumed at the outset in the most reasonable fashion possible. The accuracy of the final result is directly dependent on how closely the assumed velocity profile matches the real one.

The use of equation (5-40) is illustrated in Example 5.1.

Example 5.1

Evaluate the boundary layer thickness and the local and mean coefficients of skin friction for the case of a laminar boundary layer on a flat plate. The velocity profile may be assumed to be a cubical parabola within the boundary layer and uniform outside the boundary layer.

An expression for $v_x(y)$ must be obtained first. According to the given information, the velocity profile is of the form

$$v_x(y) = \begin{cases} a + by + cy^2 + dy^3 & 0 < y < \delta \\ v_\infty\ (\text{constant}) & \delta < y \end{cases} \qquad (5\text{-}41)$$

The constants a, b, c, and d may be evaluated by means of the following boundary conditions on the velocity:

(1) $v_x = 0$ at $y = 0$ (no slip at the wall)

(2) $v_x = v_\infty$ at $y = \delta$ (by definition of δ)

(3) $\dfrac{dv_x}{dy} = 0$ at $y = \delta$ (by continuity of the slope of the velocity profile)

(4) $\dfrac{d^2v_x}{dy^2} = 0$ at $y = 0$ (required by Newton's 2nd law; see equation

$$(5\text{-}19))$$

Applying each of these boundary conditions, the constants in equation (5-41) become

$$a = 0 \qquad b = \frac{3}{2\delta} v_\infty \qquad c = 0 \qquad d = -\frac{v_\infty}{2\delta^3}$$

and the velocity profile expression becomes

$$\frac{v_x}{v_\infty} = \frac{3}{2}\frac{y}{\delta} - \frac{1}{2}\left(\frac{y}{\delta}\right)^3 \qquad 0 \le y \le \delta \qquad (5\text{-}42)$$

$$\frac{v_x}{v_\infty} = 1 \qquad\qquad \delta \le y \qquad (5\text{-}43)$$

For the situation described, that of constant free stream velocity, we may also write $dv_\infty/dx = 0$.

With the appropriate substitutions, equation (5-40) may be written for this case as

$$\frac{3}{2}\frac{\mu}{\rho}\frac{v_\infty}{\delta} = \frac{d}{dx}\int_0^\delta v_\infty{}^2 \left[\frac{3}{2}\frac{y}{\delta} - \frac{1}{2}\left(\frac{y}{\delta}\right)^3\right]\left[1 - \frac{3}{2}\frac{y}{\delta} + \frac{1}{2}\left(\frac{y}{\delta}\right)^3\right] dy$$

which, after integrating, becomes

$$\frac{3}{2}\,\nu\,\frac{v_\infty}{\delta} = \frac{39}{280}\frac{d}{dx}(v_\infty{}^2\,\delta)$$

As noted previously, v_∞ is constant; thus the simple differential equation for δ

$$\delta\,d\delta = \frac{140}{13}\frac{\nu}{v_\infty}\,dx$$

is obtained.

Carrying out the integration, we obtain, for the boundary layer thickness,

$$\frac{\delta}{x} = 4.64\,Re_x^{-1/2} \qquad (5\text{-}44)$$

The local coefficient of skin friction is obtained as follows:

$$C_{fx} \equiv \frac{\tau_0}{\dfrac{\rho v_\infty^2}{2}} = \frac{\mu \left.\dfrac{dv_x}{dy}\right|_{y=0}}{\dfrac{\rho v_\infty^2}{2}} = \frac{2\nu}{v_\infty^2} \frac{3}{2} \frac{v_\infty}{\delta}$$

$$= 0.646 \, \mathrm{Re}_x^{-1/2} \qquad (5\text{-}45)$$

The mean coefficient of skin friction is obtained by integrating the local coefficient expression from $x = 0$ to $x = L$. The result is

$$C_{fL} = 1.292 \, \mathrm{Re}_L^{-1/2} \qquad (5\text{-}46)$$

Equations (5-44), (5-45), and (5-46) may be compared with their counterparts from an exact analysis, equations (5-30), (5-35), and (5-37), respectively. Of note is the identical form of the two sets of results. The boundary layer thickness expression is approximately 7% lower than the exact result, and the skin friction coefficients differ from the exact results by only 3%. The error is due to the assumed variation for $v_x(y)$, given by equation (5-42). The agreement is remarkably good, however.

The problem illustrated in Example 5.1 is more than an academic exercise. It has been illustrated that the approximate integral approach to boundary layer analysis produces results which compare favorably with the correct answer. For very few cases other than laminar flow over a flat plate with constant free stream velocity is an exact result known. For most physical situations, the integral approach is the only reasonable way to treat the problem.[6] With the experience gained in Example 5.1, such problems may be approached with reasonable confidence.

5.1-1.7 Turbulent Flow Considerations

Turbulent flow is encountered more frequently than laminar flow, yet it is not nearly as amenable to analytical treatment. The distinction between the two types of flow has been made several times in our work thus far, but all analysis has been with regard to laminar flow. We shall now consider some basic ideas concerning analytical modeling of turbulent flows.

The fluid and flow variables in turbulent flow vary with time. For flow in a tube, the axial velocity varies with time, as shown in Figure 5.9.

The mean value of v_x is constant, indicating steady flow in the mean; however, at any instant in time the actual velocity differs from the mean

---

[6] Numerical methods treating fluid flow and convective heat transfer problems are being developed rapidly; the technique and physical modeling involved are still quite esoteric and complex. These methods may (and probably will) replace approximate techniques in future years.

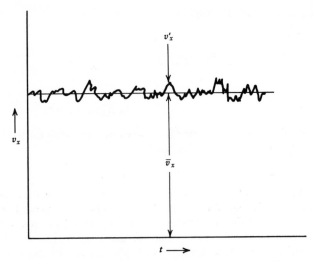

Figure 5.9   Time dependence of velocity in turbulent flow.

value by the relatively small amount v_x'. The mean, or time-average, velocity over the time interval $t_1 \leq t \leq t_2$ is represented by

$$\bar{v}_x(x, y, z) = \frac{1}{t_2 - t_1} \int_{t_1}^{t_2} v_x(x, y, z, t)\, dt \qquad (5\text{-}47)$$

In terms of $\bar{v}_x$, then, the axial velocity at any time may be written as

$$v_x(x, y, z, t) = \bar{v}_x(x, y, z) + v_x'(x, y, z, t) \qquad (5\text{-}48)$$

The time-dependent quantity on the right-hand side of equation (5-48) is called the *fluctuating velocity* and is due to the local randomness and chaotic nature of turbulent flow. These fluctuations characterize the effect on axial velocity of eddies or packets of fluid particles moving in a direction normal to that of bulk flow. The period of any velocity fluctuation is quite short. Intuitively, one should expect the time average of v_x' over any reasonable time interval to be zero. An integration of equation (5-48) in time will yield the result that

$$\bar{v}_x' = \frac{1}{t_2 - t_1} \int_{t_1}^{t_2} v_x'\, dt = 0 \qquad (5\text{-}49)$$

This result, $\bar{v}_x' = 0$, is quite reasonable since, over a long period of time, the fluctuating velocity $v_x' = v_x - \bar{v}_x$ will be positive and negative in equal amounts. A quantity which involves the square of the velocity will not behave in this same manner. Consider the time-averaged kinetic energy per unit volume expressed as

$$\overline{KE} = (1/2)\rho\overline{(v_x{}^2 + v_y{}^2 + v_z{}^2)}$$
$$= (1/2)\rho\overline{[(\bar{v}_x + v_x')^2 + (\bar{v}_y + v_y')^2 + (\bar{v}_z + v_z')^2]}$$

By expressing the average of the sum as the sum of the averages, we may also write

$$\overline{KE} = (1/2)\rho[\overline{(\bar{v}_x{}^2 + 2\bar{v}_x v_x' + v_x'{}^2)} + \overline{(\bar{v}_y{}^2 + 2\bar{v}_y v_y' + v_y'{}^2)}$$
$$+ \overline{(\bar{v}_z{}^2 + 2\bar{v}_z v_z' + v_z'{}^2)}]$$

and, since $\bar{v}_x' = \bar{v}_y' = \bar{v}_z' = 0$, according to equation (5-49), we have

$$\overline{KE} = (1/2)\rho(\bar{v}_x{}^2 + \bar{v}_y{}^2 + \bar{v}_z{}^2 + \overline{v_x'{}^2} + \overline{v_y'{}^2} + \overline{v_z'{}^2}) \qquad (5\text{-}50)$$

Thus the mean kinetic energy in a turbulent flow is greater than that for a laminar flow with the same mean velocity by the amount

$$\overline{v_x'{}^2} + \overline{v_y'{}^2} + \overline{v_z'{}^2}$$

The root-mean-square (rms) value of the velocity fluctuations is a part of a parameter designated the *intensity of turbulence*, defined as

$$I \equiv \frac{\left(\dfrac{\overline{v_x'{}^2} + \overline{v_y'{}^2} + \overline{v_z'{}^2}}{3}\right)^{1/2}}{v_{\mathrm{avg}}} \qquad (5\text{-}51)$$

where v_{avg} is the mean velocity of the flow.

We now direct our attention to the effect of these random fluctuations on pertinent flow parameters. Consider the control volume shown in Figure 5.10, located in a turbulent flow field.

The equation of motion in integral form was presented earlier as equation (2-28). This relation is repeated here for completeness.

$$\sum \mathbf{F} = \frac{\partial}{\partial t} \int_{\mathrm{cv}} \mathbf{v}\rho \, dV + \int_{\mathrm{cs}} \mathbf{v}\rho(\mathbf{v} \cdot \mathbf{n}) \, dA \qquad (2\text{-}28)$$

As a simple case, we consider the flux of x-directional momentum across the top of the element shown. At this surface there is mean flow in the x direction accompanied by fluctuating velocities v_x' and v_y'. Writing equation (2-28) for the x direction and taking a time average of each term, we obtain

$$\overline{\sum F_x} = \overline{\frac{\partial}{\partial t} \int_{\mathrm{top}} (\bar{v}_x + v_x')\rho \, dV}^{0} + \overline{\int_{\mathrm{top}} (\bar{v}_x + v_x')\rho v_y' \, dA}$$

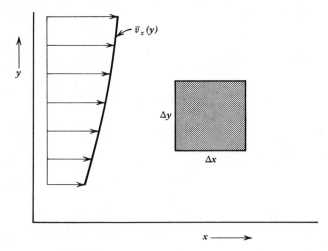

Figure 5.10   Control volume in turbulent flow.

which becomes

$$\sum F_x = \int_{\text{top}} \overset{0}{\overbrace{\bar{v}_x \rho v_y'}} \, dA + \int_{\text{top}} \overline{v_x' \rho v_y'} \, dA \tag{5-52}$$

The presence of turbulent fluctuations is thus seen to account for a flux of x-directional momentum at the top surface in the amount $\rho \overline{v_x' v_y'}$ per unit of area. This term may be thought of as the stress at the top surface due to turbulent effects, or a *turbulent shear stress*. Writing the total shear stress on the top surface as the sum of molecular (laminar) and turbulent contributions, we have

$$\tau_{yx} = \mu \frac{d\bar{v}_x}{dy} + \rho \overline{v_x' v_y'} \tag{5-53}$$

The turbulent contributions $\rho \overline{v_x' v_y'}$ is sometimes referred to as the *Reynolds stress*. There are six shear components to the complete Reynolds stress.

By analogy with the laminar flow relation, the turbulent shear stress may be written as

$$\tau_{yx}|_{\text{turbulent}} = \rho \overline{v_x' v_y'} = A_t \frac{d\bar{v}_x}{dy} \tag{5-54}$$

where A_t is a turbulent, or *eddy, viscosity*. An *eddy diffusivity of momentum* ε_m is obtained by introducing the density as

$$\varepsilon_m \equiv \frac{A_t}{\rho} \tag{5-55}$$

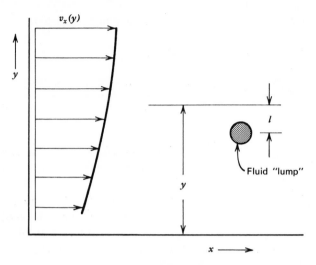

Figure 5.11 The Prandtl mixing length.

and equation (5-54) becomes

$$\tau_{yx}|_{\text{turbulent}} = \rho \varepsilon_m \frac{d\bar{v}_x}{dy} \tag{5-56}$$

It was Ludwig Prandtl,[7] in 1925, who postulated a model for the eddy diffusivity of momentum in terms of a mean free path of a turbulent eddy. This turbulent mean free path is designated the *mixing length* or, frequently, the *Prandtl mixing length*. In words, the mixing length may be thought of as the distance traveled by a group of particles, in a direction normal to that of mean flow, through which the properties remain those at the point of origin. Figure 5.11 should be observed for the following discussion.

The lump of fluid shown originates at a location $y - l$ from the x axis. It travels to its new location y through the action of the fluctuating velocity $v_{y'}$. The x velocity of the fluid lump will differ from that of its surroundings at its destination in the amount $\bar{v}_{x}|_{y-l} - \bar{v}_{x}|_{y}$. This difference in mean velocity is due to and, in fact, proportional to the fluctuating velocity $v_{y'}$. With these ideas in mind, we may write

$$v_{y}{}' = \bar{v}_{x}\big|_{y-l} - \bar{v}_{x}\big|_{y} = l \frac{d\bar{v}_x}{dy} \tag{5-57}$$

Equation (5-57) relates the fluctuating velocity component $v_{y'}$ to the mixing length and the mean velocity gradient. Assuming that x- and y-directional fluctuations are approximately equal, Prandtl was able to write,

[7] L. Prandtl, *ZAMM* **5** (1925): 136.

for the turbulent shear stress $\overline{\rho v_x' v_y'}$, given in equation (5-53),

$$\overline{\rho v_x' v_y'} = \rho l^2 \left(\frac{d\bar{v}_x}{dy}\right)^2 \tag{5-58}$$

and the eddy diffusivity, expressed in equation (5-56), becomes

$$\varepsilon_m = l^2 \frac{d\bar{v}_x}{dy} \tag{5-59}$$

One thing that should be noted from equation (5-59) is the complete dependence of the eddy diffusivity on the conditions of flow. It is thus a property of a flow rather than of a given fluid.

The utility of the mixing length hypothesis will now be demonstrated. An initial assumption that has proven reasonably accurate is that, in the vicinity of a wall, the mixing length is directly proportional to the distance from the wall. The shear stress in turbulent flow may thus be written with l replaced by Ky, K being a proportionality constant, as

$$\tau_{yx} = \rho K^2 y^2 \left(\frac{d\bar{v}_x}{dy}\right)^2 \tag{5-60}$$

Taking the square root, we have

$$\frac{d\bar{v}_x}{dy} = \frac{\sqrt{\tau_{yx}/\rho}}{Ky} \tag{5-61}$$

The quantity $\sqrt{\tau_{yx}/\rho}$ has the units of velocity and is often referred to as the *shear velocity*. Integration of equation (5-61) yields the expression

$$\bar{v}_x = \frac{1}{K} \sqrt{\frac{\tau_{yx}}{\rho}} \ln y + c \tag{5-62}$$

c being a constant of integration.

Figure 5.12 shows a comparison between equation (5-62) and some data obtained by Nikuradse[8] for flow in a smooth tube at $\mathrm{Re} = 10^6$. The agreement is seen to be quite good, in fact, remarkable considering the assumptions made which led to equation (5-62). The constant K in equation (5-62) has been found experimentally to be 0.4.

A general result of the empirical development thus far is the representation of shear stress, or momentum flux, in terms of both molecular and turbulent contributions. A complete expression is given by equation (5-63).

$$\frac{\tau_{yx}}{\rho} = (\nu + \varepsilon_m) \frac{d\bar{v}_x}{dy} \tag{5-63}$$

[8] J. Nikuradse, VDI-Forschungsheft, **356**, 1932.

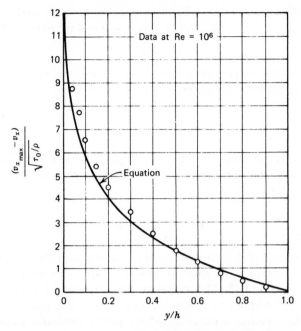

Figure 5.12 A comparison of equation (5-62) and data for flow in smooth tubes.

A usual procedure at this point is the definition of two nondimensional parameters as follows:

$$v^+ \equiv \frac{\bar{v}_x}{\sqrt{\tau_{yx}/\rho}} \tag{5-64}$$

$$y^+ \equiv \frac{\sqrt{\tau_{yx}/\rho} \; y}{\nu} \tag{5-65}$$

and the use of v^+ and y^+ to rewrite equation (5-63) in the form

$$dv^+ = \frac{dy^+}{1 + \dfrac{\varepsilon_m}{\nu}} \tag{5-66}$$

In the laminar sublayer $\varepsilon_m \simeq 0$, and equation (5-66) reduces to the simple form

$$dv^+ = dy^+$$

which may be integrated to yield

$$v^+ = y^+ + c$$

The integration constant may be evaluated, using the boundary condition that $v^+ = 0$ at $y^+ = 0$. The result is that $c = 0$ and we have, in the laminar sublayer,

$$v^+ = y^+ \qquad (5\text{-}67)$$

In a flow region where turbulence exists, $\varepsilon_m \neq 0$ and the ratio $\varepsilon_m/\nu \gg 1$, thus equation (5-66) may be written as

$$dv^+ = \frac{dy^+}{\varepsilon_m/\nu}$$

Utilizing earlier developments, the ratio ε_m/ν may be written as Ky^+. This substitution into the above equation and subsequent integration yields equation (5-62) in terms of v^+ and y^+

$$v^+ = \frac{1}{K}\ln y^+ + c \qquad (5\text{-}68)$$

Figure 5.13 is a plot of velocity data on semi-logarithmic coordinates of v^+ versus y^+. From plots such as this, one may observe the validity of equations (5-67) and (5-68) for representing turbulent velocity profiles. It is possible, in addition, to determine values for the constants in equation (5-68) and to establish a range on y^+ for which particular values of these constants apply.

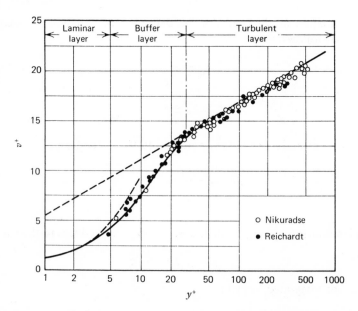

Figure 5.13   Nondimensional velocity profile for flow in tubes at high Reynolds numbers.

The flow regime is generally broken down into three parts with the following equations and ranges on y^+ pertaining:
the laminar sublayer, $0 < y^+ \leq 5$.

$$v^+ = y^+ \tag{5-67}$$

the buffer layer, $5 \leq y^+ \leq 30$

$$v^+ = 5.0 \ln y^+ - 3.05 \tag{5-69}$$

the turbulent core, $30 \leq y^+$

$$v^+ = 2.5 \ln y^+ + 5.5 \tag{5-70}$$

Equations (5-67), (5-69), and (5-70) comprise what is known as the *universal velocity profile*.

Two additional empirical results are of much use in describing turbulent flow effects. Both are attributed to Blasius. The first is a simple expression for the velocity in turbulent flow

$$\frac{\bar{v}_x}{\bar{v}_{x\ \text{max}}} = \left(\frac{y}{R}\right)^{1/7} \tag{5-71}$$

Equation (5-71) is the *one-seventh power law* for turbulent velocity profiles. The second empirical observation is that of the wall shear stress in a turbulent flow given by

$$\frac{\tau_0}{\rho} = \frac{0.0225\bar{v}_{x\ \text{max}}^2}{\left(\dfrac{\bar{v}_{x\ \text{max}}R}{\nu}\right)^{1/4}} \tag{5-72}$$

Both of the above equations are written for a circular conduit of radius R. For flow over a plane surface, these same equations apply with R replaced by δ, the boundary layer thickness.

The real utility of the Blasius observations expressed as equations (5-71) and (5-72) lies in their use in the momentum integral expression. Equation (5-40) was developed earlier and applies equally well for turbulent flow as for laminar flow. Example 5.2 illustrates the analysis of a turbulent boundary layer for flow over a flat plane surface.

Example 5.2

Evaluate the boundary layer thickness and the local and mean coefficients of skin friction for a turbulent boundary layer on a flat plate. The velocity profile is given by

$$\frac{v_x}{v_\infty} = \begin{cases} \left(\dfrac{y}{\delta}\right)^{1/7} & \text{for } 0 \leq y \leq \delta \\ 1 & \text{for } \delta \leq y \end{cases}$$

and the wall shear stress is given by equation (5-72). The free stream velocity v_∞ may be assumed constant. The integral equation in applicable form is now

$$\frac{\tau_0}{\rho} = \frac{d}{dx} \int_0^Y v_x(v_\infty - v_x)\, dy$$

Substituting the given velocity profile and equation (5-72), we have

$$0.0225\left(\frac{\nu}{v_\infty \delta}\right)^{1/4} = \frac{d}{dx} \int_0^\delta \left[\left(\frac{y}{\delta}\right)^{1/7} - \left(\frac{y}{\delta}\right)^{2/7} \right] dy$$

Integrating, then differentiating as indicated, we obtain

$$0.0225 \left(\frac{\nu}{v_\infty \delta}\right)^{1/4} = \frac{7}{62}\frac{d\delta}{dx}$$

Another integration yields the expression

$$\left(\frac{\nu}{v_\infty}\right)^{1/4} x = 3.45\delta^{5/4} + c \qquad (5\text{-}73)$$

The assumption of a turbulent boundary layer starting at $x = 0$ gives, for the boundary layer thickness,

$$\frac{\delta}{x} = 0.376\, \text{Re}_x^{-1/5} \qquad (5\text{-}74)$$

With this relation for δ substituted into equation (5-72), the local skin friction coefficient becomes

$$C_{fx} = \frac{\dfrac{\tau_0}{\rho}}{\dfrac{v_\infty{}^2}{2}}$$

$$= 0.0576\, \text{Re}_x^{-1/5} \qquad (5\text{-}75)$$

The mean coefficient of skin friction may be calculated from this expression in the usual way. The resulting equation for C_{fL} is

$$C_{fL} = 0.072\, \text{Re}_L^{-1/5} \qquad (5\text{-}76)$$

The results of Example 5.2 are, of course, approximate and limited to the range of flow where the Blasius one-seventh power law expression is reasonably valid. This restricts the use of equations (5-74), (5-75), and (5-76) to values of $\text{Re}_x < 10^7$. One should still observe the utility of the momentum integral in this case where no analytical results have been obtained as yet.

5.1-2   Combined Fluid Flow—Energy Considerations

Several basic considerations were made in Section 5.1-1 regarding viscous fluid flow and the hydrodynamic boundary layer. Each of the situations described and analyzed in the previous section has a heat transfer counterpart. Our consideration will next be given to the case where a fluid is adjacent to a wall at a different temperature. The resulting heat transfer rates will be determined, in large part, by the fluid flow phenomena discussed previously. Additional considerations regarding energy exchange and thermal properties will now be made.

The basic rate equation for convective energy exchange was expressed in Chapter 1. The relation is

$$q = hA \, \Delta T \tag{1-8}$$

where q is the rate of heat transfer between the surface and fluid, A is the area of contact, ΔT is the temperature difference between the surface and the bulk of the fluid, and h is the convective heat transfer coefficient. This equation is, basically, the defining relationship for h, and it is this coefficient or one of its associated parameters which is sought in most convective heat transfer problems. Equation (1-8) is extremely simple and, once a value of h has been determined, the heat flow rate q or heat flux q/A can be evaluated easily. The film coefficient h is related to the fluid flow mechanism, fluid properties, and flow geometry.

In any flow situation there is always a fluid layer, sometimes extremely thin, adjacent to a solid boundary wherein flow is laminar. We have already mentioned that laminar flow regions are characterized by molecular exchange of momentum. This is likewise true for energy; heat is transferred across a laminar film by molecular means (conduction); this conduction is always present in a system which is classified as convective. It is true, moreover, that the major resistance to heat transfer in a convective situation is that offered by the conduction process in the laminar film. Conduction is, thus, not only present but a major consideration in convection analysis.

It seems reasonable to associate the difficulty in transferring heat between a surface and adjacent fluid to the thickness of the laminar film; if this film is thick, then the conduction path is correspondingly long. A considerable difference in heat transfer rates exists between laminar and turbulent flow situations. In a turbulent flow there is large-scale transfer of fluid particles between regions at different temperatures. The heat transfer rate is obviously increased in the case of turbulent flow compared to that if the entire flow field is laminar.

The two general classifications of convective heat transfer are *natural*, or *free*, and *forced convection*. In natural convection, fluid motion is a result of the heat transfer. As a fluid is heated or cooled by an interaction with a solid boundary, there is an associated change in fluid density. The buoyant effects caused by this density change produce a natural circulation in which the affected fluid will move of its own accord past the solid surface; the fluid which replaces it is similarly affected by heat transfer, and the process is continued. In forced convection the fluid flows due to the influence of an external agency such as a fan or a pump. Forced convection is generally associated with greater velocities and correspondingly higher heat transfer rates than natural convection.

Thus far the term "boundary layer" has referred to the viscous effects associated with a fluid flowing past a solid surface. An additional consideration is the *thermal boundary layer*, which will be of great interest to us in heat transfer analysis.

Our sequence of coverage regarding convective heat transfer analysis and the convective heat transfer coefficient will be as follows:

(a) dimensional analysis combined with experimental results
(b) exact boundary layer analysis
(c) approximate (integral) boundary layer analysis
(d) analogy between momentum and heat transfer

5.1-2.1 Dimensional Analysis in Convective Heat Transfer

The treatment given to dimensional analyses of convection in this section is in no way intended to be exhaustive. A sufficient introduction to this technique will be included in the consideration of internal flow forced convection and external flow natural convection cases. The dimensional analysis approach will be discussed as we proceed but not considered in depth. For a more detailed treatment of dimensional analysis, the reader may consult Welty, Wicks, and Wilson[9] or Langhaar.[10]

Case 1

This situation is diagrammed in Figure 5.14. The flowing fluid and pipe wall are at different temperatures. The average velocity of flow is v, and the fluid properties of interest are the density, viscosity, heat capacity, and thermal conductivity. All of the significant variables, their symbols, and dimensional representations are listed in the table below.

[9] J. R. Welty, C. E. Wicks, and R. E. Wilson, *Fundamentals of Momentum, Heat and Mass Transfer*. (New York: John Wiley and Sons, Inc., 1969).
[10] H. L. Langhaar, *Dimensional Analysis and Theory of Models*, (New York: John Wiley and Sons, Inc., 1951).

Four dimensions are taken to be fundamental in this case: mass M, length L, time t, and temperature T. All dimensions are expressed in terms of these. The units of c_p, k, and h include a heat term; for this investigation, heat has been represented by energy with dimensions ML^2/t^2.

Variable	Symbol	Dimensions
Velocity	v	L/t
Tube diameter	D	L
Fluid density	ρ	M/L^3
viscosity	μ	M/Lt
heat capacity	c_p	L^2/t^2T
thermal conductivity	k	ML/t^3T
Heat transfer coefficient	h	M/t^3T

The Buckingham pi theorem states that the number of independent dimensionless groups i, needed to correlate n dimensional variables, is given by the expression

$$i = n - r \tag{5-77}$$

where r is the rank of a matrix having n columns and a number of rows equivalent to the number of fundamental dimensions—in the present case, four. The subject matrix is formed as follows:

	v	D	ρ	μ	c_p	k	h
M	0	0	1	1	0	1	1
L	1	1	-3	-1	2	1	0
t	-1	0	0	-1	-2	-3	-3
T	0	0	0	0	-1	-1	-1

The number in each position in the table is the exponent to which each of the dimensions must be raised to properly represent the given variable. For instance, the dimensions of k were shown to be ML/t^3T; thus the numbers in the k column are, in order, $(1, 1, -3, -1)$ since these are the powers of M, L, t, and T, respectively.

The array of numbers in this table comprises the matrix for the case under consideration. Its rank is 4, hence we must form $i = 7 - 4 = 3$ dimensionless parameters.

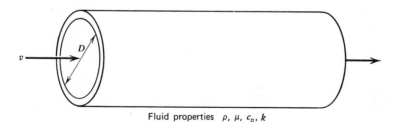

Fluid properties ρ, μ, c_p, k

Figure 5.14   Dimensional analysis parameters for forced convection in a circular conduit.

Each dimensionless parameter will be formed by combining a *core group* of r variables with one of the remaining variables not in the core. The core may include any 4 (in this case) of the variables which, among them, include all of the basic dimensions. Arbitrarily, we will choose D, ρ, μ, and k as the core. The parameters to be formed are now represented as pi groups, where

$$\pi_1 = D^a \, \rho^b \, \mu^c \, k^d \, v$$
$$\pi_2 = D^e \, \rho^f \, \mu^g \, k^h \, c_p$$
$$\pi_3 = D^i \, \rho^j \, \mu^k \, k^l \, h$$

Each pi group must be dimensionless. This is accomplished by appropriate choices of the exponents a, b, c, etc., using the very simple, mechanical technique which will now be illustrated.

Starting with π_1, we write it dimensionally as

$$1 = L^a \left(\frac{M}{L^3}\right)^b \left(\frac{M}{Lt}\right)^c \left(\frac{ML}{t^3 T}\right)^d \frac{L}{t}$$

In order that the equality be maintained, the exponents on M, L, t, and T must be the same on both sides of this expression. Noting that the exponents are all zero on the left hand side, we generate the following set of algebraic equations by equating exponents:

M: $\quad 0 = b + c + d$

L: $\quad 0 = a - 3b - c + d + 1$

t: $\quad 0 = -c - 3d - 1$

T: $\quad 0 = -d$

These equations may be solved to yield values for a, b, c, and d of 1, 1, -1, and 0, respectively. The first parameter may be written as

$$\pi_1 = \frac{D\rho v}{\mu} \equiv \mathrm{Re}_D$$

which we recognize as the *Reynolds number*.

A similar process may now be carried out for π_2.

$$1 = L^e \left(\frac{M}{L^3}\right)^f \left(\frac{M}{Lt}\right)^g \left(\frac{ML}{t^3 T}\right)^h \frac{L^2}{t^2 T}$$

M: $\quad 0 = f + g + h$

L: $\quad 0 = e - 3f - g + h + 2$

t: $\quad 0 = -g - 3h - 2$

T: $\quad 0 = -h - 1$

Values of e, f, g, and h are 0, 0, 1, and -1, respectively, and we have

$$\pi_2 = \frac{\mu c_p}{k} \equiv \mathrm{Pr}$$

The dimensionless parameter that results is designated Pr, the *Prandtl number*, which, we may note, is equal to the ratio of the molecular diffusivities of momentum and heat. This important parameter is defined according to

$$\Pr \equiv \frac{\alpha}{\nu} = \frac{\mu c_p}{k} = \nu / \alpha \tag{5-78}$$

The third parameter is now formed in the same manner as before.

$$1 = L^i \left(\frac{M}{L^3}\right)^j \left(\frac{M}{Lt}\right)^k \left(\frac{ML}{t^3 T}\right)^l \frac{M}{t^3 T}$$

$$M: \quad 0 = j + k + 1 + 1$$

$$L: \quad 0 = i - 3j - k + 1$$

$$t: \quad 0 = -k - 3l - 3$$

$$T: \quad 0 = -1 - 1$$

Solving for i, j, k, and l, we obtain 1, 0, 0, and -1, respectively, and we may write

$$\pi_3 = \frac{hD}{k} \equiv \text{Nu}_D$$

The parameter which resulted this time is designated Nu, the *Nusselt number*, which is one form of a nondimensional heat transfer coefficient. This parameter will be seen numerous times; its definition is

$$\text{Nu}_L \equiv \frac{hL}{k} \tag{5-79}$$

where the length L takes on various values, depending on system geometry. The significant length used in the Nusselt number is the same as that used in the Reynolds number.

The dimensional analysis of this internal flow case has led to a relationship of the form

$$\text{Nu} = f(\text{Re, Pr}) \tag{5-80}$$

which is common for forced convection heat transfer.

Had we chosen a different core group in our dimensional analysis, for instance D, ρ, μ, c_p, the pi groups formed would have been Re, Pr, and a nondimensional form of the heat transfer coefficient which is designated St, the *Stanton number*. The Stanton number may be formed by dividing Nu by the product RePr. It is defined as

$$\text{St} \equiv \frac{\text{Nu}}{\text{RePr}} = \frac{h}{\rho v c_p} \tag{5-81}$$

Using the Stanton number instead of Nu, we may write an alternate expression to equation (5-80) as

$$\text{St} = f(\text{Re, Pr}) \tag{5-82}$$

which is another popular form for correlating forced convection heat transfer data.

Dimensional analysis has indicated a way to reduce the seven variables significant to this situation to three dimensionless parameters. One must now obtain experimental data for this case in order to determine the functional relationships between parameters expressed in equations (5-80) and (5-82).

Case 2. Natural Convection Adjacent to a Heated Vertical Plate

Figure 5.15 is a representation of this situation. There is no specified velocity in this case; flow is a result of energy transfer between the plate at temperature T_0 and the fluid at ambient temperature T_∞. Fluid properties of interest are ρ, μ, c_p, k, and β. The last property listed is the coefficient of thermal expansion, which is used to represent the variation in fluid density with temperature, according to

$$\rho = \rho_0(1 + \beta \, \Delta T) \tag{5-83}$$

where ρ_0 is a reference density inside the heated layer and ΔT is the temperature difference between the fluid at the plate surface and that far removed from the plate.

The buoyant force per unit volume, F_B may be written as

$$F_B = (\rho - \rho_0)g$$

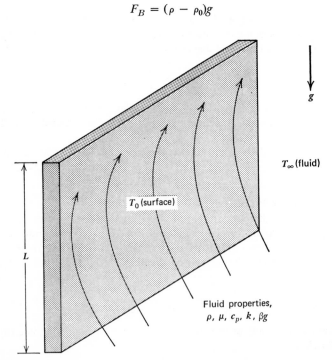

Figure 5.15 Dimensional analysis parameters for natural convection adjacent to a heated vertical plate.

and, with the substitution of equation (5-83),

$$F_B = \rho_0 \beta g \, \Delta T \qquad (5\text{-}84)$$

As a consequence of equation (5-84), the variables β, g, and ΔT should be included in a dimensional analysis of natural convection. Since β and g appear together in the buoyant force expression, they will be combined and treated as a single variable in our dimensional analysis.

The variables pertinent to this natural convection case are listed below, along with the symbols and dimensional representation for each.

Variable	Symbol	Dimension
Height	L	L
Temperature difference	ΔT	T
Fluid coefficient of thermal expansion	βg	L/Tt^2
Fluid density	ρ	M/L^3
viscosity	μ	M/Lt
heat capacity	c_p	$L^2/t^2 T$
thermal conductivity	k	$ML/t^3 T$
Heat transfer coefficient	h	$M/t^3 T$

Application of the Buckingham pi theorem will show that four dimensionless pi groups must be formed. If we designate as a core group the variables L, ρ, μ, and k, the pi groups are

$$\pi_1 = L^a \, \rho^b \, \mu^c \, k^d \, \Delta T$$

$$\pi_2 = L^e \, \rho^f \, \mu^g \, k^h \, \beta g$$

$$\pi_3 = L^i \, \rho^j \, \mu^k \, k^l \, c_p$$

$$\pi_4 = L^m \, \rho^n \, \mu^o \, k^p \, h$$

We shall not go through the mechanical process of evaluating the exponents, a, b, c, etc., needed to make the pi groups dimensionless. The results of this procedure yield the following:

$$\pi_1 = \frac{L^2 \, \rho^2 \, k \, \Delta T}{\mu^3}$$

$$\pi_2 = \frac{L \mu \, \beta g}{k}$$

$$\pi_3 = \frac{\mu c_p}{k} \equiv \mathrm{Pr}$$

$$\pi_4 = \frac{hL}{k} \equiv \mathrm{Nu}_L$$

The last two groups are recognized as the Prandtl and Nusselt numbers, respectively.

Analysis and experimental work have shown that the first two groups always appear together as a single dimensionless group. The parameter so formed is

$$\pi_1 \pi_2 = \frac{L^2 \rho^2 k \Delta T}{\mu^3} \frac{L \mu \beta g}{k}$$

$$= \frac{\rho^2 \beta g}{\mu^2} L^3 \Delta T$$

The resulting parameter is the *Grashof* number. It is designated Gr and defined as

$$\text{Gr} \equiv \frac{\rho^2 \beta g}{\mu^2} L^3 \Delta T \qquad (5\text{-}85)$$

The dimensional analysis in this case has shown that natural convection heat transfer data may be represented in nondimensional form as

$$\text{Nu} = f(\text{Gr}, \text{Pr}) \qquad (5\text{-}86)$$

Equation (5-86) for natural convection is very similar to equation (5-80), which applies to forced convection. Fluid velocity is represented non-dimensionally by the Reynolds number, which appears in forced convection analyses. In natural convection the flow is produced by buoyant effects resulting from a temperature difference. These effects are included in the Grashof number, and this parameter replaces the Reynolds number in the case of natural convection. The Stanton number has no significance in the case of natural convection.

A discussion of empirical correlations of heat transfer data in forms suggested by dimensional analysis is included in Section 5.3. Presently we shall examine some analytical approaches to the thermal boundary layer. The reader will observe how many of the dimensionless parameters developed in this section will arise quite naturally from our analysis.

5.1-2.2   Exact Analysis of the Thermal Boundary Layer

Much of the discussion pertinent to the thermal boundary layer is identical to that presented in Section 5.1-1.4 for the hydrodynamic boundary layer. The reader is referred to this earlier section for details. The present discussion will avoid unnecessary repetition.

The heat transfer situation directly amenable to analysis is that of external flow of a fluid parallel to an isothermal flat plate. In the region where boundary layer flow is laminar, both momentum and heat transfer are via molecular exchange. The thermal boundary layer for this situation is shown in Figure 5.16.

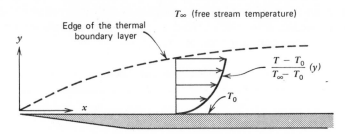

Figure 5.16 The thermal boundary layer for laminar flow over an isothermal plate.

Equations to be solved for this situation include the two-dimensional, incompressible form of the continuity equation

$$\frac{\partial v_x}{\partial x} + \frac{\partial v_y}{\partial y} = 0 \qquad (5\text{-}18)$$

the equation of motion with appropriate boundary layer modifications

$$v_x \frac{\partial v_x}{\partial x} + v_y \frac{\partial v_x}{\partial y} = \nu \frac{\partial^2 v_x}{\partial y^2} \qquad (5\text{-}19)$$

and the applicable form of the first law of thermodynamics (the energy equation).

For steady, incompressible, isobaric, two-dimensional, constant property flow without energy generation or viscous dissipation, the differential form of the energy equation, given by equation (2-49), reduces to

$$\rho c_p \left(v_x \frac{\partial T}{\partial x} + v_y \frac{\partial T}{\partial y} \right) = k \left(\frac{\partial^2 T}{\partial x^2} + \frac{\partial^2 T}{\partial y^2} \right) \qquad (5\text{-}87)$$

A boundary layer assumption relating second derivatives of T in order of magnitude is

$$\frac{\partial^2 T}{\partial y^2} \gg \frac{\partial^2 T}{\partial x^2}$$

thus equation (5-87) can be simplified and rearranged as

$$v_x \frac{\partial T}{\partial x} + v_y \frac{\partial T}{\partial y} = \alpha \frac{\partial^2 T}{\partial y^2} \qquad (5\text{-}88)$$

This expression, and equations (5-18) and (5-19), comprise the set to be solved in the present case.

Equations (5-18) and (5-19) were originally solved by Blasius to yield the results discussed in Section 5.1-1.4. The applicable velocity boundary conditions in dimensionless form were

$$\frac{v_x}{v_\infty} = \frac{v_y}{v_\infty} = 0 \quad \text{at } y = 0$$

$$\frac{v_x}{v_\infty} = 1 \quad \text{at } y = \infty$$

The differential equation for energy, equation (5-88), is obviously similar to equation (5-19). This similarity suggests the possibility of applying the Blasius solution or a similar approach to the thermal boundary layer.

We may apply the Blasius solution directly to the present case if the equations of energy and momentum can be made equivalent and if the boundary conditions are the same in both cases. These conditions are satisfied provided that

1. The coefficients of the second-order terms are equal; this requires that $v = \alpha$ or $Pr = 1$.
2. The nondimensional forms of the dependent variables are equivalent; this is accomplished by treating velocity as v_x/v_∞ and temperature as $(T - T_0)/(T_\infty - T_0)$. The boundary conditions may now be written as

$$\frac{v_x}{v_\infty} = \frac{T - T_0}{T_\infty - T_0} = 0 \quad \text{at } y = 0$$

$$\frac{v_x}{v_\infty} = \frac{T - T_0}{T_\infty - T_0} = 1 \quad \text{at } y = \infty$$

With these modifications one may perform the same transformations in the case of the energy equation as was done earlier for the equation of motion. Equation (5-28) now applies for both the momentum and energy exchange. This relation and the associated parameters are repeated here to aid in the present discussion. The equation is

$$f''' + \frac{1}{2} ff'' = 0 \tag{5-28}$$

where the parameters involved are

$$f'(\eta) = \frac{v_x}{v_\infty} = \frac{T - T_0}{T_\infty - T_0} \tag{5-89}$$

and

$$\eta \equiv y \left(\frac{v_\infty}{\nu x} \right)^{1/2} = \frac{y}{x} \left(\frac{x v_\infty}{\nu} \right)^{1/2} = \frac{y}{x} Re_x^{1/2} \tag{5-90}$$

Values of f, f', and f'' are presented for various values of η in Table 5.1.

Using the Blasius result, we may write

$$\frac{df'}{d\eta}(0) = f''(0) = \frac{d\left(\dfrac{v_x}{v_\infty}\right)}{d\left(\dfrac{y}{x}\mathrm{Re}_x^{1/2}\right)\bigg|_{y=0}}$$

$$= \frac{d\left(\dfrac{T - T_0}{T_\infty - T_0}\right)}{d\left(\dfrac{y}{x}\mathrm{Re}_x^{1/2}\right)\bigg|_{y=0}} \tag{5-91}$$

The above relations indicate that the velocity and temperature profiles are identical in the laminar boundary layer for the conditions specified; this is a direct consequence of specifying Pr = 1. A corresponding result is that the thermal and hydrodynamic boundary layers are of equal thickness. It is significant that the Prandtl numbers for most gases are sufficiently close to unity that the two boundary layers correspond very closely.

The heat flux at the surface may now be written as

$$\frac{q_y}{A} = h_x(T_0 - T_\infty) = -k\frac{\partial T}{\partial y}\bigg|_{y=0} \tag{5-92}$$

The surface temperature gradient may be developed from equation (5-91) according to

$$\frac{\partial T}{\partial y}\bigg|_{y=0} = (T_\infty - T_0)\left(\frac{0.332}{x}\mathrm{Re}_x^{1/2}\right) \tag{5-93}$$

which may be substituted into equation (5-92) to yield

$$\frac{q_y}{A} = h_x(T_0 - T_\infty) = -k(T_\infty - T_0)\left(\frac{0.332}{x}\mathrm{Re}_x^{1/2}\right)$$

Solving for h_x, we obtain

$$h_x = 0.332\frac{k}{x}\mathrm{Re}_x^{1/2} \tag{5-94}$$

or, finally, in dimensionless form

$$\mathrm{Nu}_x = \frac{h_x x}{k} = 0.332\mathrm{Re}_x^{1/2} \tag{5-95}$$

One should keep in mind the fact that equation (5-95) applies specifically to cases where Pr = 1. A more complete analysis would allow for other

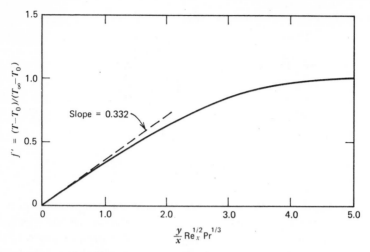

Figure 5.17 Nondimensional temperature profile in laminar flow over a flat plate.

values of the Prandtl number. Pohlhausen[11] was able to relate the hydrodynamic and thermal boundary layer thicknesses to the Prandtl number as

$$\frac{\delta}{\delta_t} = Pr^{1/3} \tag{5-96}$$

This function, $Pr^{1/3}$, included in the parameter η, allows the solution for the thermal boundary layer to be extended to include Pr values other than unity. A plot of dimensionless temperature versus the redefined η is shown in Figure 5.17. With the redefined η the heat flux at the surface expressed as before leads to the following expression for h_x:

$$h_x = 0.332 \frac{k}{x} Re_x^{1/2} Pr^{1/3} \tag{5-97}$$

and for Nu_x, we have

$$Nu_x \equiv \frac{h_x x}{k} = 0.332 Re_x^{1/2} Pr^{1/3} \tag{5-98}$$

Equations (5-97) and (5-98) may be used for the Prandtl number range of $0.6 < Pr < 50$.

These equations relate local values of h and Nu, i.e., those applying at a specific value of x. In engineering practice we are more interested in an average, or mean, value of h or Nu applying over a plate of some length L and width

[11] E. Pohlhausen, *ZAMM* **1** (1921): 115.

W. A mean value for these parameters may be obtained from local values by the process of integration. For a plate having the dimensions specified above, we may write

$$q_y = h_L A(T_0 - T_\infty) = \int_A h_x(T_0 - T_\infty)\, dA$$

Equation (5-97) may be used to express h_x in the integral; continuing, we have

$$h_L = \frac{1}{A} \int_A h_x\, dA$$

$$= \frac{1}{WL} \int_0^L 0.332 \frac{k}{x} \operatorname{Re}_x^{1/2} \operatorname{Pr}^{1/3} W\, dx$$

$$= \frac{0.332k}{L}\left(\frac{v_\infty}{\nu}\right)^{1/2} \operatorname{Pr}^{1/3} \int_0^L [x^{-1/2}]\, dx$$

$$= 0.664k\left(\frac{v_\infty}{\nu L}\right)^{1/2} \operatorname{Pr}^{1/3} \tag{5-99}$$

In terms of the Nusselt number this becomes

$$\operatorname{Nu}_L \equiv \frac{h_L L}{k} = 0.664 \operatorname{Re}_L^{1/2} \operatorname{Pr}^{1/3} \tag{5-100}$$

thus, in a similar fashion to the mean skin friction coefficient, the mean Nusselt number over a flat plate of length L has a value equal to twice that of the local Nusselt number evaluated at $x = L$.

In all of the preceding equations it is recommended that fluid properties be evaluated at the *film temperature* T_f, defined as

$$T_f \equiv \frac{T_0 + T_\infty}{2} \tag{5-101}$$

5.1-2.3 *Approximate Integral Analysis of the Thermal Boundary Layer*

As with the preceding section, the approach to the thermal boundary layer to be considered next has a counterpart in the hydrodynamic case. The development of this section will parallel the integral analysis of the hydrodynamic boundary layer considered in Section 5.1-1.6. The rationale behind an integral analysis is in the very limited scope of available analytical techniques. Exact analyses are, at present, limited to laminar flow and very simple geometries. Any situation of greater complexity will require an integral or numerical approach.

The control volume to be considered for purposes of illustration is shown in Figure 5.18. The applicable form of the integral expression for the first law of thermodynamics is repeated here for clarity. For the present case, each term in this relation is evaluated as follows:

$$\frac{\delta Q}{dt} = -k\,\Delta x\,\frac{\partial T}{\partial y}\bigg|_{y=0}$$

$$\frac{\delta W_s}{dt} = \frac{\delta W_\mu}{dt} = 0$$

$$\int_{cs}\left(e + \frac{P}{\rho}\right)\rho(\mathbf{v}\cdot\mathbf{n})\,dA = \int_0^Y \left(\frac{v^2}{2} + gy + u + \frac{P}{\rho}\right)\rho v_x\,dy\bigg|_{x+\Delta x}$$

$$-\int_0^Y\left(\frac{v^2}{2} + gy + u + \frac{P}{\rho}\right)\rho v_x\,dy\bigg|_x$$

$$-\frac{d}{dx}\int_0^Y \rho v_x\left(\frac{v^2}{2} + gy + u + \frac{P}{\rho}\right)\bigg|_\infty dy\,\Delta x$$

$$\int_{cv}\frac{\partial}{\partial t}e\rho\,dV = 0$$

In the absence of gravitational effects the convective energy flux terms may be written

$$\frac{v^2}{2} + u + \frac{P}{\rho} \equiv h_0 \simeq c_p T$$

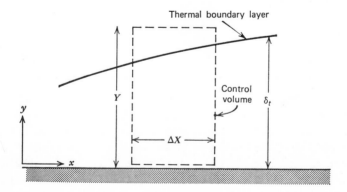

Figure 5.18 Control volume for integral analysis of the thermal boundary layer.

where h_0 and T are stagnation enthalpy and temperature, respectively. With this substitution the energy expression may now be written as

$$-k\,\Delta x\,\frac{\partial T}{\partial y}\bigg|_{y=0} = \int_0^Y \rho v_x c_p T\,dy\bigg|_{x+\Delta x} - \int_0^Y \rho v_x c_p T\,dy\bigg|_x$$

$$- \rho c_p\,\Delta x\,\frac{d}{dx}\int_0^Y v_x T_\infty\,dy \quad (5\text{-}102)$$

In equation (5-102) the free stream stagnation temperature is designated T_∞. The product ρc_p has been taken outside the integral; each property is evaluated at "average" conditions. The usual procedure will now be carried out; each term is divided by Δx, and the expression is evaluated in the limit as $\Delta x \to 0$. The result is

$$\frac{k}{\rho c_p}\frac{\partial T}{\partial y}\bigg|_{y=0} = \frac{d}{dx}\int_0^Y v_x(T_\infty - T)\,dy \quad (5\text{-}103)$$

This is the integral form of the energy equation, which is a counterpart to equation (5-40) for momentum. The procedure to solve equation (5-103) is to assume functions for $v_x(y)$ and $T(y)$ which may be substituted into the integral term and the entire expression then solved for the desired quantities. Note that both velocity and temperature profiles must be assumed in this case in contrast to the momentum integral equation where only velocity profiles were required.

The boundary conditions to be satisfied by an assumed temperature profile include

(1) $T - T_s = 0$ at $y = 0$

(2) $T - T_s = T_\infty - T_s$ at $y = \delta_t$

(3) $\dfrac{\partial}{\partial y}(T - T_s) = 0$ at $y = \delta_t$

(4) $\dfrac{\partial^2(T - T_s)}{\partial y^2} = 0$ at $y = 0$ [see equation (5-88)]

In Section 5.1-1.6 a cubical parabola was assumed as the shape of the velocity profile. We shall here assume a similar form for the temperature distribution in the thermal boundary layer. Doing so, the temperature profile has the form

$$T - T_s = \begin{cases} A + By + Cy^2 + Dy^3 & 0 < y < \delta_t \\ T_\infty - T_s & \delta_t < y \end{cases} \quad (5\text{-}104)$$

An application of the boundary conditions yields the following expression for $T - T_s$:

$$\frac{T - T_s}{T_\infty - T_s} = \frac{3}{2}\frac{y}{\delta_t} - \frac{1}{2}\left(\frac{y}{\delta_t}\right)^3 \tag{5-105}$$

The corresponding velocity expression, obtained earlier, is

$$\frac{v_x}{v_\infty} = \frac{3}{2}\frac{y}{\delta} - \frac{1}{2}\left(\frac{y}{\delta}\right)^3 \tag{5-42}$$

The substitution of equations (5-42) and (5-104) into the integral expression and subsequent solution, assuming $\delta = \delta_t$, will yield

$$\mathrm{Nu}_x = 0.36\,\mathrm{Re}_x^{1/2}\,\mathrm{Pr}^{1/3} \tag{5-106}$$

a result which is identical in form but approximately 8% larger than that of exact analysis reported in equation (5-98).

The conclusions stated earlier for momentum integral analysis apply here as well. In many cases of practical interest, this is the only reasonable approach to be used. The achieved result, equation (5-106), is sufficiently close to that of an exact analysis to give one confidence in applying this technique to problems for which no analytical result has been found.

5.1-2.4   Energy Transfer in Turbulent Flow

Certain turbulent flow concepts and models were introduced in Section 5.1-1.7. Most of the ideas presented in this earlier section related to velocities and momentum exchange. The discussion to be presented here pertains to the case of temperatures and the transport of thermal energy. The reader is referred to the earlier discussion for many of the details which apply to the present development.

Depicted in Figure 5.19 is a temperature profile in turbulent flow. The nomenclature and concepts are similar to those used in Figures 5.10 and 5.11.

A "lump" of fluid as shown, originating at a position $y - l$ from the x axis, will move to a new location y, due to the presence of a fluctuating velocity v_y', which is characteristic of turbulent flow. The distance l, moved by the "lump" of fluid normal to the direction of mean flow through which the fluid properties remain those at the point of origin, has previously been designated the Prandtl mixing length. The instantaneous rate of energy transport in the y direction is

$$(\rho v_y')(c_p T)$$

where the temperature at $y - l$ is composed of mean and fluctuating components; this may be expressed as

$$T = \bar{T} + T' \tag{5-107}$$

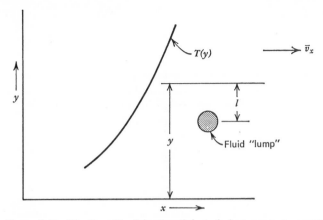

Figure 5.19   The Prandtl mixing length in turbulent energy transport.

with the meaning of each term evident from the sketch of temperature variation with time at a point shown in Figure 5.20. Completely analogous to a fluctuating velocity component is the temperature fluctuation written as

$$T' \simeq -l\frac{d\bar{T}}{dy} \tag{5-108}$$

With these ideas in mind, the y-directional heat transfer per unit area may be written

$$\frac{q_y}{A} = \rho c_p v_y'(\bar{T} + T')$$

A time average of this expression will yield the energy transport due to turbulent effects; we obtain the expression

$$\frac{q_y}{A}\bigg|_{\text{turbulent}} = \rho c_p (\overline{v_y'\bar{T}} + \overline{v_y'T'})$$

$$0$$

$$= \rho c_p \overline{v_y'T'} \tag{5-109}$$

By substituting equation (5-108) for T', we have

$$\frac{q_y}{A}\bigg|_{\text{turbulent}} = -\rho c_p \overline{v_y'l}\frac{d\bar{T}}{dy} \tag{5-110}$$

The term $\overline{v_y'l}$ is seen to play the role in turbulent flow that α, the molecular diffusivity of heat, does in laminar flow. For this reason we designate the

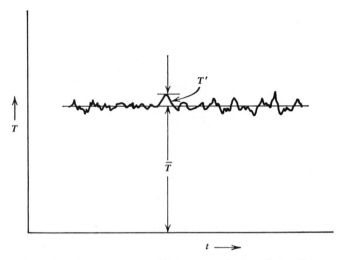

Figure 5.20   Temperature-time history at a point in turbulent flow.

quantity ε_H, the *eddy diffusivity of heat*, defined as

$$\varepsilon_H \equiv \overline{v'l} \tag{5-111}$$

and write the turbulent energy flux expression as

$$\frac{q_y}{A}\bigg|_{\text{turbulent}} = -\rho c_p \varepsilon_H \frac{d\bar{T}}{dy} \tag{5-112}$$

A complete expression for heat flux, including both molecular and turbulent effects, is

$$\frac{q_y}{A} = -\rho c_p (\alpha + \varepsilon_H) \frac{d\bar{T}}{dy} \tag{5-113}$$

The reader will recall that the ratio of the molecular diffusivities of momentum and heat, ν/α, was designated the *Prandtl number*. In analogous fashion a *turbulent Prandtl number* may be defined by the ratio of the eddy diffusivities

$$\text{Pr}_t \equiv \frac{\varepsilon_m}{\varepsilon_H} \tag{5-114}$$

5.1-2.5   Energy and Momentum Transfer Analogies

In the discussions and analyses thus far many similarities in mechanism and in analytical techniques have been noted between momentum and energy transfer. In the present section we shall use this analogous behavior to obtain applicable heat transfer parameters from their analogous momentum transfer counterparts.

The Reynolds Analogy. The similarities between momentum and heat transfer were noted first by Osborne Reynolds in 1874.[12] The first quantitative relationship between the two was made by Reynolds in 1883.[13]

It was noted in the previous section that a Prandtl number exists in both turbulent and laminar flow. The turbulent parameter is the ratio of ε_m to ε_H, both of these eddy diffusivities having been developed equal to $\overline{lv_x'}$. Assuming that all of the hypotheses upon which the mixing length model is based are correct, then the turbulent Prandtl number, being the ratio of like quantities, should be unity. Jenkins[14] has calculated the ratio of $\varepsilon_H/\varepsilon_m$ for various values of the molecular Prandtl number and turbulence intensity. His results indicate that $\varepsilon_H/\varepsilon_m$ approaches unity for large values of ε_m/ν for all values of Pr. In a fully turbulent regime it is quite reasonable to assume a value of 1 for Pr_t.

In a case where $\mathrm{Pr} = 1$, either fully turbulent flow for any fluid or laminar flow of a fluid for which $\nu/\alpha \simeq 1$, the dimensionless velocity and temperature gradients are related according to

$$\frac{d}{dy}\left(\frac{v_x}{v_\infty}\right)\bigg|_{y=0} = \frac{d}{dy}\left(\frac{T - T_0}{T_\infty - T_0}\right)\bigg|_{y=0} \tag{5-115}$$

With $\mathrm{Pr} = 1$ we may write $k = \mu c_p$, and equation (5-125) may be altered in the form

$$\mu c_p \frac{d}{dy}\left(\frac{v_x}{v_\infty}\right)\bigg|_{y=0} = k \frac{d}{dy}\left(\frac{T - T_0}{T_\infty - T_0}\right)\bigg|_{y=0}$$

which may be written

$$\frac{\mu c_p}{v_\infty}\frac{dv_x}{dy}\bigg|_{y=0} = -\frac{k}{T_0 - T_\infty}\frac{d(T - T_0)}{dy}\bigg|_{y=0} \tag{5-116}$$

By definition of the convective heat transfer coefficient h, we have

$$\frac{q_y}{A} = h(T_0 - T_\infty) = -k\frac{d}{dy}(T - T_0)\bigg|_{y=0}$$

or

$$h = -\frac{k}{T_0 - T_\infty}\frac{d}{dy}(T - T_0)\bigg|_{y=0}$$

[12] O. Reynolds, *Proc. Manchester Lit. Phil. Soc.* **14** (1874): 7.
[13] O. Reynolds, *Trans. Roy. Soc. (London)* **174A** (1883): 935.
[14] R. Jenkins, *Ht. Trans Fl. Mech. Inst.* Stanford, Calif.: Stanford Univ. Press, (1951): 147.

Equation (5-116) may now be written as

$$h = \frac{\mu c_p}{v_\infty} \frac{dv_x}{dy}\bigg|_{y=0} \tag{5-117}$$

We next recall the definition of the skin friction coefficient

$$C_f \equiv \frac{F/A}{\dfrac{\rho v_\infty^2}{2}} = \frac{\tau_0}{\dfrac{\rho v_\infty^2}{2}} \tag{5-4}$$

or

$$C_f = \frac{\mu \dfrac{dv_x}{dy}\bigg|_{y=0}}{\dfrac{\rho v_\infty^2}{2}}$$

Making the appropriate substitution in equation (5-117), we obtain

$$h = (\rho v_\infty c_p) \frac{C_f}{2}$$

or, in dimensionless form,

$$\frac{h}{\rho c_p v_\infty} \equiv \mathrm{St} = \frac{C_f}{2} \tag{5-118}$$

Equation (5-118) is *Reynolds analogy* between energy and momentum transfer. This simple expression allows the coefficient h to be determined from the fluid flow parameter C_f. Reynolds analogy is valid, provided that (1) $\mathrm{Pr} = 1$; this has been discussed previously; and (2) the drag forces are wholly viscous in nature; this is obvious in that C_f is used. If the drag is not wholly viscous, then equation (5-5) does not hold and Reynolds analogy is not valid. Flow situations which satisfy the requirement of no form drag are flow in a closed conduit or any exterior flow in which no boundary layer separation occurs.

The Prandtl Analogy. The restriction on the Reynolds analogy that $\mathrm{Pr} = 1$ is a severe limitation, and various techniques have been used to generate an expression which might be valid over a reasonable span of Prandtl numbers.

In Section 5.1-2.4 it was noted that the effective Prandtl number in turbulent flow is the ratio of eddy diffusivities and that this ratio, based on the mixing length hypothesis, is approximately 1. Thus in a region where flow is fully turbulent, the Reynolds analogy is valid. It is true, however, that in any convective heat transfer situation, a boundary layer exists at the solid-fluid interface where molecular transfer occurs and molecular rather than eddy properties predominate. Some allowance must be made for this region where the molecular Prandtl number is significant and may have values considerably different from unity.

As a first look at such an analysis let us divide our considerations into two parts; the first will be in the region where flow is laminar and turbulent effects are absent, and the second will be in a region of fully turbulent flow where molecular effects are insignificant. These two flow regimes will be joined at some hypothetical distance from the solid boundary where $y = \xi$.

Within the region of laminar flow the applicable momentum and energy transport equations are

$$\frac{\tau}{\rho} = \nu \frac{dv_x}{dy}$$

$$\frac{q_y}{A} = -\rho c_p \alpha \frac{dT}{dy}$$

Separating variables and integrating between $y = 0$ and $y = \xi$, we have in both cases

$$\int_0^{v_\xi} dv_x = \frac{\tau}{\rho\nu} \int_0^\xi dy$$

$$\int_{T_0}^{T_\xi} dT = -\frac{q_y}{A\rho c_p \alpha} \int_0^\xi dy$$

The velocity and temperature differences across the laminar layer become

$$v_\xi = \frac{\tau}{\rho\nu} \xi \tag{5-119}$$

and

$$T_\xi - T_0 = -\frac{q_y}{A\rho c_p \alpha} \xi \tag{5-120}$$

The hypothetical distance ξ may be eliminated between these two equations to yield

$$\frac{\rho\nu v_\xi}{\tau} = \frac{A\rho c_p \alpha}{q_y} (T_0 - T_\xi) \tag{5-121}$$

We now direct our attention to the region of fully turbulent flow where Reynolds analogy applies. Equation (5-118) may be written as

$$\frac{h}{\rho c_p (v_\infty - v_\xi)} = \frac{C_f}{2} \tag{5-118}$$

and, expressing h and C_f in terms of their definitions, we obtain

$$\frac{q_y/A}{\rho c_p (v_\infty - v_\xi)(T_\xi - T_\infty)} = \frac{\tau}{\rho (v_\infty - v_\xi)^2}$$

A simple rearrangement of this expression yields a modified form of Reynolds analogy in the region $\xi \le y \le \infty$.

$$\frac{\rho (v_\infty - v_\xi)}{\tau} = \frac{A \rho c_p}{q_y} (T_\xi - T_\infty) \qquad (5\text{-}122)$$

We now eliminate T_ξ between equations (5-121) and (5-122) with the following result:

$$\frac{\rho}{\tau}\left[v_\infty + v \left(\frac{\nu}{\alpha} - 1 \right) \right] = \frac{A \rho c_p}{q_y} (T_0 - T_\infty) \qquad (5\text{-}123)$$

At this point the definitions of C_f and h are reintroduced, these forms being

$$C_f = \frac{\tau}{\dfrac{\rho v_\infty^2}{2}}$$

and

$$h = \frac{q_y/A}{T_0 - T_\infty}$$

and equation (5-123) becomes

$$\frac{v_\infty + v_\xi \left(\dfrac{\nu}{\alpha} - 1 \right)}{\dfrac{C_f}{2} v_\infty^2} = \frac{\rho c_p}{h}$$

Inverting both sides of this expression and dividing through by v_∞, we obtain the nondimensional form

$$\frac{h}{\rho c_p v_\infty} = \frac{\dfrac{C_f}{2}}{1 + \dfrac{v_\xi}{v_\infty}\left(\dfrac{\nu}{\alpha} - 1 \right)} \qquad (5\text{-}124)$$

In equation (5-124) the familiar dimensionless forms $\mathrm{St} \equiv h/\rho c_p v_\infty$ and $\mathrm{Pr} \equiv \nu/\alpha$ are apparent. It is also clear that, in the case where $\mathrm{Pr} \equiv \nu/\alpha = 1$, this relation reduces to equation (5-118), the Reynolds analogy. For a **Prandtl** number not equal to unity, the Stanton number is a function of C_f,

Pr, and the velocity ratio v_ξ/v_∞. This latter quantity is inconvenient and we may eliminate it as follows.

Equation (5-67) expressed the condition at the edge of the laminar sub-layer as

$$v^+ = y^+ = 5 \tag{5-67}$$

The definition of v^+ may be introduced to yield

$$v^+ \equiv \frac{v_\xi}{\sqrt{\tau/\rho}} = 5$$

along with that of the skin friction coefficient

$$C_f = \frac{\tau}{\dfrac{\rho v_\infty{}^2}{2}}$$

from which we may write

$$\left(\frac{\tau}{\rho}\right)^{1/2} = v_\infty \left(\frac{C_f}{2}\right)^{1/2}$$

This may now be combined with the previous expression to yield

$$\frac{v_\xi}{v_\infty} = 5\left(\frac{C_f}{2}\right)^{1/2} \tag{5-125}$$

and, with equation (5-125) substituted in appropriate fashion, equation (5-124) becomes

$$\mathrm{St} = \frac{C_f/2}{1 + 5\left(\dfrac{C_f}{2}\right)^{1/2}(\mathrm{Pr} - 1)} \tag{5-126}$$

The above relation is the *Prandtl analogy*. It is written in the form $\mathrm{St} = f(C_f, \mathrm{Pr})$, thus is relatively easy to use. It is applicable to a heat transfer situation, provided that (1) no boundary layer separation exists, and (2) the Prandtl number is in the moderate range, viz., $0.5 < \mathrm{Pr} < 30$.

The Prandtl analogy is an obvious improvement on the Reynolds analogy since it applies to cases in which $\mathrm{Pr} \neq 1$. It is still approximate, however, in that the assumption was made that at $y^+ = 5$ there is a sudden transition from completely laminar to completely turbulent flow. A more realistic model would include considerations of the laminar sublayer, the buffer layer, and the turbulent core, as described and analyzed in Section 5.1-1.7. Such a consideration was presented by T. von Kármán, and his result, the *von Kármán*[15] *analogy,* was

$$St = \cfrac{C_f/2}{1 + 5\left(\cfrac{C_f}{2}\right)^{1/2}\{Pr - 1 + \ln[1 + \tfrac{1}{6}(5\,Pr - 5)]\}} \tag{5-127}$$

The von Kármán analogy is subject to the same restrictions as the Prandtl analogy. As with the Prandtl analogy, equation (5-127) becomes the Reynolds analogy for $Pr = 1$.

Several refinements on the Prandtl and von Kármán analogies, as well as other analogies based on different turbulent models, are presented in Kays.[16]

The Colburn Analogy. A. P. Colburn,[17] in 1933, presented an analogical form which differs from those mentioned previously. We shall now examine the situation where the Colburn analogy has some analytical basis.

The reader may recall the discussion in Section 5.1-2.2 where, for a laminar boundary layer on a flat plate, the exact result is

$$Nu_x = 0.332\,Re_x^{1/2}\,Pr^{1/3} \tag{5-98}$$

and, in Section 5.1-1.5, the result was obtained that

$$C_{fx} = 0.664\,Re_x^{-1/2} \tag{5-33}$$

If both sides of equation (5-98) are divided by the product $Re_x Pr^{1/3}$, and equation (5-33) is applied to the result, we obtain

$$\frac{Nu_x}{Re_x\,Pr^{1/3}} = \frac{0.332}{Re_x^{1/2}} = \frac{C_{fx}}{2} \tag{5-128}$$

Multiplying and dividing the left-hand side of this expression by $Pr^{2/3}$, we have

$$\frac{Nu_x}{Re_x\,Pr^{1/3}}\,\frac{Pr^{2/3}}{Pr^{2/3}} = \frac{Nu_x}{Re\,Pr}\,Pr^{2/3} = St\,Pr^{2/3}$$

and equation (5-128) becomes

$$St\,Pr^{2/3} = \frac{C_f}{2} \tag{5-129}$$

Equation (5-129) is the *Colburn analogy* expression and is seen to be exact for the laminar external boundary layer. Colburn showed this relation to be

[15] T. von Kármán, *Trans. ASME* **61** (1939): 705.
[16] Kays, Chapter 9.
[17] A. P. Colburn, *Trans. AIChE* **29** (1933): 174.

extremely effective in correlating data for a wide range of convective heat transfer situations, including various geometries and flow conditions. He concluded that this expression predicts reliable values for the Stanton number, provided that (1) there is no boundary layer separation, and (2) the Prandtl number is within the range $0.5 < \text{Pr} < 50$. As with the other analogies presented, the Colburn analogy reduces to Reynolds analogy for $\text{Pr} = 1$.

Each analogy expression presented in this section has been in the form $\text{St} = f(C_f, \text{Pr})$. The requirement that $\text{Pr} = 1$ makes the Reynolds analogy useful only as a rough approximation in all but a very few practical cases. The Prandtl and von Kármán expressions are more useful but also more complex than the Reynolds analogy expression. The Colburn analogy includes the best combination of attributes; it is valid over a range of Prandtl numbers, and is a quite simple expression.

An additional remark concerning the Prandtl number is appropriate at this point. The range of Prandtl numbers for which the expressions presented thus far are valid is between 0.5 and 50. This range is adequate for gases and most commonly used liquids. Liquids whose Prandtl numbers lie outside this range vary from the liquid metals, typical values of Pr being 0.01 to 0.03, to the heavy oils, which have Pr values of 10^5 and higher.

5.1-2.6 Flow in a Tube; Solutions for Constant Wall Heat Flux and Constant Wall Surface Temperature

The basic problem for which the analog expressions from the previous section are useful is that of flow in a closed conduit. Conduit flow, the most common type being in a tube or pipe, is wholly viscous in nature, thus satisfying the requirement on all analog expressions that no boundary layer separation exist. In practice, almost any variations or combinations of boundary conditions may exist at a tube surface. The limiting cases are those of constant heat flux and constant surface temperature. The first condition can be achieved experimentally by electrical heating through a heater sheath or by winding a tube uniformly with a heating wire or ribbon. The constant temperature condition is most nearly realized in practice by allowing condensation of a saturated vapor on the outside surface of the tube. Expressions which pertain to each of these conditions will now be developed.

Specified Wall Heat Flux. Our considerations will be based on the control volume Δx thick and bounded by the tube walls, as shown in Figure 5.21.

A first law of thermodynamics analysis for a steady-state condition will yield

$$q_1 - q_2 + q_3 = 0$$

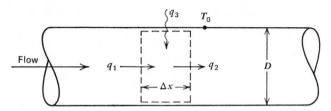

Figure 5.21   Control volume for analysis of heat transfer in tube flow with specified wall heat flux.

where q_1 and q_2 represent the energy transported in and out of the control volume by fluid flow and q_3 is that transferred to the control volume by the condition at the wall. These terms may be evaluated individually as

$$q_1 = \rho \, \frac{\pi D^2}{4} \, v_x c_p T \big|_x$$

$$q_2 = \rho \, \frac{\pi D^2}{4} \, v_x c_p T \big|_{x+\Delta x}$$

$$q_3 = \frac{q_r}{A} \, \pi D \, \Delta x$$

Making these substitutions into the first law expression, we have

$$\rho \, \frac{\pi D^2}{4} \, c_p (v_x T \big|_{x+\Delta x} \, - \, v_x T \big|_x) - \frac{q_r}{A} \, \pi D \, \Delta x = 0$$

For fully developed flow, $v_x \neq v_x(x)$; thus our expression may be rearranged to the form

$$\rho \, \frac{\pi D^2}{4} \, c_p v_x \, \frac{T \big|_{x+\Delta x} - T \big|_x}{\Delta x} - \frac{q_r}{A} \, \pi D = 0$$

Taking the limit as $\Delta x \to 0$, we have

$$\rho c_p v_x \, \frac{D}{4} \frac{dT}{dx} - \frac{q_r}{A} = 0$$

The temperature gradient may be solved for as

$$\frac{dT}{dx} = \frac{1}{\rho c_p v_x \, \dfrac{D}{4}} \, \frac{q_r}{A} \tag{5-130}$$

and, after separating variables and integrating, we obtain an expression for the mean fluid temperature as a function of position along the tube

$$T - T_e = \frac{1}{\rho c_p v_x \dfrac{D}{4}} \int_0^x \frac{q_r}{A} \, dx \qquad (5\text{-}131)$$

which, for the case of constant wall heat flux, is

$$T - T_e = 4 \frac{q_r/A}{\rho c_p v_x} \frac{x}{D} \qquad (5\text{-}132)$$

At the entrance, $x = 0$, the temperature is T_e.

Equation (5-131) may be used with any known function for wall heat flux. A very complete analysis of heat transfer in a tube with various distributions of surface heat flux is given by Kays.[18]

One quantity that is often of interest to an engineer is the surface temperature of a tube in which a coolant flows. In this case, with the heat flux known and the mean temperature available from equation (5-132), the surface temperature can be found from

$$T_0 = T_\infty + \frac{q_r/A}{h} \qquad (5\text{-}133)$$

The convective heat transfer coefficient may be evaluated, using an applicable analog expression from the previous section.

Example 5.3

The heat flux along the axis of a tube varies sinusoidally according to the expression

$$\frac{q}{A} = 600 + 1800 \sin \frac{\pi x}{L}$$

where q/A is in Btu/hr-ft^2. The tube length is 10 ft and its diameter is 1 in. Show, as a function of x,

(a) the local heat flux.
(b) the mean fluid temperature.
(c) the local surface temperature.

The fully developed condition may be assumed valid such that h_x is constant. The flowing fluid is water with an entering velocity of 1 ft/sec and a temperature of 50°F.

[18] Kays, Chapter 10.

Heat flux levels may be calculated readily from the given expression. Substituting the properties which apply, we have

$$\frac{q}{A} = 600 + 1800 \sin \pi \frac{x}{10}$$

where x is in feet.

The mean fluid temperature is expressed by equation (5-131). After inserting the proper quantities, the mean temperature expression is

$$T_x - 50 = \left[\frac{1}{\left(62.4 \frac{\text{lb}_m}{\text{ft}^3}\right)\left(1 \frac{\text{Btu}}{\text{lb}_m\text{-}^\circ\text{F}}\right)\left(1 \frac{\text{ft}}{\text{sec}}\right)\left(\frac{1\ \text{ft}}{48}\right)\left(\frac{3600\ \text{sec}}{\text{hr}}\right)} \right]$$

$$\left[\int_0^x \left(600 + 1800 \sin \frac{\pi x}{L}\right) dx \frac{\text{Btu}}{\text{hr-ft}} \right]$$

$$= \left(\frac{1}{4680}\right)\left[600x - 1800 \frac{L}{\pi} \cos \frac{\pi x}{L} \right]_0^x$$

$$T_x - 50 = 0.128 \left[x - \frac{30}{\pi} \cos \frac{\pi x}{L} \right]_0^x$$

$$= 0.128 \left[x + 9.54 \left(1 - \cos \frac{\pi x}{L}\right) \right]$$

Values of q_x/A and T_x are tabulated below and plotted in Figure 5.22. It is seen that the fluid temperature changes by less than 4°F throughout the 10 feet of pipe length.

The tube surface temperature will be determined from equation (5-133); the Colburn analogy will be used to predict a value of h.

First the Reynolds number is evaluated as

$$\text{Re} = \frac{Dv_x}{\nu} = \frac{(1/12\ \text{ft})\left(1 \frac{\text{ft}}{\text{sec}}\right)}{1.25 \times 10^{-5}\ \text{ft}^2/\text{sec}} = 6.66 \times 10^3$$

An average film temperature of 55°F was assumed in reading ν from the tables in the appendix. With this value for Re, assuming a smooth 1-inch tube, Figure 5.6 may be used to obtain C_f; the value which applies is 0.0084. At 55°F, Pr for water is 8.2. The Stanton number may now be determined as

$$\text{StPr}^{2/3} = \frac{C_f}{2}$$

$$\text{St} = \frac{0.0084}{2} (8.2)^{-2/3} = 0.00103$$

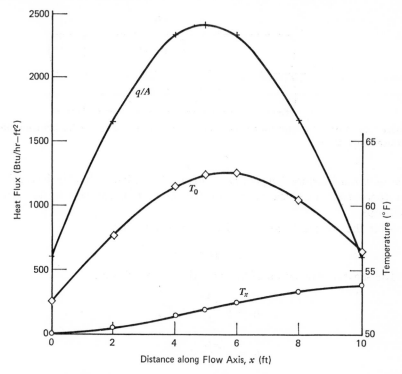

Figure 5.22 Heat flux, fluid temperature, and wall surface temperature for water flowing in a one-inch-diameter tube.

The heat transfer coefficient is now obtained as

$$h = \text{St } \rho c_p v_x = (1.03 \times 10^{-3})\left(62.4 \frac{\text{lb}_m}{\text{ft}^3}\right)\left(\frac{1 \text{ Btu}}{\text{lb}_m\text{-}^\circ\text{F}}\right)\left(\frac{1 \text{ ft}}{\text{sec}}\right)\left(\frac{3600 \text{ sec}}{\text{hr}}\right)$$

$$= 231 \text{ Btu/hr-ft}^2\text{-}^\circ\text{F}$$

We may now use this value for h in equation (5-133) to obtain the tube surface temperatures. The values for T_0 are tabulated and plotted along with q/A and T_x.

x	q/A	T_x	T_0
0	600	50	52.6
2	1658	50.5	57.7
4	2312	51.4	61.4
5	2400	51.9	62.2
6	2312	52.4	62.4
8	1658	53.2	60.4
10	600	53.7	56.3

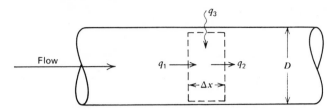

Figure 5.23   Control volume for analysis of heat transfer in tube flow with a constant tube surface temperature.

Constant Wall Surface Temperature.   Figure 5.23 will be helpful in developing an appropriate analytical expression for the case of constant surface temperature.

As in the previously considered case, at steady state

$$q_1 - q_2 + q_3 = 0$$

The separate evaluation of these q's yields the following:

$$q_1 = \rho \, \frac{\pi D^2}{4} \, c_p v_x T \big|_x$$

$$q_2 = \rho \, \frac{\pi D^2}{4} \, c_p v_x T \big|_{x+\Delta x}$$

$$q_3 = h \pi D \, \Delta x (T_0 - T)$$

Appropriate substitution into the first law expression gives

$$\rho \, \frac{\pi D^2}{4} \, c_p (v_x T \big|_{x+\Delta x} - v_x T \big|_x) - h \pi D \, \Delta x (T_0 - T) = 0$$

Once more considering flow to be fully developed where $v_x \neq v_x(x)$, dividing by Δx and rearranging, we obtain

$$\rho c_p v_x \, \frac{D}{4} \, \frac{T \big|_{x+\Delta x} - T \big|_x}{\Delta x} + h(T - T_0) = 0$$

In the limit as $\Delta x \to 0$, this becomes

$$\rho c_p v_x \, \frac{D}{4} \, \frac{dT}{dx} + h(T - T_0) = 0$$

and the temperature gradient may be written as

$$\frac{dT}{dx} = -\frac{h}{\rho c_p v_x} \, \frac{4}{D} \, (T - T_0) \tag{5-134}$$

We now separate variables and integrate from the entrance to some position x along the flow axis as follows:

$$\int_{T_e-T_0}^{T-T_0} \frac{dT}{T - T_0} = -\frac{h}{\rho c_p v_x} \frac{4}{D} \int_0^x dx$$

$$\ln \frac{T - T_0}{T_e - T_0} = -\frac{h}{\rho c_p v_x} 4 \frac{x}{D}$$

(5-135)

The temperature variation with x may now be expressed as

$$\frac{T_e - T_0}{T_f - T_0} = e^{-\text{St } 4(x/D)}$$

(5-136)

It is necessary that x/D be greater than 60 in order that the assumption of fully developed conditions be valid (specifically $h \neq h(x)$).

Notice that in equation (5-136) the Stanton number arose quite naturally from our analysis. The final determination of $T(x)$ requires that St be evaluated. Some of the analog expressions presented in the previous section may be used for this purpose.

The use of equation (5-136) and the analogies presented in Section 5.1-2.5 will be illustrated in Example 5.4.

Example 5.4

Water flowing at 10 ft/sec enters a 1-inch-diameter tube at a temperature of 50°F. Saturated pentane vapor at 96°F is condensing on the outside tube wall. This temperature may be assumed constant along the inside surface of the tube. Determine the water exit temperature and the total heat transfer over 10 feet of tube length, using (a) the Reynolds analogy, (b) the Prandtl analogy, (c) the von Kármán analogy, and (d) the Colburn analogy.

Each of these analog expressions will involve the evaluation of C_f from Figure 5.6 after Re has been calculated. Reynolds number includes ν, which is a strong function of temperature. All expressions, with the exception of Reynolds analogy, involve Pr, which is also temperature dependent. The conclusion which may be reached at this point is that, before the problem may be solved for the exit temperature, a good guess for exit temperature must be made. A trial-and-error procedure must be used to solve this and similar problems. The procedure is depicted in the flow chart of Figure 5.24.

The use of this algorithm is considerably simplified in this example since the temperature extremes are within 46°F of each other. The initial assumption will be made that $T_{\text{exit}} = 78°F$. The average fluid temperature would thus be 64°F, and the average film temperature is 80°F. Fluid properties will initially be evaluated at 80°F; Reynolds number is

$$\text{Re} = \frac{Dv}{\nu} = \frac{(1/12 \text{ ft})(10 \text{ ft/sec})}{(0.929 \times 10^{-5} \text{ ft}^2/\text{sec})} = 8.97 \times 10^4$$

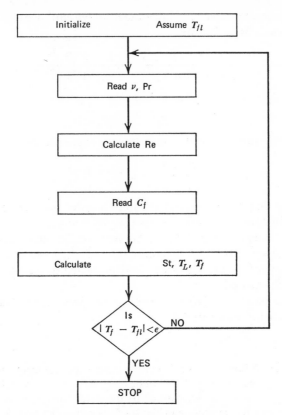

Figure 5.24 Flow chart for evaluating exit fluid temperature in a tube with constant wall temperature.

For smooth tubing, the corresponding value of C_f, from Figure 5.6, is 0.0046. By each of the four analogies we now have, for the Stanton number,

(a) Reynolds analogy

$$St = \frac{C_f}{2} = 0.0023$$

(b) Prandtl analogy

$$St = \frac{C_f/2}{1 + 5\sqrt{\dfrac{C_f}{2}}(Pr - 1)}$$

$$= \frac{0.0023}{1 + 5(0.048)(4.89)} = 0.00106$$

(c)   von Kármán analogy

$$St = \frac{C_f/2}{1 + 5\sqrt{\dfrac{C_i}{2}}\left\{Pr - 1 + \ln\left[1 + \dfrac{1}{6}(5Pr - 5)\right]\right\}}$$

$$= \frac{0.0023}{1 + 5(0.048)[(4.89) + \ln (5.08)]} = 0.000897$$

(d)   Colburn analogy

$$StPr^{2/3} = \frac{C_f}{2}$$

$$St = 0.0023(5.89)^{-2/3} = 0.000705$$

Substituting each of these values for St, in turn, into equation (5-136) we have, for T_L,

(a) $T_L = 65.3°F$
(b) $T_L = 77.6°F$
(c) $T_L = 80.0°F$
(d) $T_L = 82.7°F$

   The results of the Prandtl, von Kármán, and Colburn analogies yielded reasonably consistent results. The Reynolds analogy result is considerably different from the rest, as would be expected by the value of Pr = 5.87. The resulting exit temperature is sufficiently close to the one assumed that no additional trials are necessary.

   The heat transfer may be calculated simply from the temperature increase of the water, according to
$$q = \rho A v c_p (T_L - T_e)$$

The results of this example are summarized in the table below.

	T_{exit} (°F)	q (Btu/sec)
Reynolds analogy	65.3	52
Prandtl analogy	77.6	93.9
von Kármán analogy	80.0	102.0
Colburn analogy	82.7	111.2

5.2   NATURAL CONVECTION: THEORETICAL AND EXPERIMENTAL CONSIDERATIONS

Much of the theory of the previous section developed specifically for forced convection is valid for natural convection as well. Some obvious differences exist, however, and these will be brought out presently. Some limited theory will be presented for natural convection, along with accepted empirical results sufficient to solve most problems one may encounter involving heat transfer by the mechanism of natural convection.

It has been mentioned that the natural convection phenomenon involves heat exchange between a fluid and an adjacent boundary where fluid motion occurs due to the density differences which result from the energy exchange. The thermal and hydrodynamic boundary layers in natural convection are essentially of the same thickness since velocity gradients are produced by temperature gradients. Of major importance are the orientation and geometry of the solid boundary. The considerations to follow will be presented according to geometry.

5.2-1 Convection in Fluids Adjacent to a Single Vertical Plane Wall

This geometry and the associated terminology are represented in Figure 5.25. The governing differential equations which apply in the boundary layer region are as follows:
continuity

$$\frac{\partial v_x}{\partial x} + \frac{\partial v_y}{\partial y} = 0 \qquad (5\text{-}137)$$

equation of motion

$$\rho v_x \frac{\partial v_x}{\partial x} + \rho v_y \frac{\partial v_x}{\partial y} = \mu \frac{\partial^2 v_x}{\partial y^2} + (\rho_\infty - \rho)g \qquad (5\text{-}138)$$

energy equation

$$v_x \frac{\partial T}{\partial x} + v_y \frac{\partial T}{\partial y} = \alpha \frac{\partial^2 T}{\partial y^2} \qquad (5\text{-}139)$$

In this set of equations the usual boundary layer approximations are made, including the assumption that flow is incompressible; density disparity effects

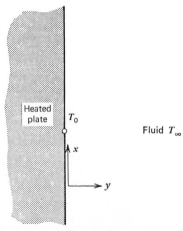

Figure 5.25 Coordinates for treatment of natural convection adjacent to a vertical wall.

are present only in the buoyancy term, which is in the equation of motion, equation (5-138). This treatment of density comprises the so-called *Boussinesq approximation*.

Boundary conditions which apply to this problem are

$$T(x, y) = T_0 \quad \text{at } (x, 0)$$
$$T(x, y) = T_\infty \quad \text{at } (x, \infty)$$
$$T(x, y) = T_\infty \quad \text{at } (0, y)$$
$$v_x(x, y) = 0 \quad \text{at } (x, 0)$$
$$v_x(x, y) = 0 \quad \text{at } (x, \infty)$$

Solutions to these equations with various simplifying assumptions have been achieved by Lorenz[19] and by Schmidt and Beckmann.[20] The Lorenz solution was based on the assumption that v_x and T were functions of y only. Schmidt and Beckmann measured both velocity and temperature profiles within the boundary layer and found these profiles to vary significantly at different locations along the plate. The Schmidt and Beckmann data for a heated 12.5-centimeters high plate are shown in Figures 5.26 and 5.27.

Similarity solutions to equations (5-137), (5-138), and (5-139) have been achieved by E. Pohlhausen[21] and by Ostrach.[22] For details of the analysis, the reader is referred to Schlichting.[23] For a fluid with $\text{Pr} = 0.733$, the results for the heated vertical plate are

$$\text{Nu}_x = 0.359 \, \text{Gr}_x^{1/4} \tag{5-140}$$

where Gr_x is the local Grashof number, defined as

$$\text{Gr}_x \equiv \frac{g x^3 (T_0 - T_\infty)}{\nu^2 T_\infty} \tag{5-141}$$

The mean Nusselt number may be determined in usual fashion from the expression for the local parameter; the result is

$$\text{Nu}_L = 0.478 \, \text{Gr}_L^{1/4} \tag{5-142}$$

The results of Pohlhausen for $\text{Pr} = 0.733$ were extended to apply to larger values of the Prandtl number by Schuh.[24] The appropriate values of Nu_L are listed in Table 5.2 for Pr up to 1000.

[19] L. Lorenz, *Wiedemann Ann. d. Phys.* **13** (1881): 582.
[20] E. Schmidt and W. Beckmann, *Tech. Mech. u. Thermodynamik* **1** (1930): 341, 391.
[21] E. Pohlhausen, 115.
[22] S. Ostrach, *NACA Report 1111* (1953).
[23] Schlichting, p. 332.
[24] H. Schuh, unpublished; see Schlichting, p. 335.

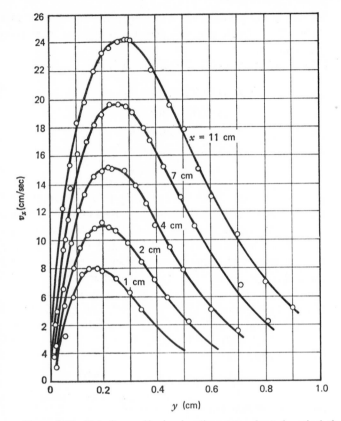

Figure 5.26   Velocity profiles in air adjacent to a heated vertical plate.

In Figure 5.28 the data of Eckert and Jackson[25] are shown as a plot of Nu_L versus GrPr for both vertical plates and cylinders. A change in the data is evident at $\mathrm{GrPr} \cong 10^9$. This is attributed to a transition from laminar to turbulent flow in the boundary layer. The correlating relationships suggested by Eckert and Jackson are

$$\mathrm{Nu}_L = 0.555(\mathrm{GrPr})^{1/4} \quad \text{for } \mathrm{GrPr} < 10^9 \tag{5-143}$$

and

$$\mathrm{Nu}_L = 0.0210(\mathrm{GrPr})^{2/5} \quad \text{for } \mathrm{GrPr} > 10^9 \tag{5-144}$$

An integral analysis of the laminar boundary layer case is presented by Eckert[26] for a constant wall temperature. A duplication of this work is

[25] E. R. G. Eckert and T. W. Jackson, *NACA Rept. 1015* (1951).
[26] E. R. G. Eckert, *Introduction to the Transfer of Heat and Mass* (New York: McGraw-Hill, 1951).

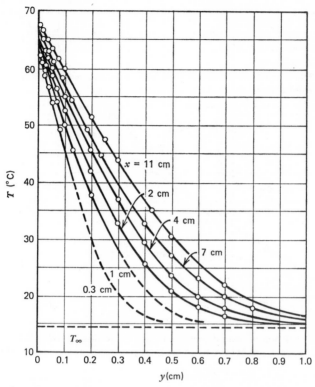

Figure 5.27 Temperature profiles in air adjacent to a vertical heated plate.

Table 5.2 Values of Nu_m for Natural Convection Adjacent to a Heated Vertical Plate

Pr	$Nu_L/Gr_L^{1/4}$	$Nu_L/Gr_L^{1/4}Pr^{1/4}$
0.73	0.478	0.517
10	1.09	0.612
100	2.06	0.652
1000	3.67	0.653

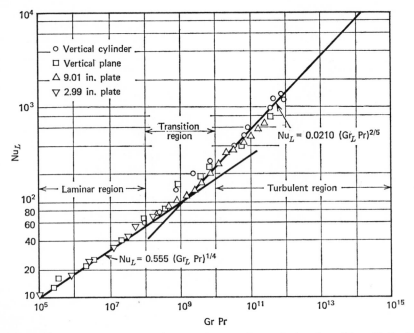

Figure 5.28 Correlation of natural convection data for vertical surfaces (From E. R. G. Eckert and T. W. Jackson, NACA RFM 50 D25, July 1950. By permission of the publishers.)

suggested as an exercise at the end of this chapter. The resulting expression for the local Nusselt number is

$$\mathrm{Nu}_x = 0.508 \frac{\mathrm{Pr}^{1/2}\,\mathrm{Gr}_x^{1/4}}{(0.952 + \mathrm{Pr})^{1/4}} \tag{5-145}$$

and the mean Nusselt number for a plate of height L is given by

$$\mathrm{Nu}_L = 0.678 \frac{\mathrm{Pr}^{1/2}\,\mathrm{Gr}_L^{1/4}}{(0.952 + \mathrm{Pr})^{1/4}} \tag{5-146}$$

The case of laminar natural convection adjacent to a vertical plate with uniform surface heat flux was solved by Sparrow and Gregg.[27]

White[28] has investigated this problem for the case of low Prandtl number fluids for both the constant temperature and constant heat flux surface conditions.

[27] E. M. Sparrow and J. L. Gregg, *Trans ASME* **78** (1956): 435.

[28] D. H. White, "An Experimental Investigation of Natural Convection Heat Transfer from Vertical Flat Plates in Mercury" (Ph.D. thesis, Oregon State University, 1971).

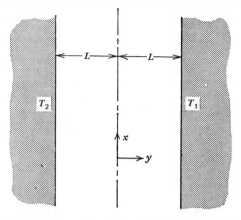

Figure 5.29 Natural convection between parallel vertical plane walls.

5.2-2 Natural Convection in Vertical Channels

A simple analytic solution is possible in the case of two vertical plane walls, one heated and one cooled with negligible end effects. The system is shown in Figure 5.29 with pertinent nomenclature. The heated wall is at a constant temperature T_1, and the cooled wall is at constant temperature T_2. With end effects neglected, the average temperature T_{avg} is $(T_1 + T_2)/2$.

The applicable form of the first law, assuming steady, laminar flow and no end effects, is

$$\frac{d^2T}{dy^2} = 0 \tag{5-147}$$

This expression may be separated and integrated twice to yield, for the temperature variation,

$$T = c_1 y + c_2$$

Applying the boundary conditions that $T(L) = T_1$ and $T(-L) = T_2$, we obtain

$$T = \frac{\Delta T}{2} \frac{y}{L} + T_{\text{avg}} \tag{5-148}$$

where ΔT is the difference $T_1 - T_2$.

The equation of motion, assuming flow to be steady, laminar, and one-dimensional, and neglecting end effects, becomes

$$\mu \frac{d^2 v_x}{dy^2} = \frac{dp}{dx} + \rho g \tag{5-149}$$

The density ρ may be expanded in a Taylor series about the average temperature, according to

$$\rho = \rho_{\text{avg}} + \frac{\partial \rho}{\partial T}\bigg|_{T_{\text{avg}}} (T - T_{\text{avg}}) + \cdots$$

Expressing the density derivative in terms of the coefficient of thermal expansion β, and neglecting higher-order terms, we may write

$$\rho = \rho_{\text{avg}} - \rho\beta|_{T_{\text{avg}}} (T - T_{\text{avg}}) \qquad (5\text{-}150)$$

For the case where pressure is hydrostatic only, the pressure gradient is written

$$\frac{dp}{dy} = -\rho g \qquad (5\text{-}151)$$

Equations (5-150) and (5-151) may be substituted into equation (5-149) to yield

$$\mu \frac{d^2 v_x}{dy^2} = -\rho\beta g(T - T_{\text{avg}}) \qquad (5\text{-}152)$$

which merely expresses the equality between viscous and buoyant forces.
Equation (5-148) may now be substituted, giving

$$\mu \frac{d^2 v_x}{dy^2} = -\rho\beta g \frac{\Delta T}{2} \frac{y}{L} \qquad (5\text{-}153)$$

This relation may now be separated and integrated twice to yield

$$v_x = -\frac{\rho\beta g}{\mu} \frac{\Delta T}{12} \frac{y^3}{L} + c_1 y + c_2$$

The boundary conditions which apply are $v_x(\pm L) = 0$; when these are substituted, the velocity profile expression becomes

$$v_x = \frac{\rho\beta g \,\Delta T L^2}{12\mu}\left[\frac{y}{L} - \left(\frac{y}{L}\right)^3\right] \qquad (5\text{-}154)$$

When dimensionless velocity and length, defined as

$$v_x{}^* \equiv \frac{L v_x \rho}{\mu}$$

and

$$y^* \equiv y/L$$

are substituted, the velocity expression becomes

$$v_x{}^* = \frac{1}{12}\left[\frac{\rho^2 \beta g L^3 \, \Delta T}{\mu^2}\right](y^* - y^{*3})$$

or

$$v_x{}^* = \frac{\text{Gr}}{12}(y^* - y^{*3}) \qquad (5\text{-}155)$$

where the bracketed term is observed to be the Grashof number.

The temperature and velocity variation between the two parallel vertical plates are shown in Figure 5.30. The profiles are graphical representations of equations (5-148) for temperature and (5-154) for velocity.

Several investigators have concerned themselves with the case of fluids within rectangular enclosures of height H and width L for one vertical wall heated and one cooled. Correlations for heat transfer in these cases are normally expressed in the form Nu = Nu(Gr, Pr, H/L). A summary of correlations for this geometry has been presented by Pagnani.[29] An empirical correlation by Jakob[30] of experimental data, primarily that of Mull and Reiher,[31] included three regions where the dominant effects were conduction, laminar flow, and turbulent flow, respectively. The correlations and corresponding Grashof numbers are

conduction: $\text{Gr}_L < 2 \times 10^3$ $\text{Nu}_L = 1$ (5-156)

laminar flow: $2 \times 10^3 \leq \text{Gr}_L \leq 2 \times 10^5$ $\text{Nu}_L = 0.18 \, \text{Gr}^{1/4}\left(\dfrac{H}{L}\right)^{-1/9}$

(5-157)

turbulent flow: $2 \times 10^5 \leq \text{Gr}_L \leq 2 \times 10^7$ $\text{Nu}_L = 0.065 \, \text{Gr}^{1/3}\left(\dfrac{H}{L}\right)^{-1/9}$

(5-158)

For open channels between parallel vertical plane walls, Elenbaas[32] has developed the plot shown in Figure 5.31. The values in this plot are valid for laminar flow only.

Heat transfer data along with velocity profiles were obtained by Colwell[33] for the case of natural convection with mercury in a vertical channel with

[29] B. R. Pagnani, "An Explicit Finite-Difference Solution for Natural Convection in Air in Rectangular Enclosures" (Ph.D. thesis, Oregon State University, 1968).

[30] M. Jakob, *Trans. ASME* **68** (1946): 189.

[31] W. Mull and H. Reiher, *Beihefte Z. Gesundheits-ingeneure* **28** (1930): 126.

[32] W. Elenbaas, *Phillips Research Rept. 3*, N. V. Phillips Gloeilamp-enfabrieken, Eindhoben, Neth. (1948).

[33] R. G. Colwell, "Experimental Investigation of Natural Convection of Mercury in an Open, Uniformly Heated, Vertical Channel" (Ph.D. thesis, Oregon State University, 1974).

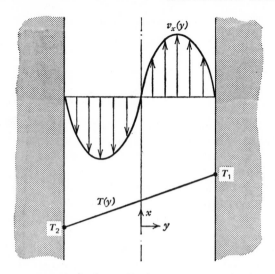

Figure 5.30 Temperature and velocity profiles in natural convection between two vertical plane walls.

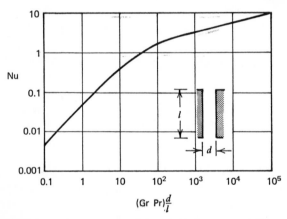

Figure 5.31 Correlation for laminar flow natural convection between two vertical plane walls.

constant heat flux boundaries. Colwell and Welty[34] have presented correlations for laminar natural convection flows in a vertical open-ended channel obtained for a range of channel widths. Figure 5.32 shows the data and

[34] R. G. Colwell and J. R. Welty, "An Experimental Study of Natural Convection with Low Prandtl Number Fluids in a Vertical Channel with Uniform Wall Heat Flux," ASME paper 73-HT-52, presented at the 14th National Heat Transfer Conference, Atlanta, Georgia, August 1973.

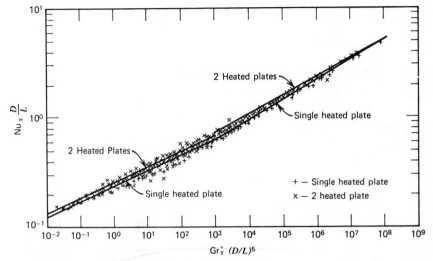

Figure 5.32 Generalized heat transfer correlations for natural convection with mercury in a vertical channel (R. G. Colwell and J. R. Welty, ASME paper 73-HT-52, presented at the 14th National Heat Transfer Conference, Atlanta, Georgia, August 1973).

generalized correlations for the cases where one channel wall was heated and the other insulated, and where both walls were uniformly and symmetrically heated. The fluid involved was mercury with $Pr = 0.022$. The behavior exhibited by mercury is typical of that to be expected of other low-Prandtl-number fluids (liquid metals).

Correlations of the data, including both wall conditions, integrated over the channel height were obtained in the form

$$\text{Nu}_D = 0.438\left(\text{Gr}_D*\frac{D}{L}\right)^{0.141} \qquad 10^{-2} < \text{Gr}_D*\frac{D}{L} \le 10^3 \quad (5\text{-}159)$$

$$\text{Nu}_D = 1.16 + 0.269\left(\text{Gr}_D*\frac{D}{L}\right)^{0.180} \qquad 10^{-3} \le \text{Gr}_D*\frac{D}{L} \le 10^9 \quad (5\text{-}160)$$

where

$$\text{Nu}_D = \frac{hD}{k}, \quad \text{the mean Nusselt number}$$

$$D = \text{channel width}$$

$$L = \text{channel height}$$

$$\text{Gr}_D* = \frac{\beta gq/A}{\nu^2 k}\, D^4, \quad \text{the modified Grashof number}$$

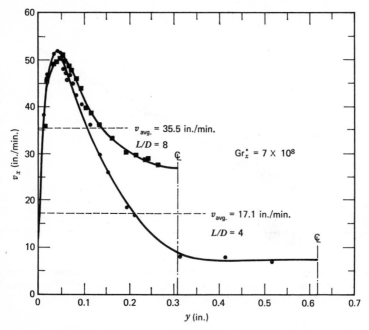

Figure 5.33 Velocity profiles for natural convection with mercury in a vertical channel, $x = 1.5$ in. (R. G. Colwell and J. R. Welty, ASME paper 73-HT-52, presented at the 14th National Heat Transfer Conference, Atlanta, Georgia, August 1973).

Figure 5.33 shows two velocity profiles obtained at a distance of 1.5 inches up from the bottom of the channel at two different channel widths, both walls heated. The local modified Grashof number Gr_x^* was the same in both cases. It is of interest to note that, for dynamically similar conditions, a decrease in channel width causes an increase in flow rate. The implication here is that viscous forces do not play a large role, at least for the channel widths considered here, 0.6175 in. and 1.235 in., respectively.

Other data presented by Colwell and Welty showed that, for a given heat flux, the wall temperature decreased as the channel became narrower. This is in contrast to the behavior of air, water, and other fluids with moderate to high Prandtl numbers. It was thus established that, in a low-Prandtl-number fluid in natural (buoyant) flow, flow is essentially inviscid. The region of high shear stress is in close proximity with the source of buoyancy, and only when the channel is made very narrow does wall friction affect the flow in an appreciable way. When frictional forces do come into play, a subsequent decrease in channel width will result in a wall temperature increase.

5.2-3   Natural Convection for Horizontal Surfaces

Horizontal Cylinders. Natural convection data for the case of heated horizontal cylinders in both liquids and gases have been correlated by McAdams,[35] as shown in Figure 5.34. The values of Nu_D recommended for various $Gr_f Pr_f$ products are shown on this plot; in the range $10^4 < Gr_D Pr < 10^9$ McAdams suggests the correlation

$$Nu_D = 0.53(Gr_D Pr)^{1/4} \tag{5-161}$$

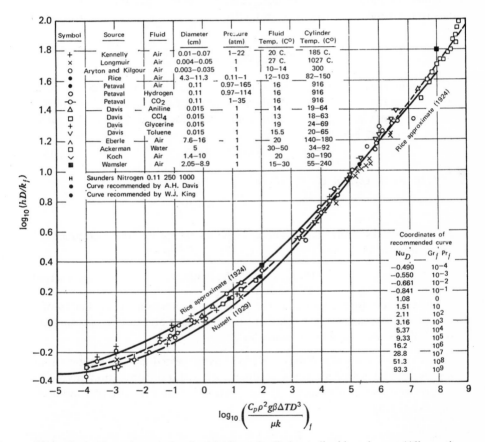

The vertical axis is $\log_{10}(hD/k_f)$ and the horizontal axis is $\log_{10}\left(\dfrac{C_p \rho^2 g\beta\Delta T D^3}{\mu k}\right)_f$

Table within figure:

Symbol	Source	Fluid	Diameter (cm)	Pressure (atm)	Fluid Temp. (C°)	Cylinder Temp. (C°)
+	Kennelly	Air	0.01–0.07	1–22	20 C.	185 C.
×	Longmuir	Air	0.004–0.05	1	27 C.	1027 C.
○	Aryton and Kilgour	Air	0.003–0.035	1	10–14	300
●	Rice	Air	4.3–11.3	0.11–1	12–103	82–150
●	Petaval	Air	0.11	0.97–165	16	916
○	Petaval	Hydrogen	0.11	0.97–114	16	916
-○-	Petaval	CO₂	0.11	1–35	16	916
△	Davis	Aniline	0.015	1	14	19–64
□	Davis	CCl₄	0.015	1	13	18–63
+	Davis	Glycerine	0.015	1	19	24–69
∨	Davis	Toluene	0.015	1	15.5	20–65
∧	Eberle	Air	7.6–16	–	20	140–180
□	Ackerman	Water	5	1	30–50	34–92
∨	Koch	Air	1.4–10	1	20	30–190
■	Wamsler	Air	2.05–8.9	1	15–30	55–240
H	Saunders	Nitrogen	0.11	250	1000	
●	Curve recommended by A.H. Davis					
●	Curve recommended by W.J. King					

Coordinates of recommended curve

Nu_D	$Gr_f Pr_f$
−0.490	10^{-4}
−0.550	10^{-3}
−0.661	10^{-2}
−0.841	10^{-1}
1.08	0
1.51	10
2.11	10^2
3.16	10^3
5.37	10^4
9.33	10^5
16.2	10^6
28.8	10^7
51.3	10^8
93.3	10^9

Figure 5.34 Natural convection from horizontal cylinders to liquids and gases. (All quantities with subscript f are to be evaluated at the film temperature.) [From W. H. McAdams, *Heat Transmission*, 3rd ed. (New York: McGraw-Hill Book Company, 1954), p. 176. By permission of the publishers.]

[35] W. H. McAdams, *Heat Transmission*, 3rd ed. (New York: McGraw-Hill, 1954).

When the cylinder diameter becomes small, as in the case of a wire, the Grashof number becomes very small. For cases where $Gr_D Pr < 10^4$, Elenbaas[36] has derived the equation

$$Nu^3 e^{-6/Nu} = \frac{Gr_D Pr}{235} \qquad (5\text{-}162)$$

which has agreed well with experimental data.

In the case of streamline flow in natural convection of both metallic and nonmetallic fluids adjacent to horizontal cylinders larger than wires, Hsu[37] has recommended a correlation in the form

$$Nu_D = 0.53 \left(\frac{Pr}{0.952 + Pr} Gr_D Pr \right)^{1/4} \qquad (5\text{-}163)$$

Horizontal Plane Surfaces. McAdams[38] has recommended the following expressions for the case of natural convection adjacent to horizontal plates. For hot plates facing upward or cold plates facing down, in the range $10^5 < Gr_L Pr < 2 \times 10^7$,

$$Nu_L = 0.54 (Gr_L Pr)^{1/4} \qquad (5\text{-}164)$$

and, in the range $2 \times 10^7 < Gr_L Pr < 3 \times 10^{10}$,

$$Nu_L = 0.14 (Gr_L Pr)^{1/3} \qquad (5\text{-}165)$$

With hot plates facing down and cold plates facing up, the recommended expression, in the range $3 \times 10^5 < Gr_L Pr < 10^{10}$, is

$$Nu_L = 0.27 (Gr_L Pr)^{1/4} \qquad (5\text{-}166)$$

The characteristic length L, in the preceding expressions, is the length of a side of a square surface, the mean of the dimensions of a rectangular surface, or 0.9 times the diameter of a circular area.

Spheres and Rectangular Solids. King[39] has suggested using the expressions for horizontal cylinders when dealing with natural convection involving spheres and rectangular solids. His suggestion includes the use of a modified significant length L, determined from

$$\frac{1}{L} = \frac{1}{L_{horiz}} + \frac{1}{L_{vert}} \qquad (5\text{-}167)$$

[36] W. Elenbaas, *J. Appl. Phys.* **19** (194): 1148; *Phillips Res. Dept.* **3** (1948): 338 and 450.
[37] S. T. Hsu, *Engineering Heat Transmission*, 3rd ed. (New York: McGraw-Hill, 1954).
[38] McAdams, op. cit.
[39] W. J. King, *Mech. Engr.* **54** (1932): 347.

where L_{horiz} and L_{vert} are the horizontal and vertical dimensions, respectively, of the solid. It is obvious that L for a sphere becomes $D/2$.

The manner of determining heat transfer information, using correlations of the type presented thus far, will be illustrated in Example 5.5.

Example 5.5

Five-hundred-pound saturated steam is transported through 2-inch schedule-80 steel pipe with 1-1/2-inch 85% magnesia insulation on the outside. It is desired to know the rate of heat loss per foot from the insulated pipe if the steam line is (a) horizontal and (b) vertical, with a run of 20 ft. The pipe is surrounded by 40°F still air. The saturation temperature of 500 psi steam is 467°F. Necessary data for this problem are available in Appendices A-1 and F-1. For present purposes, this information is compiled below.

$$
\begin{array}{lll}
\text{2-inch schedule-80 pipe} & \text{ID} = 1.939 \text{ in.} \\
& \text{OD} = 2.375 \text{ in.} \\[6pt]
\text{1-1/2-inch insulation layer} & \text{ID} = 2.375 \text{ in.} \\
& \text{OD} = 5.375 \text{ in.} \\[6pt]
\text{mild steel} & k \cong 23 \text{ Btu/hr-ft-}°\text{F} \\[6pt]
85\% \text{ magnesia} & k = 0.041 \text{ Btu/hr-ft-}°\text{F}
\end{array}
$$

The thermal resistance per foot of length of each part of the heat transfer path is determined as follows:

Inside wall (condensing steam)

$$R_t = \frac{1}{h_i A_i} \quad (h_i \text{ is large, thus this term will be neglected})$$

Steel pipe (conduction)

$$R_t = \frac{\ln (D_o/D_i)}{2\pi k L}$$

$$R_t = \frac{\ln (2.375/1.939)}{2\pi (23)(1)} = 0.00141 \; \frac{\text{hr-}°\text{F}}{\text{Btu}}$$

Insulating layer (conduction)

$$R_t = \frac{\ln (D_o/D_i)}{2\pi k L}$$

$$= \frac{\ln(5.375/2.375)}{2\pi (0.041)(1)} = 3.16 \frac{\text{hr-}°\text{F}}{\text{Btu}}$$

Outside surface (natural convection)

$$R_t = \frac{1}{h_o A_o} = \frac{1}{h_o(\pi)\left(\dfrac{5.375}{12}\right)(1)} = \frac{0.71}{h_o} \frac{\text{hr-}^\circ\text{F}}{\text{Btu}}$$

Between the steam and outside air, the heat loss per foot of pipe is

$$q = \frac{\Delta T}{R_t} = \frac{427}{3.161 + 0.71/h_o} \tag{I}$$

This heat loss can also be expressed in terms of the conduction between inner and outer surfaces as

$$q = \frac{467 - T_o}{3.161} \tag{II}$$

or, in terms of the convection at the outer surface, in the form

$$q = \frac{h_o}{0.71}(T_o - 40) \tag{III}$$

In both of these latter expressions, T_o is the temperature at the outside surface of the insulation.

The reader should note that the heat transfer coefficient is a function of the unknown temperature T_o. The problem now becomes one of trial and error in which the surface temperature is assumed initially and then checked. The problem will be completed when assumed and calculated values of T_o agree.

We know that, in natural convection, the heat transfer coefficient is relatively small. For a value of $h_o = 10$ Btu/hr-ft²-°F, corresponding values of q and T_o are approximately 132 Btu/hr and 49°F, respectively. An h of 1 Btu/hr-ft²-°F will yield corresponding values of 110 Btu/hr and 118°F, respectively. The range of values which might be considered in the assumption of T_o is greatly reduced by this sort of preliminary calculation.

We shall choose $T_o = 100$°F for our initial assumption; corresponding values of Gr_D and Pr are

$$\text{Gr}_D = \frac{\beta g}{\nu^2} D^3 \Delta T$$

$$= \left(2.29 \times 10^6 \frac{1}{^\circ\text{F-ft}^3}\right)\left(\frac{5.375}{12}\text{ ft}\right)^3 (60^\circ\text{F})$$

$$= 12.35 \times 10^6$$

$$\text{Pr} = 0.710$$

A film temperature of 70°F was used in determining the needed properties of air.

According to equation (5-161), the value of Nu_D applying to the case of a horizontal steam line is

$$\text{Nu}_D = 0.53[(12.35 \times 10^6)(0.71)]^{1/4}$$

$$= 28.8$$

and the heat transfer coefficient is

$$h_o = \frac{k}{D} \text{Nu} = \frac{(0.0149 \text{ Btu/hr-ft-}°\text{F})(28.8)}{5.375/12 \text{ ft}}$$

$$= 0.959 \text{ Btu/hr-ft}^2\text{-}°\text{F}$$

The heat flow rate and outside surface temperature which correspond to this value of h are

$$q = 109.5 \text{ Btu/hr per ft}$$

and

$$T_o = 121°\text{F}$$

With this value of T_o used in a second series of calculations, we obtain, for the horizontal case,

$$\text{Gr}_D = 15.05 \times 10^6$$
$$\text{Pr} = 0.708$$
$$\text{Nu}_D = 30.3$$
$$h = 1.03$$
$$q = 111 \text{ Btu/hr}$$
$$T_o = 116.5$$

It is clear that one more calculation will produce a refined value of T_o; however, the heat loss value will be changed a very small amount; thus, the answer to (a) of this problem is $q = 111$ Btu/hr per foot of pipe.

In the case of a vertical pipe the significant length in Gr_L and Nu_L becomes the height rather than the diameter. Equations (I), (II), and (III) still apply.

If an initial value of 120°F is assumed for T_o, the important parameters are determined as follows:

$$\text{Gr}_L = \frac{\beta g}{\nu^2} L^3 \Delta T$$

$$\text{Gr}_L = \left(2.09 \times 10^6 \frac{1}{°\text{F-ft}^3}\right)(20 \text{ ft})^3(80°\text{F})$$

$$= 1.338 \times 10^{12}$$

$$\text{Pr} = 0.708$$

The Nusselt number may now be determined from equation (5-135) or read from Figure 5.29. Using equation (5-135), we obtain

$$\text{Nu}_L = 0.0210(\text{Gr}_L\text{Pr})^{2/5}$$
$$= 0.0210[(1.338 \times 10^{12})(0.708)]^{2/5}$$
$$= 1300$$

and, from this, the heat transfer coefficient is evaluated as

$$h = \frac{k}{L} \mathrm{Nu}_L$$

$$= \frac{0.0152 \text{ Btu/hr-ft-}°F}{20 \text{ ft}} (1300)$$

$$= 0.988 \text{ Btu/hr-ft}^2\text{-}°F.$$

Values for q and T_o may now be determined from expressions (I) and (III). They are

$$q = \frac{427}{3.161 + \dfrac{0.71}{0.988}} = 110 \text{ Btu/hr}$$

$$T_o = 40 + \frac{0.71}{0.988} (q) = 119°F$$

This agreement with the assumed value for T_o of 120°F is excellent; there is no need for additional calculation. The answer to (b) of this problem, for a vertical steam line, is $q = 110$ Btu/hr per foot. No appreciable difference in heat loss exists, for this case, between a vertical and a horizontal configuration.

5.2-4 Simplified Expressions for Natural Convection in Air

In a decided majority of actual cases, the fluid involved in a natural convection situation is air at atmospheric pressure. Since this is true, the preceding expressions have been modified to apply specifically to air and a simplified expression for the heat transfer coefficient written in the form

$$h = A\left(\frac{\Delta T}{L}\right)^b \tag{5-168}$$

where A and b are constants, depending on geometry and flow conditions, and L is the significant length, also a function of geometry and flow.

Table 5.3 lists the values suggested by McAdams[40] for various orientations, geometries, and flow conditions as indicated by the magnitude of the product GrPr. Values of h determined, using constants from Table 5.3 in equation (5-168), have the dimensions of Btu/hr-ft^2-°F. The temperature difference referred to is that between the surface and the bulk air with units of °F.

Example 5.6 illustrates the savings to be realized when equation (5-168) is used rather than one of the more involved correlations in earlier sections.

[40] McAdams, op. cit.

Table 5.3   Constants to Be Used in Equation (5-168) for Natural Convection in Air

Geometry	Applicable range	A	b	L
Vertical surfaces (planes and cylinders)	$10^4 < Gr_L Pr < 10^9$	0.29	1/4	height
	$10^9 < Gr_L Pr < 10^{12}$	0.19	1/3	1
Horizontal cylinders	$10^3 < Gr_D Pr < 10^9$	0.27	1/4	diameter
	$10^9 < Gr_D Pr < 10^{12}$	0.18	1/3	1
Horizontal planes hot plates facing up or cold plates facing down	$10^5 < Gr_L Pr < 2 \times 10^7$	0.27	1/4	length of side
	$2 \times 10^7 < Gr_L Pr < 3 \times 10^{10}$	0.22	1/3	1
cold plates facing up or hot plates facing down	$3 \times 10^5 < Gr_L Pr < 3 \times 10^{10}$	0.12	1/4	length of side

Example 5.6

For the steam line described in Example 5.5, with all specifications and conditions the same, determine the heat loss per foot from the insulated pipe to 40°F air for (a) horizontal and (b) vertical orientations by means of the simplified relationships for air given by equation (5-168).

The overall heat transfer analysis already performed still applies. Expressions (I), (II), and (III) from Example 5.5 are repeated below for reference:

$$q = \frac{427}{3.161 + 0.71/h_o} \tag{I}$$

$$q = \frac{467 - T_o}{3.161} \tag{II}$$

$$q = \frac{h_o}{0.71}(T_o - 40) \tag{III}$$

Once again it is necessary to calculate h_o before completing the solution for q. The convenience associated with the use of equation (5-168) lies primarily in the minimal trial and error involved and the decreased burden of calculating and manipulating cumbersome dimensionless parameters.

We shall first consider the horizontal configuration. With the assumption that $Gr_D Pr < 10^9$, h_o may be written in the form of equation (5-168) as

$$h_o = 0.27 \left(\frac{\Delta T}{5.375/12} \right)^{1/4} = 0.33 \, \Delta T^{1/4}$$

The heat flow rate is now written in the form of both expressions (I) and (III) as

$$q = \frac{427}{3.161 + 2.15/\Delta T^{1/4}} = 0.465 \, \Delta T^{5/4}$$

Solving this equality quite easily, we obtain the result that $\Delta T \simeq 80°F$. The corresponding value of q is 111.8 Btu/hr, which compares extremely well with the value of 111 Btu/hr obtained in the previous example by a much more laborious process.

The remaining step is to verify, using the result for ΔT, that $Gr_D Pr$ falls within the range specified in Table 5.3.

For the vertical configuration, since the significant length is 20 ft, the higher range of $Gr_L Pr$ will be assumed insofar as values read from Table 5.3 are concerned. The expression for h_o thus becomes

$$h_o = 0.19(\Delta T)^{1/3}$$

Equating expressions (I) and (III), with this value of h_o incorporated, we obtain

$$q = \frac{427}{3.161 + 3.74/\Delta T^{1/3}} = 0.268 \, \Delta T^{4/3}$$

The value which satisfies this expression is $\Delta T \simeq 90°F$, and the heat flow rate which corresponds is approximately 107 Btu/hr per foot. In Example 5.5 the result for the vertical case was 110 Btu/hr per foot.

We may quickly ascertain that $Gr_L Pr$ for this vertical case lies within the range $10^9 < Gr_L Pr < 10^{12}$ specified in Table 5.3.

The foregoing example has illustrated the great utility afforded by equation (5-168) when it applies. The reader is reminded that this simplified expression applies only when the fluid involved is air at atmospheric pressure.

5.3 FORCED CONVECTION: THEORETICAL AND EMPIRICAL CONSIDERATIONS

Forced convection heat transfer is encountered very frequently in engineering practice. One of the more easily controlled parameters in a heat transfer situation involving a fluid is the rate of flow of the fluid through a pipe or past a surface; such situations are classified as forced convection.

Some basic ideas in forced convection were considered at length in Section 5.1. We shall use some of the information developed in the earlier section while extending the coverage of forced convection considerably in the present section. A distinction will be made, in the discussions to follow, between internal and external flow.

5.3-1 Forced Convection in Internal Flow

A distinction will be made between laminar and turbulent flows. The very different nature of these two types of flow was discussed at length in the early part of this chapter.

5.3-1.1 Laminar Flow inside Pipes and Ducts

For the case of laminar pipe flow with radial symmetry, the equation of motion for steady, fully developed flow reduces to

$$\frac{\mu}{r} \frac{d}{dr}\left(r \frac{dv_x}{dr}\right) - \frac{dP}{dx} = 0 \qquad (5\text{-}169)$$

which may be separated and integrated twice to yield

$$v_x = \frac{r^2}{4\mu} \frac{dP}{dx} + c_1 \ln r + c_2$$

The boundary conditions necessary to evaluate the integration constants c_1 and c_2 are

$$\frac{dv_x}{dr} = 0 \qquad \text{at} \quad r = 0$$

and

$$v_x = 0 \qquad \text{at} \quad r = R$$

Substituting these boundary conditions into the velocity profile expression, we obtain

$$v_x = -\frac{dP}{dx} \frac{R^2}{4\mu}\left[1 - \left(\frac{r}{R}\right)^2\right] \qquad (5\text{-}170)$$

which is a symmetric parabola about the pipe centerline. Since $v_x = v_{\max}$ at $r = 0$, the coefficient $-(dP/dx)(R^2/4\mu)$ is equal to the maximum velocity. An alternate form of equation (5-176) may thus be written as

$$v_x = v_{\max}\left[1 - \left(\frac{r}{R}\right)^2\right] \qquad (5\text{-}171)$$

An average velocity can be determined by a simple application of the equation of continuity to yield the expression

$$v_{\mathrm{avg}} = \frac{1}{A} \int_A v_x \, dA \qquad (5\text{-}172)$$

If equation (5-171) is substituted for v_x, the result is obtained that

$$v_{\text{avg}} = \frac{v_{\text{max}}}{2} \tag{5-173}$$

thus an expression for the fully developed laminar velocity profile in a circular pipe equivalent to those already introduced is

$$v_x = 2v_{\text{avg}}\left[1 - \left(\frac{r}{R}\right)^2\right] \tag{5-174}$$

The applicable form of the energy equation which applies to steady, laminar, fully developed flow with symmetric heat transfer is

$$\frac{k}{r}\frac{\partial}{\partial r}\left(r\frac{\partial T}{\partial r}\right) + \frac{\partial^2 T}{\partial x^2} = \rho c v_x \frac{\partial T}{\partial x} \tag{5-175}$$

For the case where axial conduction is negligible compared to that in the radial direction, $\partial^2 T/\partial x^2 = 0$; and the expression which results is

$$\frac{1}{r}\frac{\partial}{\partial r}\left(r\frac{\partial T}{\partial r}\right) = \frac{v_x}{\alpha}\frac{\partial T}{\partial x} \tag{5-176}$$

where the ratio of physical properties $k/\rho c$ has been replaced by α, the thermal diffusivity

For a *fully developed temperature profile*, the requirement is that the generalized temperature profile does not vary along the flow axis. The generalized profile is the ratio of temperatures $(T_o - T)/(T_o - T_m)$, where T_o is the surface temperature, T is the fluid temperature at a point, and T_m is the *mixed-mean*, or *mixing-cup*, *temperature*, defined as

$$T_m = \frac{1}{A v_{\text{avg}}} \int_A v_x T \, dA \tag{5-177}$$

A fully developed temperature profile now requires that

$$\frac{\partial}{\partial x}\left(\frac{T_o - T}{T_o - T_m}\right) = 0 \tag{5-178}$$

Differentiating this expression and solving for $\partial T/\partial x$, we obtain

$$\frac{\partial T}{\partial x} = \frac{dT_o}{dx} + \frac{T_o - T}{T_o - T_m}\left(\frac{dT_m}{dx} - \frac{dT_o}{dx}\right) \tag{5-179}$$

This result may be substituted into the right side of the energy expression, equation (5-176). Two special cases will be considered: constant wall heat flux and constant wall surface temperature.

Constant Wall Heat Flux. One exercise at the end of this chapter involves showing that, for a fully developed temperature profile, $h \neq h(x)$. Assuming that this can be shown, we may write, for the case of constant heat flux where

$$\frac{q_o}{A} = h(T_o - T_m) = \text{constant}$$

the result that $T_o - T_m = \text{constant}$. Equation (5-179) thus reduces to the form

$$\frac{\partial T}{\partial x} = \frac{dT_o}{dx} = \frac{dT_m}{dx} \tag{5-180}$$

and the energy equation can be written as

$$\frac{1}{r}\frac{d}{dr}\left(r\frac{dT}{dr}\right) = \frac{v_x}{\alpha}\frac{dT_m}{dx} \tag{5-181}$$

The surface and mean fluid temperatures both vary linearly in this case; Figure 5.35 shows this distribution.

Equation (5-181) can be solved, using the boundary conditions $T(R) = T_o$, $(dT/dr)(0) = 0$ to obtain an expression for $T(r)$. Subsequent manipulation will produce the Nusselt number. This calculation is left as an exercise at the end of the chapter; the result of such a computation is

$$\text{Nu}_D = \frac{hD}{k} = 4.364 \tag{5-182}$$

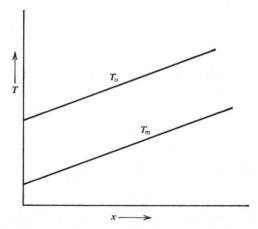

Figure 5.35 Axial variation of wall and mean fluid temperature for steady, fully developed laminar flow with constant wall heat flux.

Table 5.4   Nusselt Numbers in the
Entrance Region of a
Circular Pipe with
Laminar Flow and
Constant Wall Heat
Flux

$\dfrac{x/R}{\text{RePr}}$	Nu_x
0	∞
0.002	12.00
0.004	9.93
0.010	7.49
0.020	6.14
0.040	5.19
0.100	4.51
∞	4.36

A thorough examination of the constant wall heat flux case, including end effects, is presented by Kays.[41] Table 5.4 lists the Nusselt number for various values of the dimensionless parameter $(x/R)/\text{RePr}$, from 0 to ∞. A fully developed condition is approached for values of this parameter greater than 0.100.

Figure 5.36 shows a representative variation in mean fluid temperature along the axis of flow for the constant heat flux condition. A constant

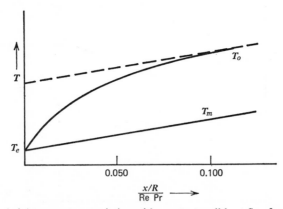

Figure 5.36   Axial temperature variation with constant wall heat flux for steady, laminar pipe flows.

[41] Kays, op. cit.

difference between wall temperature and mean fluid temperature is achieved for $(x/R)/RePr$ greater than 0.100.

Constant Wall Temperature. The variation in mean fluid temperature along the flow axis for a constant wall temperature and constant h was developed in Section 5.1-2.6, with the result given by equation (5-136) repeated here for reference:

$$\frac{T_{\ell} - T_o}{T_{\ell}' - T_o} = e^{-(h/\rho c_p v_x)(4x/D)} \tag{5-136}$$

The Nusselt number for a fully developed temperature profile in the case of constant wall temperature will involve equations (5-174), (5-176), (5-177), and (5-179) as previously developed. For constant T_o, the axial derivative of T, $\partial T/\partial x$, as expressed in equation (5-179), becomes

$$\frac{\partial T}{\partial x} = \frac{T_o - T}{T_o - T_m}\frac{dT_m}{dx} \tag{5-183}$$

Substitution of equation (5-183) for $\partial T/\partial x$ and equation (5-174) for $v_x(r)$ into the energy expression, equation (5-181), yields

$$\frac{1}{r}\frac{\partial}{\partial r}\left(r\frac{\partial T}{\partial r}\right) = \frac{2v_{\mathrm{avg}}}{\alpha}\left[1 - \left(\frac{r}{R}\right)^2\right]\left[\frac{T_o - T}{T_o - T_m}\right]\frac{dT_m}{dx} \tag{5-184}$$

This expression is more difficult to solve than was the case for constant wall heat flux and is beyond the scope of this text. The result, expressed as Nu_x, obtained by iterative means is

$$\mathrm{Nu}_D = 3.658 \tag{5-185}$$

which, the reader should note, is approximately 16% less than the result for the constant wall heat flux case.

The complete problem for constant wall temperature, including end effects, was first considered by Graetz.[42] This problem was solved in most complete fashion by Sellars, Tribus, and Klein.[43] Values of the local Nusselt number are listed in Table 5.5 for $(x/R)/RePr$ from 0 to ∞. A fully developed condition is reached for $(x/R)/RePr$ greater than 0.10.

Seider and Tate[44] have correlated experimental data for laminar tube flow with constant wall temperature. Their reasonably well-accepted expression is

$$\mathrm{Nu}_D = 1.86\left(RePr\frac{D}{L}\right)^{1/3}\left(\frac{\mu_m}{\mu_o}\right)^{0.14} \tag{5-186}$$

[42] L. Graetz, *Ann. Phys. u. Chem.* **25** (1885): 337.
[43] J. R. Sellars, M. Tribus, and J. S. Klein, *Trans. ASME* **78** (1956): 441.
[44] E. N. Seider and G. E. Tate, *Ind. Eng. Chem.* **28** (1936): 1429.

Table 5.5 Nusselt Numbers in the Entrance Region of a Circular Pipe with Laminar Flow and Constant Wall Temperature

$\dfrac{x/R}{\mathrm{Re}\mathrm{Pr}}$	Nu_x
0	∞
0.001	12.86
0.004	7.91
0.01	5.99
0.04	4.18
0.08	3.79
0.10	3.71
∞	3.66

where all fluid properties are evaluated at the average of the mean fluid temperature except μ_o, which is the viscosity evaluated at the wall temperature T_o.

A comparison between the Seider-Tate equation and the theoretical results of Graetz is shown in Figure 5.37.

Example 5.7

A 1-inch, 14-BWG copper tube is to be used to heat hydraulic fluid (MIL-M-5606) from 60°F to 150°F. The outside tube surface is wound uniformly with an electric

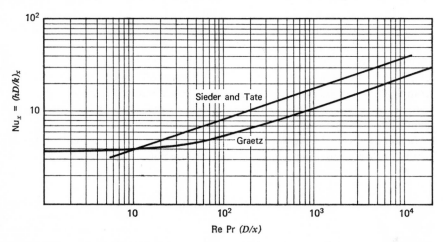

Figure 5.37 A comparison between empirical and analytical results for laminar, forced convection in pipes.

strip heater to provide a uniform wall heat flux. The hydraulic fluid flows at 10 ft/min. Determine (a) the required heat flux to produce the specified temperature change and (b) the surface temperature at the exit of the 10-foot-long tube.

The heat flux determination is straightforward. The energy transferred to the hydraulic fluid is

$$q = \rho A v_{avg} c_p (T_L - T_e)$$

$$= \left(50.9 \frac{\text{lb}_m}{\text{ft}^3}\right)(0.00379\ \text{ft}^2)\left(10\ \frac{\text{ft}}{\text{min}}\right)\left(0.502 \frac{\text{Btu}}{\text{lb}_m\text{-}°\text{F}}\right)$$

$$\times (150 - 60)°\text{F}\left(\frac{60\ \text{min}}{\text{hr}}\right)$$

$$= 5230\ \text{Btu/hr}$$

The outside surface area in 10 ft of 1-inch 14-BWG tubing is

$$A = \pi D L = \pi \left(\frac{1}{12}\ \text{ft}\right)(10\ \text{ft}) = 2.62\ \text{ft}^2$$

and the required wall heat flux is

$$\frac{q}{A} = \frac{5230\ \text{Btu/hr}}{2.62\ \text{ft}^2} = 2000\ \frac{\text{Btu}}{\text{hr-ft}^2}$$

The wall surface temperature at the exit is determined from

$$\frac{q}{A} = h(T_o|_L - T_m)$$

which requires that h be determined.

Reynolds number for this case is evaluated as

$$\text{Re}_D = \frac{Dv}{\nu} = \frac{\left(\frac{0.834}{12}\ \text{ft}\right)\left(\frac{10}{60}\ \text{ft/sec}\right)}{10.2 \times 10^{-5}\ \text{ft}^2/\text{sec}} = 113.6$$

which is in the laminar range.

The parameter $(x/R)/\text{RePr}$ is now determined as

$$\frac{x/R}{\text{RePr}} = \frac{10\ \text{ft}\left/\dfrac{0.834}{2 \times 12}\ \text{ft}\right.}{113.6(130)} = 0.0195$$

Interpolating the values in Table 5.4, we obtain, as an approximate value for the local Nusselt number,

$$\text{Nu}_x = 6.3$$

from which

$$h_x = \frac{k}{D} \mathrm{Nu}_x$$

$$= \frac{0.0685 \text{ Btu/hr-ft-}^\circ\text{F}}{0.834/12 \text{ ft}} \quad (6.3)$$

$$= 6.21 \text{ Btu/hr-ft}^2\text{-}^\circ\text{F}$$

The surface temperature at the exit is now evaluated from

$$T_o|_L = T_m + \frac{q/A}{h_x}$$

$$= 150 + \frac{2000 \text{ Btu/hr-ft}^2}{6.21 \text{ Btu/hr-ft}^2\text{-}^\circ\text{F}} = 472^\circ\text{F}$$

5.3-1.2.   Turbulent Flow inside Pipes and Ducts

No simple analytical solution exists for the case of heat transfer in turbulent pipe flow. The techniques for predicting heat transfer rates in this case are the analogy expressions considered in Section 5.1-2.5 and correlations of experimental data. The reader is referred to the earlier section for appropriate analog expressions. We shall now consider some of the most used empirical expressions.

An often used correlation for turbulent pipe flow is the Dittus-Boelter[45] equation

$$\mathrm{Nu}_D = 0.023 \, \mathrm{Re}_D^{0.8} \mathrm{Pr}^n \quad (5\text{-}187)$$

where

(1) $n = 0.3$ if the fluid is being cooled
      $= 0.4$ if the fluid is being heated
(2)   all fluid properties are evaluated at the average mean fluid temperature
(3) $\mathrm{Re}_D > 10^4$
(4) $0.7 < \mathrm{Pr} < 100$
(5) $L/D > 60$

A similar correlation was suggested by Colburn,[46] who used a Stanton number rather than Nu_D and a constant exponent on Pr,

$$\mathrm{St} = 0.023 \, \mathrm{Re}_D^{-0.2} \mathrm{Pr}^{-2/3} \quad (5\text{-}188)$$

[45] F. W. Dittus and L. M. K. Boelter, *Univ. of California Publ. Eng.* 2 (1930): 443.
[46] Colburn, op. cit.

where

(1) St is evaluated at the average mean fluid temperature
(2) Re_D and Pr are evaluated at the average film temperature
(3) $Re_D > 10^4$
(4) $0.7 < Pr < 160$
(5) $L/D > 60$

McAdams[47] modified the Colburn expression, using a Seider and Tate[48] type of viscosity correction term, to achieve an expression which applies over a much larger Prandtl number range; his expression is

$$St = 0.023\ Re_D^{-0.2}Pr^{-2/3}\left(\frac{\mu_m}{\mu_o}\right)^{0.14} \tag{5-189}$$

where

(1) all fluid properties are evaluated at the average mean temperature except μ_o, which is evaluated at T_o
(2) $Re_D > 10^4$
(3) $0.7 < Pr < 17,000$
(4) $L/D > 60$

A typical problem employing an empirical expression for h is presented below.

Example 5.8

Water at 50°F enters a 1/2-inch schedule-40 pipe flowing at 60 gal/min. The pipe wall is maintained at 210°F by condensing steam. For 5-foot-long tube, find the exit water temperature and the total heat transferred.

The exit water temperature will be determined by means of equation (5-136)

$$\frac{T_L - T_o}{T_e - T_o} = e^{-St4L/D} \tag{5-136}$$

and the heat transfer can be determined, using the first law of thermodynamics once T_L is known, according to

$$q = \rho A v_{avg} c_p (T_L - T_e)$$

The problem now becomes one of determining the Stanton number which is needed to solve for T_L. A complication exists in that the fluid properties included in any Stanton number correlation are temperature dependent, and we do not know, as yet, what value the exit temperature has. A trial-and-error procedure, similar to that diagrammed in Figure 5.24, must be employed. An initial guess will

[47] McAdams, op. cit.
[48] Seider and Tate, op. cit.

be made for T_L; St will then be evaluated, and T_L solved for from equation (5-136). The problem will be completed when assumed and calculated values of T_L agree. The velocity is evaluated as

$$v_m = \frac{\left(\dfrac{60 \text{ gal}}{\min}\right)\left(\dfrac{\min}{60 \text{ sec}}\right)\left(\dfrac{\text{ft}^3}{7.48 \text{ gal}}\right)}{\dfrac{\pi}{4}\left(\dfrac{0.622}{12}\text{ ft}\right)^2} = 63.4 \text{ ft/sec}$$

Initial assumptions are made that $T_L = 60°F$ and the corresponding average mean and film temperatures are 55°F and 132.5°F, respectively. Using the film temperature to evaluate ν, we have, for Re_D,

$$\text{Re}_D = \frac{Dv}{\nu} = \frac{\left(\dfrac{0.622}{12}\text{ ft}\right)(63.4 \text{ ft/sec})}{0.57 \times 10^{-5} \text{ ft}^2/\text{sec}}$$

$$= 5.76 \times 10^5$$

thus flow is definitely turbulent.

Using the Colburn correlation, equation (5-188), we calculate the Stanton number as

$$\text{St} = 0.023(5.76 \times 10^5)^{-0.2}(3.35)^{-2/3}$$
$$= 7.24 \times 10^{-4}$$

The exit temperature is calculated as

$$T_L = 210 - (210 - 50)\exp\left[-(7.24 \times 10^{-4})\left(\frac{4 \times 60}{0.622}\right)\right]$$

$$= 210 - 160\,e^{-0.279}$$

$$= 210 - 121 = 91°F$$

A second series of calculations, assuming that $T_L = 91°F$, yields the following values:

$$T_{mavg} = 70.5°F, \qquad T_f = 140.3°F$$

$$\text{Re}_D = \frac{\left(\dfrac{0.622}{12}\text{ ft}\right)(63.4 \text{ ft/sec})}{0.525 \times 10^{-5} \text{ ft}^2/\text{sec}} = 6.26 \times 10^5$$

$$\text{Pr} = 3.05$$

$$\text{St} = 0.023(6.25 \times 10^5)^{-0.2}(3.05)^{-2/3}$$

$$= 7.58 \times 10^{-4}$$

$$T_L = 210 - 160\,e^{-0.292} = 90.6°F$$

which is close enough to the assumed value of T_0 for our purposes. The heat transferred is

$$q = \left(60 \frac{gal}{min}\right) \left(\frac{60\,min}{hr}\right) \left(\frac{ft^3}{7.48\,gal}\right) \left(62.2 \frac{lb_m}{ft^3}\right)$$

$$(1\,Btu/lb_m\text{-}°F)(90 - 50)°F$$

$$= 12 \times 10^5\,Btu/hr$$

The answers to this problem are thus 90°F for the exiting fluid temperature and a total heat transfer of 12×10^5 Btu/hr.

5.3-2 Forced Convection for External Flow

Numerous situations in engineering practice deal with the flow of a fluid over the exterior surface of a solid. The shapes of greatest interest are cylinders and spheres; heat transfer between these surfaces and fluids in cross flow is frequently encountered.

Some of the phenomena associated with the flow of fluids over plane and bluff bodies, described in Section 5.1, should be referred to at this point. The reader will recall that the phenomenon of boundary layer separation may be encountered in external flow; this situation occurs in nearly all practical cases involving bluff bodies.

5.3-2.1 Heat Transfer for Cylinders in Crossflow

Single Cylinders. The work of Eckert and Soehngen[49] and of Giedt[50] is pertinent to the case of fluids flowing normal to the axis of single cylinders. The Eckert and Soehngen values of local Nusselt numbers around the periphery of a cylinder for Reynolds numbers varying between 20 and 600 are shown in Figure 5.38. Giedt investigated much higher Reynolds number cases; his results are shown in Figure 5.39.

The Nusselt number is seen to vary smoothly from a relatively large value near the stagnation point. At the lower Reynolds numbers the Nusselt number decreases regularly around the surface from the stagnation point to the point where boundary layer separation occurs; at this location the Nusselt number increases slightly. At higher Reynolds numbers this same behavior occurs near the stagnation point; however, the Nusselt number experiences two sudden increases, one at the separation point and one where the boundary layer undergoes a transition from laminar to turbulent flow.

In Figure 5.39, at Reynolds numbers below 10^5, the laminar boundary layer is observed to undergo separation at a location approximately 80°

[49] E. R. G. Eckert and E. Soehngen, *Trans. ASME* **74** (1952): 343.
[50] W. H. Giedt, *Trans. ASME* **71** (1949).

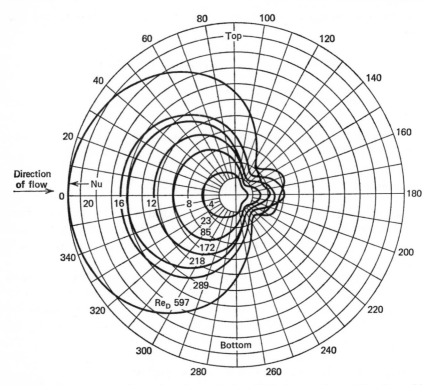

Figure 5.38 Local Nusselt numbers for a single cylinder in crossflow at low Reynolds numbers. [From E. R. Eckert and E. Soehngen, *Trans. A.S.M.E.* 74 (1952): 346. By permission of the publishers.]

from the stagnation point with no large change in local Nusselt number. For $Re_D > 1.5 \times 10^5$, a portion of the boundary layer is in turbulent flow. For this case, the separation point moves past 90°, and less of the surface is engulfed in the separated wake. In the region where flow is turbulent, the local Nusselt number reaches a maximum due to the greater conductance of the turbulent boundary layer.

While plots such as Figures 5.38 and 5.39 are helpful in predicting local conditions around a cylinder in nonisothermal crossflow, one is usually more interested in describing the total heat transfer, which includes all local effects around the cylinder. Obviously, the actual conditions are such that analysis is of minimal value; thus empirical correlations must be used to predict average values of the heat transfer coefficient. McAdams[51] has plotted the data of 13 different investigators as Nu_D versus Re_D. Excellent correlation has been found; this plot is shown in Figure 5.40.

[51] McAdams, op. cit.

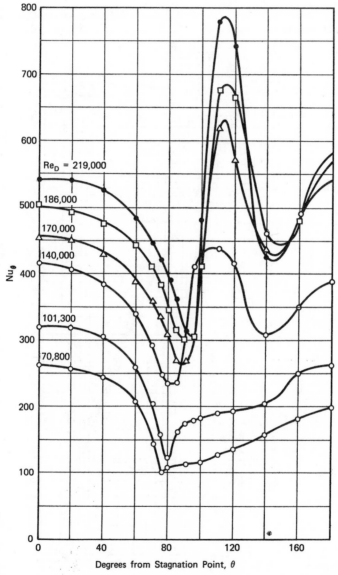

Figure 5.39 Local Nusselt numbers about a single cylinder in crossflow at high Reynolds numbers. [From W. H. Giedt, *Trans. A.S.M.E.* 71 (1949): 378. By permission of the publishers.]

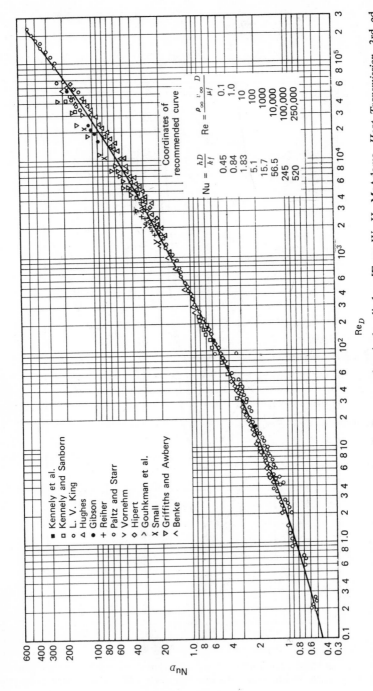

Figure 5.40 Mean Nusselt numbers versus Re_D for flow normal to single cylinders. [From W. H. McAdams, *Heat Transmission*, 3rd ed. (New York: McGraw-Hill Book Company, 1954), p. 259. By permission of the publishers.]

The following labels appear within the figure:

- Kennely et al.
- Kennely and Sanborn
- L. V. King
- Hughes
- Gibson
- Reiher
- Paltz and Starr
- Vornehm
- Hipert
- Gouhkman et al.
- Small
- Griffiths and Awbery
- Benke

Coordinates of recommended curve

$Nu = \dfrac{hD}{k_f}$	$Re = \dfrac{\rho_\infty \, v_\infty \, D}{\mu_f}$
0.45	0.1
0.84	1.0
1.83	10
5.1	100
15.7	1000
56.5	10,000
245	100,000
520	250,000

Table 5.6   Values of the Constants B and n to Be Used
in Equation (5-190)

Re_D	B	n
0.4–4	0.891	0.330
4–40	0.821	0.385
40–4,000	0.615	0.466
4,000–40,000	0.174	0.618
40,000–400,000	0.0239	0.805

The values shown in this figure are expressed empirically in the form

$$Nu_D = B\,Re^n \qquad (5\text{-}190)$$

with values of the constants B and n given in Table 5.6 for different Reynolds numbers. Fluid properties involved in equation (5-190) should be evaluated at the film temperature. Values of h determined from Figure 5.40 or equation (5-190) apply to gases. When used for liquids in external flow the right-hand side of equation (5-190) should be modified by adding the term $(1.1\,Pr^{1/3})$.

Banks of Cylinders in Crossflow.   Circular cylinders are often arranged in bundles or banks to accomplish a desired total heat transfer in a relatively small space. The tube arrangement in many heat exchangers is a good example of this.

When arranged in banks, the heat transfer for each cylinder involves some of the considerations just discussed; however, there is considerable interaction between one cylinder and the others adjacent to it. Additional factors which must be considered are tube arrangement, tube spacing, and orientation relative to fluid flow.

Figure 5.41 shows the work of Bergelin, Colburn, and Hull[52] for laminar flow with $1 < Re < 1000$. The parameters plotted include fluid properties, with the exception of μ_o, evaluated at the average mean fluid temperature; μ_o is the viscosity at the wall temperature. The length parameter D_{eq}, used in calculating Reynolds number, is an equivalent diameter of a tube bundle, defined as

$$D_{eq} = \frac{4(S_L S_T - \pi D^2/4)}{\pi D} \qquad (5\text{-}191)$$

where S_L is the center-to-center distance between tubes *in the direction of flow*, S_T is the center-to-center distance between tubes *normal to the flow*, and D is the tube OD.

[52] O. P. Bergelin, A. P. Colburn, and H. L. Hull, *Univ. of Delaware Engr. Exp. Sta. Bulletin No. 2* (1950).

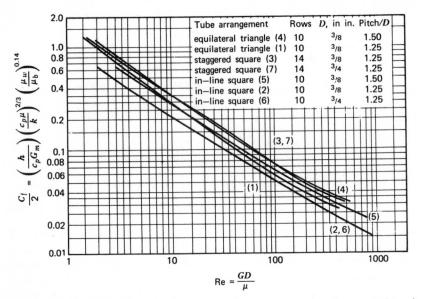

$$C_f = \left(\frac{h}{c_p G_m}\right)\left(\frac{c_p \mu}{k}\right)^{2/3}\left(\frac{\mu_w}{\mu_b}\right)^{0.14}$$

Tube arrangement	Rows	D, in in.	Pitch/D
equilateral triangle (4)	10	$3/8$	1.50
equilateral triangle (1)	10	$3/8$	1.25
staggered square (3)	14	$3/8$	1.25
staggered square (7)	14	$3/4$	1.25
in–line square (5)	10	$3/8$	1.50
in–line square (2)	10	$3/8$	1.25
in–line square (6)	10	$3/4$	1.25

$$Re = \frac{GD}{\mu}$$

Figure 5.41   Heat transfer coefficient versus Re_D for liquids in laminar flow normal to tube bundles. [From O. P. Bergelin, A. P. Colburn, and H. L. Hull, University of Delaware, *Engineering Experiment Station Bulletin No.* 2 (1950). By permission of the publishers.]

For values of Re_D up to 10^4, Bergelin, Brown, and Doberstein[53] have presented plots for both heat transfer and pressure drop determination. Figure 5.42 shows the curves for evaluating h and f in various tube arrangements. Reynolds number is calculated, using D_t, the tube diameter, in this figure.

It is apparent that almost limitless combinations of tube spacing and arrangement can exist. No single plot could show all such possibilities. Probably the most fruitful source for information of this sort, including data for gases flowing through tube bundles, is the book, *Compact Heat Exchangers*, by Kays and London.[54] Heat transfer coefficients for the shell side of shell and tube heat exchangers remain difficult to predict; fortunately, in practice, the controlling thermal resistance is usually that on the inside tube wall, which is relatively easy to evaluate.

5.3-2.2   Heat Transfer for Flow about Spheres

Local heat transfer phenomena about a spherical surface occur in similar fashion t'o those about cylinders, as discussed earlier. The work of Cary[55] is

[53] O. P. Bergelin, G. A. Brown, and S. C. Doberstein, *Trans. ASME* 74 (1952): 1958.
[54] W. M. Kays and A. L. London, *Compact Heat Exchangers*, (New York: McGraw-Hill, 1958).
[55] J. R. Cary, *Trans. ASME* 75 (1953): 483.

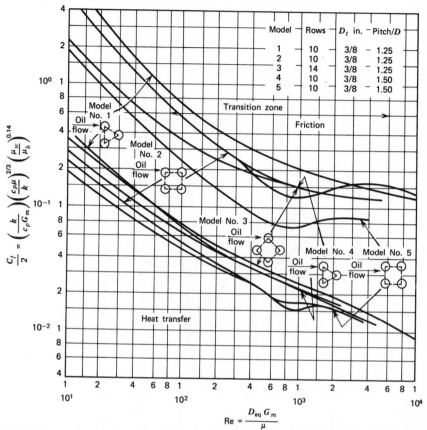

Figure 5.42 Heat transfer and pressure drop parameters for liquids flowing through tube bundles for Reynolds numbers to 10^4. [From O. P. Bergelin, G. A. Brown, and S. C. Doberstein, *Trans. A.S.M.E.* 74 (1952): 1958. By permission of the publishers.]

shown in Figure 5.43 for local heat transfer coefficients with flow past a spherical surface.

A correlation of the work of several investigators for flow past spheres has been presented by McAdams. Figure 5.44 shows McAdams' plot of Nu_D versus Re_D.

Suggested empirical equations for flow past spheres are the following:

for liquids with $1 < \mathrm{Re}_D < 70,000$

$$\mathrm{Nu}_D = 2.0 + 0.60 \, \mathrm{Re}_D^{1/2} \mathrm{Pr}^{1/3} \tag{5-192}$$

for air with $20 < \mathrm{Re}_D < 150,000$

$$\mathrm{Nu}_D = 0.33 \, \mathrm{Re}_D^{0.6} \tag{5-193}$$

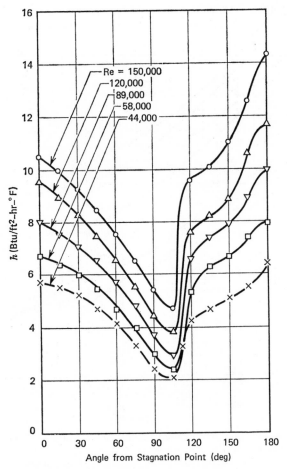

Figure 5.43 Local heat transfer coefficients for flow past a sphere. [From J. R. Cary, *Trans. A.S.M.E.* 75 (1953): 485. By permission of the publishers.]

for gases other than air, with $1 < \mathrm{Re}_D < 25$

$$\mathrm{St} = \frac{2.2}{\mathrm{Re}_D} + 0.48\ \mathrm{Re}_D^{-1/2} \qquad (5\text{-}194)$$

and with $25 < \mathrm{Re}_D < 150{,}000$

$$\mathrm{Nu}_D = 0.37\ \mathrm{Re}_D^{0.6}\mathrm{Pr}^{1/3} \qquad (5\text{-}195)$$

In all of the preceding equations, the film temperature should be used for fluid property evaluation.

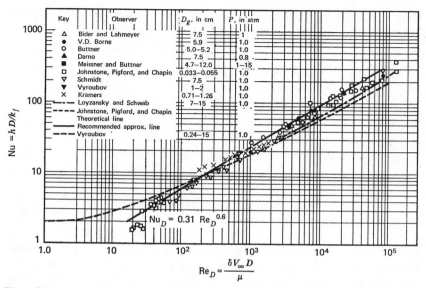

Figure 5.44 Mean Nusselt numbers for air flow past a sphere. [From W. H. McAdams, *Heat Transmission*, 3rd ed. (New York: McGraw-Hill Book Company, 1954), p. 266. By permission of the publishers.]

5.3-2.3 Heat Transfer with Flow about Plane Surfaces

In the case of flow parallel to a straight plane wall, the analytical results of Section 5.1-2 apply. The appropriate equations will be repeated here to make the present section complete.

For laminar boundary layer flow on an isothermal flat plate, the local and mean Nusselt numbers are given by

$$Nu_x = 0.332 \ Re_x^{1/2} Pr^{1/3} \tag{5-98}$$

and

$$Nu_L = 0.664 \ Re_L^{1/2} Pr^{1/3} \tag{5-100}$$

respectively, with the film temperature used for property evaluation. These equations are valid for fluids with Prandtl numbers in the range $0.6 < Pr < 50$.

For situations where no boundary layer separation exists, the analogies between momentum and heat transfer from Section 5.1-2.5 apply. The combined attributes of simplicity and reasonable accuracy make the Colburn analogy the most convenient to use of those presented. The Colburn analogy expression is

$$StPr^{2/3} = \frac{C_f}{2} \tag{5-129}$$

With this analog expression and equations (5-75) and (5-76) for the turbulent boundary layer over a flat plate, the heat transfer coefficient in a turbulent boundary layer may be expressed in nondimensional form as

$$\text{Nu}_x = 0.0288 \, \text{Re}_x^{4/5} \text{Pr}^{1/3} \tag{5-196}$$

and

$$\text{Nu}_L = 0.036 \, \text{Re}_L^{4/5} \text{Pr}^{1/3} \tag{5-197}$$

When considering a surface over which boundary layer flow is both laminar and turbulent, a combination of equations (5-100) and (5-197) must be used. Transition from laminar to turbulent flow within the boundary layer occurs where the local Reynolds number Re_x achieves a value of approximately 10^6.

Numerous variations on wall conditions are possible. Kays[56] discusses the cases of arbitrarily specified surface temperatures and wall heat fluxes. A specific result of interest in Kays' discussion applies to the case of a step change in wall temperature above the free stream value at the leading edge, followed by a linear increase in wall temperature from the leading edge. The constants a and b refer to the step change and incremental variation in wall temperature, respectively; thus

$$T_o = T_\infty + a + bx \tag{5-198}$$

$$\frac{dT_o}{dx} = b \tag{5-199}$$

The temperature profile along the direction of flow is shown in Figure 5.45. The heat flux, for this case, becomes

$$\frac{q}{A} = 0.332 \frac{k}{x} \text{Re}_x^{1/2} \text{Pr}^{1/3}(a + 1.612bx) \tag{5-200}$$

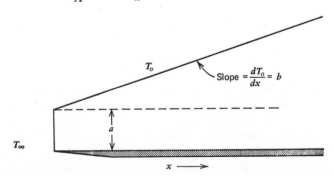

Figure 5.45 Temperature variation along a flat plate with a step increase a at the leading edge and a linear increase bx from the leading edge.

[56] Kays, pp. 218–222.

and the local Nusselt number is

$$\text{Nu}_x = \frac{0.332 \, \text{Re}_x^{1/2} \text{Pr}^{1/3}(a + 1.612bx)}{a + bx} \tag{5-201}$$

The constant wall temperature case is achieved when $b = 0$ in these expressions. For the constant $a = 0$, the situation is one with a linear increase in surface temperature from a value of T_∞ at the leading edge. The result is 61 % higher than for the situation where the wall temperature is constant. The need to consider wall temperature variation in predicting or evaluating heat transfer rates is thus most obvious.

5.4 HEAT TRANSFER WITH A CHANGE IN PHASE

The phenomena of boiling and condensation are of considerable practical importance because they may involve very large heat fluxes with accompanying temperature differences which are quite small.

Although boiling and condensation are not convective processes in the same sense as those previously discussed, the heat transfer coefficient h is used to describe the heat flux as a function of the temperature difference between the solid surface and adjacent fluid. The added considerations of surface characteristics, surface tension, latent heats of phase change, and other mechanisms involved with phase changes make empirical correlations more complex and analysis more difficult than in simple convective situations involving single-phase fluids.

Boiling and condensation involve changes between liquid and vapor phases of a fluid. These phenomena will be considered separately in the sections to follow.

5.4-1 Boiling Heat Transfer

In the boiling phenomenon the addition of thermal energy to a saturated liquid results in a change from liquid to vapor phase at the same temperature. Relatively large amounts of heat can be transferred when boiling occurs; thus in certain circumstances this may be a most important heat transfer mechanism. In other situations the formation of vapor from a liquid may be the desired result of a heat exchange process; the need to know heat transfer rates and controlling mechanisms is vital to a well-designed and efficiently operating process.

When boiling occurs on a heated surface submerged in a liquid at rest, the situation is classified as *pool boiling*. If the fluid is moving past the hot surface, *flow boiling* occurs.

5.4-1.1 The Regimes of Boiling; the Boiling Curve

A well-used and most descriptive approach to understanding the mechanisms of boiling heat transfer is the "boiling curve" shown in Figure 5.46. The plot is one of heat flux versus the temperature difference between a hot surface and an adjacent saturated liquid. The boiling surface may be thought of as an electrically heated wire; thus the heat flux is easily controlled by the voltage drop across the wire of fixed resistance. The heat flux curve is separated into six regimes for purposes of discussion. The values shown for $T_o - T_{sat}$ are representative of saturated water at atmospheric pressure. Different fluids will behave in the same way; however, the temperature differences may be other than those used here for discussion purposes.

Regime I. For a wire surface temperature which is a very few degrees above the adjacent saturated liquid, the energy exchange mechanism is one of natural convection. Heat transfer occurs via natural convection currents which transport superheated liquid from the hot surface to the free liquid surface, where evaporation takes place.

Regime II. As the wire surface temperature increases, bubbles of water vapor form on the wire surface. When the bubbles reach a size sufficient for buoyant forces to overcome surface tension forces, they break off, rise through the cooler liquid, and condense before reaching the free liquid

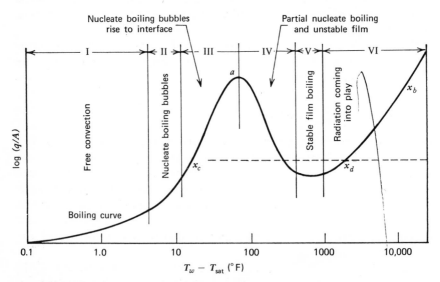

Figure 5.46 Heat flux versus temperature difference for pool boiling of water on a horizontal wire at atmospheric pressure.

surface. Bubbles of vapor are observed to form at favored locations on the wire surface; these locations are designated *active sites* or *nucleation sites*.

Regime III. When the power level imposed on the wire is increased such that the surface temperature rises still higher, the bubbles form, break off, and rise with increasing frequency and intensity. The bubbles will now reach the free surface and expel vapor formed at the wire surface directly into the surrounding atmosphere. In this regime the wire surface is never completely covered with vapor, and bubbles still form at preferred locations on the wire. The type of boiling occurring is termed *nucleate boiling*.

One should note the relatively high heat flux values possible in the nucleate boiling regime. With water, heat fluxes on the order of 10^5 Btu/hr-ft^2 are possible for a temperature difference of approximately 100°F. The equivalent value of h is thus 1000 Btu/hr-ft^2-°F. This high value of h possible with a relatively small temperature difference makes the nucleate boiling regime one of considerable practical importance.

The reason for the large heat transfer rates in the nucleate regime should be considered briefly. The major contribution to heat transfer is the violent agitation of the liquid as bubbles form, break away from the solid surface, and rise rapidly through the liquid. The circulation of the liquid past the hot surface, a portion of which is not covered by vapor, is largely responsible for the high heat transfer rates. The energy involved in forming vapor bubbles which subsequently rise through the liquid and enter the air is significant but not the major part of the total nucleate boiling heat transfer process

As the temperature difference increases, while in the nucleate range, the rate of heat flux increase is quite rapid. The process of bubble formation and breakaway at the heated surface increases in frequency, and more of the surface becomes "active." As more of the surface becomes covered with vapor, less is in contact with the liquid which is swept by; thus at ever-larger heat fluxes, the heat transfer potential of the nucleate boiling process approaches a maximum. Eventually the point is reached where an increase in surface temperature will cause enough of the hot surface to be covered with vapor so that the heat flux potential will decrease. The point of maximum heat flux, designated a in Figure 5.46, is termed the *burnout point* for reasons we will discuss presently. The burnout point represents the upper end of Regime III. In water, at atmospheric pressure, burnout occurs for ΔT slightly above 100°F and a heat flux on the order of 5×10^5 Btu/hr-ft^2.

Regime IV. At temperature differences past the burnout point, more of the parent surface becomes covered by the vapor film as ΔT increases. Heat transfer through the film is first by conduction, then by direct transport as vapor in the film is "torn" away from the film and carried to the surface by buoyant forces and by the agitation of the liquid caused by this movement of vapor bubbles. Much of the heated surface is covered continuously by the vapor film in this region, with ever-decreasing amounts of the surface exposed directly to saturated liquid. The heat transfer potential thus decreases continuously for larger ΔT's until the entire surface is covered continuously by the vapor film.

When the surface is covered by vapor, the phenomenon which occurs is designated *film boiling*. Regime IV is often termed the regime of *transition boiling*, where the phenomena of nucleate and film boiling both occur.

Near the burnout point more of the surface experiences nucleate boiling, while the proportion of surface experiencing film boiling increases as ΔT increases. For reasons yet to be discussed, stable operation in Regime IV cannot be maintained for electrically heated surfaces.

Regime V. This regime is designated the *stable film boiling regime* and is characterized by a minimum in the boiling curve shown in Figure 5.46. In water, at atmospheric pressure, stable film boiling is achieved at ΔT values near 500°F

Regime VI. This regime is characterized by very large temperature differences between the surface and saturated liquid. Boiling is still of the film type; however, the temperature differences are so large that radiant heat transfer becomes significant—in fact, controlling—and the heat flux curve rises once more with increasing ΔT.

5.4-1.2 Burnout

The nature of an electrically heated surface on which boiling occurs will now be considered further. At a heat flux level corresponding to point b in Figure 5.46, a small change in heat flux will cause the surface temperature to vary slightly, and the operating condition will change such that the new point is still in the nucleate boiling regime.

If, however, the burnout point is reached, a slight increase in heat flux will cause the surface temperature to rise as before; the boiling curve shows that an increase in ΔT above burnout is accompanied by a *decrease* in heat transfer capability of the surface. The result is a continued increase in surface temperature, and a continued decrease in heat transfer—an ever-worsening situation. This process would continue until point c is reached in Regime VI. At this point the required surface temperature is so large that the material would long since have melted and broken the electrical heating circuit. The term "burnout" is thus observed to be very descriptive of the physical process which will occur. For obvious reasons the transition boiling regime can never be achieved for stable operation of an electrically heated surface.

Stable operation at any point on the boiling curve can be achieved if ΔT, rather than heat flux, is controlled. This has been accomplished by having the surface heated by condensing fluid vapors. A description of such a system and its necessary considerations is given by Aoki.[57]

It is worthy of note here that, in boiling heat transfer, a somewhat anomalous behavior is encountered. In general, it is expected that heat transfer will increase continuously with ever-larger values of ΔT. We have seen that this is not true when boiling occurs due to the formation of an insulating vapor film.

5.4-1.3   Correlations of Boiling Heat Transfer Data

In this section some of the better accepted correlations of experimental data for boiling heat transfer are presented.

Rohsenow[58] has developed the following empirical expression for stable nucleate pool boiling, based upon the data obtained by Addoms.[59] A plot of Addoms' data is shown as Figure 5.47.

$$\frac{c_{pL}(T_o - T_{\text{sat}})}{h_{fg}} = c_{sf}\left[\frac{q/A}{\mu_L h_{fg}}\sqrt{\frac{\sigma}{g(\rho_L - \rho_v)}}\right]^{1/3} \Pr_L^{1.7} \qquad (5\text{-}202)$$

In equation (5-202) the terms have their usual meanings; subscripts L and v refer to the saturated liquid and saturated vapor, respectively. The term c_{sf} is an empirical constant whose value depends on the particular combination of fluid and heated surface materials involved in a boiling situation. The line in Figure 5.47 is for a value of c_{sf} equal to 0.013. Table 5.7, adopted from Rohsenow and Choi,[60] gives values of c_{sf} which pertain to various fluid/surface combinations.

For the point of maximum heat flux, Rohsenow and Griffith[61] have correlated data in the form

$$\frac{(q/A)_{\text{max}}}{\rho_v h_{fg}} = 143g^{1/4}\left(\frac{\rho_L - \rho_v}{\rho_v}\right)^{0.6} \qquad (5\text{-}203)$$

where g is gravitational acceleration in G's; the subscripts L and v refer to liquid and vapor properties, respectively; and the remaining terms have their usual meanings.

[57] T. Aoki, "An Experimental Study of Transition Pool Boiling" (Ph.D. thesis, Oregon State University, 1970).

[58] W. H. Rohsenow, Trans. A.S.M.E. 74 (1952): 969.

[59] J. N. Addoms, D.Sc. Thesis, Chemical Engineering Department, Massachusetts Institute of Technology, June 1948.

[60] W. M. Rohsenow and H. Choi, Heat, Mass, and Momentum Transfer, (Englewood Cliffs, N.J.: Prentice-Hall, 1961).

[61] W. M. Rohsenow and P. Griffith, AIChE-ASME Heat Transfer Symposium, Louisville, Ky., 1955.

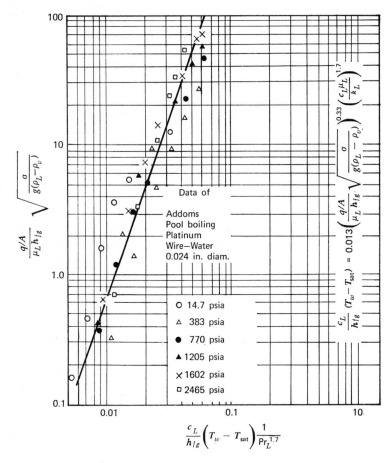

Figure 5.47 Data for stable nucleate pool boiling. [From W. M. Rohsenow and H. Choi, *Heat, Mass, and Momentum Transfer* (Englewood Cliffs, N.J.: Prentice-Hall, Inc., 1961), p. 224. By permission of the publishers.]

In the case of *stable film pool boiling*, the following expressions are recommended. For horizontal tubes, Bromley[62] suggests the equation

$$h = 0.62 \left[\frac{k_v^3 \rho_v (\rho_L - \rho_v) g(h_{fg} + 0.4 c_{pv}\, \Delta T)}{D_o \mu_v (T_o - T_{\text{sat}})} \right]^{1/4} \quad (5\text{-}204)$$

where $\Delta T = T_o - T_{\text{sat}}$ and D_o is the outside diameter of the tube.

[62] L. A. Bromley, *Chem. Engr. Progr.* **46** (May 1950): 5, 221; Bromley et al., *Ind. Engr. Chem.* **45** (1953); 2639.

Table 5.7   Values of c_{sf} to Be Used in
Equation (5-199)

Fluid/Surface Combination	c_{sf}
water/nickel	0.006
water/platinum	0.013
water/copper	0.013
water/brass	0.006
CCl_4/copper	0.013
benzene/chromium	0.010
n-pentane/chromium	0.015
ethanol/chromium	0.0027
isopropyl alcohol/copper	0.0025
35 % KOH/copper	0.0054
50 % KOH/copper	0.0027
n-butyl alcohol/copper	0.0030

For a horizontal plane surface, Berenson[63] has suggested a modified form of equation (5-204) with tube diameter D_o, replaced by $[\sigma/g(\rho_L - \rho_v)]^{1/2}$. The resulting expression is

$$h = 0.425 \left[\frac{k_{vf}^3 \rho_{vf}(\rho_L - \rho_v)g(h_{fg} + 0.4c_{pv}\,\Delta T)}{\mu_{vf}(T_o - T_{\text{sat}})\sqrt{\sigma/g(\rho_L - \rho_v)}} \right]^{1/4} \qquad (5\text{-}205)$$

where k_{vf}, ρ_{vf}, and μ_{vf} are to be evaluated at the film temperature.

For a vertical tube, experimental data were correlated by Hsu and Westwater[64] by

$$h = 0.0020 \ \text{Re}^{0.6} \left[\frac{\mu_v^2}{g\rho_v(\rho_L - \rho_v)k_v^3} \right]^{-1/3} \qquad (5\text{-}206)$$

where

$$\text{Re} = \frac{4w}{\pi D_o \mu_v} \qquad (5\text{-}207)$$

w being the vapor flow rate in lb_m/hr at the upper end of the tube. For like conditions, Hsu[65] states that heat transfer rates are higher for film boiling with vertical tubes than for horizontal tubes.

For boiling combined with convection, as in the case of flow boiling, the heat fluxes associated with each mechanism may be added simply as

$$\left.\frac{q}{A}\right|_{\substack{\text{convection} \\ \text{and boiling}}} = \left.\frac{q}{A}\right|_{\text{convection}} + \left.\frac{q}{A}\right|_{\text{boiling}} \qquad (5\text{-}208)$$

[63] P. Berenson, *AIChE Paper No.* 18, Heat Transfer Conference, Buffalo, N.Y. (1969).
[64] Y. Y. Hsu and J. W. Westwater, *AIChE Jour.* **4** (1958): 59.
[65] S. T. Hsu, *Engineering Heat Transfer*, (Princeton. N.J.: Van Nostrand, 1963).

With radiation considered in a boiling situation, the combined effect may be determined according to

$$h_{\text{total}} = h_b \left(\frac{h_b}{h_{\text{total}}} \right)^{1/2} + h_r \qquad (5\text{-}209)$$

where h_{total} is the total heat transfer coefficient, and h_b and h_r are the boiling and radiation heat transfer coefficients, respectively. The radiation coefficient h_r will be discussed in Chapter 6.

5.4-2   Heat Transfer with Condensing Vapors

We now consider the opposite effect to boiling, specifically the heat transfer associated with a change from vapor to the liquid phase. When a surface is maintained at a temperature below the saturation temperature of an adjacent vapor, *condensation* will occur; the liquid which is condensed will collect on a flat horizontal surface or flow under the influence of gravity if the surface and its orientation will allow this to occur.

Under most circumstances the liquid condensate wets the surface, spreads out, and forms a film over the entire surface. This type of condensation is known as *film condensation*. When the surface is not wetted by the liquid, condensate forms in droplets which will run down an inclined surface, coalescing with other droplets which they touch. This is *dropwise* condensation. When filmwise condensation occurs, the surface is completely covered by a liquid film. Heat transfer necessary for condensation to occur at the liquid-vapor interface must then be by conduction through the liquid film. As the film thickness increases, the heat transfer rate, hence the rate of condensation, will decrease. With dropwise condensation, on the other hand, a portion of the cool surface is always in contact with vapor and is not subject to the insulating effect of a liquid layer. Dropwise condensation, therefore, is associated with higher heat transfer rates than is filmwise condensation. Dropwise condensation is difficult to achieve and maintain for any extended period of time; thus equipment involving the condensation phenomenon is normally designed on the basis that filmwise condensation will occur.

5.4-2.1   *Film Condensation; the Nusselt Film Model*

The problem of a pure vapor condensing as a film on a plane vertical wall was analyzed by Nusselt[66] in 1916. His result is still valid, and the development is helpful in understanding the mechanism involved. His approach will be duplicated in the following analysis.

[66] W. Nusselt, *Zeitschr, d. ver. deutsch. Ing.* **60** (1916): 514.

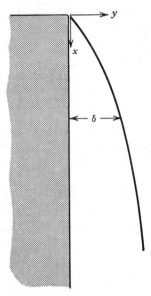

Figure 5.48   Film condensation on a plane vertical wall.

Terms used in the development to follow are related to the sketch shown in Figure 5.48. The condensate film thickness δ increases with x from a value of 0 at $x = 0$.

Our approach will involve equating the energy transfer associated with the condensation process to that which can be conducted through a liquid film in laminar flow. A heat transfer coefficient may be obtained, using the defining relationship in the form

$$h = \frac{q/A}{T_{\text{sat}} - T_o} = \frac{k}{\delta}$$ (5-210)

where T_{sat} and T_0 refer to the saturation temperature of the fluid at the prevailing pressure and the wall temperature, respectively.

The rate of liquid condensed will be determined from an analysis of the fluid flow situation; we shall evaluate the velocity profile, the rate of flow, and the change in liquid flow rate in turn.

With the assumption of laminar flow, the incompressible form of the x-directional Navier-Stokes equation applies. Under steady, two-dimensional flow conditions, assuming the film to grow relatively slowly, equation (2-66) reduces to

$$0 = -\rho g - \frac{dP}{dx} + \mu \frac{d^2 v_x}{dy^2}$$ (5-211)

The pressure term may be eliminated since the liquid-vapor interface is maintained at constant pressure. The remaining expression can be separated and integrated twice, and the boundary conditions $v_x = 0$ at $y = 0$ and $(dv_x/dy) = 0$ at $y = \delta$ applied, yielding, for v_x,

$$v_x = \frac{\rho g \delta^2}{\mu}\left[\frac{y}{\delta} - \frac{1}{2}\left(\frac{y}{\delta}\right)^2\right]$$ (5-212)

To include the possibility that the vapor density ρ_v may be significant with respect to the liquid density ρ_L (as in the case of high system pressure), equation (5-212) is modified in the form

$$v_x = \frac{(\rho_L - \rho_v)g\delta^2}{\mu}\left[\frac{y}{\delta} - \frac{1}{2}\left(\frac{y}{\delta}\right)^2\right]$$ (5-213)

The volume flow rate Γ, per unit width of film, is calculated from this expression for v_x as follows:

$$\Gamma = \int_A v_x \, dA = \int_0^\delta v_x(1) \, dy$$

For any $x > 0$, we have

$$\Gamma(x) = \frac{(\rho_L - \rho_v)g \, \delta(x)^3}{3\mu}$$ (5-214)

The rate at which the liquid flow rate changes may now be expressed as

$$d\Gamma = \frac{(\rho_L - \rho_v)g \, \delta^2 \, d\delta}{\mu}$$ (5-215)

The energy associated with this rate of condensation includes the heat of condensation plus the heat released as the saturated liquid cools to the average temperature of the liquid film. This heat flux is expressed as

$$\frac{q_y}{A} = \rho_L \frac{d\Gamma}{dx}\left[h_{fg} + \frac{1}{\rho_L \Gamma}\int_0^\delta \rho_L v_x c_{pL}(T_{sat} - T) \, dy\right]$$ (5-216)

A linear temperature profile, expressed as $T = T_o + (T_{sat} - T_o)(y/\delta)$, will yield

$$\frac{q_y}{A} = \rho_L \frac{d\Gamma}{dx}[h_{fg} + (3/8)c_{pL}(T_{sat} - T_o)]$$ (5-217)

The foregoing heat flux must transfer by conduction (since flow is laminar) through the liquid film of thickness δ, according to

$$\frac{q_y}{A} = \frac{k_L}{\delta}(T_{sat} - T_o)$$ (5-218)

Equations (5-217) and (5-218) may now be equated, with $d\Gamma$ expressed as given in equation (5-215), giving

$$\frac{(\rho_L - \rho_v)g\delta^2}{\mu} \, d\delta = \frac{k_L(T_{sat} - T_o)}{\rho_L \, \delta[h_{fg} + (3/8)c_{pL}(T_{sat} - T_o)]}$$

This expression may be simplified and solved for the film thickness δ as

$$\delta = \left\{\frac{4k_L\mu(T_{sat} - T_o)x}{\rho_L g(\rho_L - \rho_v)[h_{fg} + (3/8)c_{pL}(T_{sat} - T_o)]}\right\}^{1/4} \tag{5-219}$$

Evaluating h_x as suggested in equation (5-20), we obtain

$$h_x = \left\{\frac{\rho_L g k_L^3(\rho_L - \rho_v)[h_{fg} + (3/8)c_{pL}(T_{sat} - T_o)]}{4\mu(T_{sat} - T_o)x}\right\}^{1/4} \tag{5-220}$$

A mean heat transfer coefficient for a surface of length L is given by

$$h_L = \frac{1}{L}\int_0^L h_x \, dx$$

$$= 0.943 \, \frac{\rho_L g k_L^3(\rho_L - \rho_v)[h_{fg} + (3/8)c_{pL}(T_{sat} - T_o)]}{L\mu(T_{sat} - T_o)} \tag{5-221}$$

For fluids having $Pr > 0.5$ and a value of $c_{pL}(T_{sat} - T_o)/h_{fg} < 1.0$, Rohsenow[67] suggests replacing the term $[h_{fg} + 3/8c_{pL}(T_{sat} - T_o)]$ by $[h_{fg} + 0.68c_{pL}(T_{sat} - T_o)]$, his expression agreeing well with experimental data. The recommended expression for fluids condensing on a plane wall inclined at an angle θ from the horizontal is

$$h_L = 0.943\left[\frac{\rho_L g \sin\theta k_L^3(\rho_L - \rho_v)[h_{fg} + 0.68c_{pL}(T_{sat} - T_o)]}{L\mu(T_{sat} - T_o)}\right] \tag{5-222}$$

5.4-2.2 Film Condensation; Turbulent Flow Considerations

Flow in the condensate film will become turbulent when the film thickness becomes appreciable; i.e., when condensation rates are large or when the cool surface is long. In such cases, the laminar assumption, inherent in all expressions developed thus far, is no longer valid.

The criterion for turbulent flow is, as one would expect, the Reynolds number. The form taken by Re in this case is

$$Re = \frac{D_{eq}v_{avg}\rho}{\mu} = \frac{4A}{P} P \frac{\Gamma}{\rho A} \frac{\rho}{\mu} = \frac{4\Gamma}{\mu} \tag{5-223}$$

[67] W. M. Rohsenow, *Trans. A.S.M.E.* **78** (1956): 1645.

where Γ, as before, is the mass flow rate of condensate per unit width of film. For a vertical tube, Γ is the total mass flow rate divided by πD. The critical Reynolds number in the case of a condensate film is 1800.

An analytical procedure for cases where a condensate film is in turbulent flow has been suggested by Rohsenow and Choi.[68] The heat transfer coefficient associated with condensation on vertical surfaces is shown in Figure 5.49 as a function both of film Reynolds number and of the magnitude of the shear stress at the vapor-condensate interface. In this figure the heat

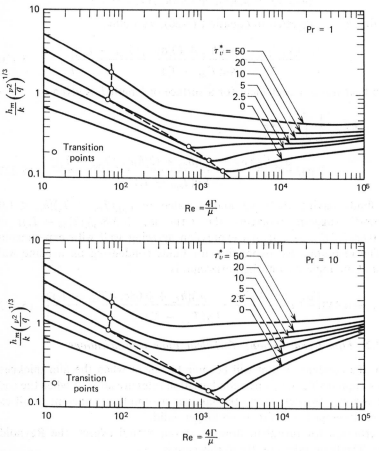

Figure 5.49 Heat transfer coefficient for turbulent flow of a condensate film on vertical surfaces. [From W. M. Rohsenow and H. Choi, *Heat, Mass, and Momentum Transfer* (Englewood Cliffs, N.J.: Prentice-Hall, Inc., 1961). By permission of the publishers.]

[68] Rohsenow and Choi, op. cit.

transfer coefficient can be determined for values of Pr of 1 and 10. The parameter τ_v^* represents the interfacial shear in nondimensional form; it is defined as

$$\tau_v^* = \frac{\tau_v}{g(\rho_L - \rho_v)(v^2/g)^{1/3}} \qquad (5\text{-}224)$$

In equation (5-224) τ_v is the average shear stress at the interface. It is calculated from

$$\tau_v = \frac{fG_{v\,\mathrm{avg}}^2}{2\rho_v} \qquad (5\text{-}225)$$

with the friction coefficient f determined from the plot given as Figure 5.50. The average mass velocity of the vapor, $G_{v\,\mathrm{avg}}$, is determined from conditions at the top and bottom of the condensing surface, according to

$$G_{v\,\mathrm{avg}} = 0.4(G_{v\,\mathrm{top}} + 1.5G_{v\,\mathrm{bottom}}) \qquad (5\text{-}226)$$

The use of these figures and equations is illustrated in Example 5.9.

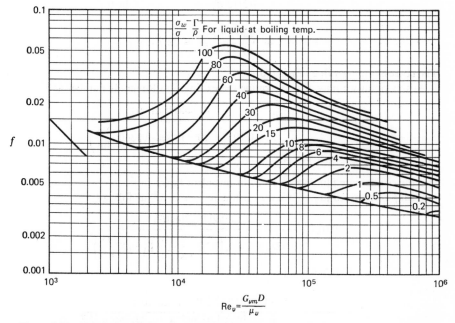

Figure 5.50 Friction coefficient for a gas flowing parallel to a liquid layer on a wall. (From O. P. Bergelin, P. K. Kegel, F. G. Carpenter, and C. Gozley, Heat Transfer and Fluid Mechanics Institute, Berkeley, Calif., 1949. By permission of the publishers.)

Example 5.9

It is desired to condense 30 lb/min of saturated steam at 250°F, using a 2-inch-ID tube whose surface is maintained at 150°F. What is the minimum length of tube required?

With a tube of minimum length, $G_{v\ \text{bottom}} = 0$, and

$$G_{v\ \text{avg}} = \frac{0.4(30\ \text{lb/min})(60\ \text{min/hr})}{\pi/4(2/12\ \text{ft})^2}$$

$$= 33{,}000\ \text{lb/hr-ft}^2$$

The vapor Reynolds number may now be calculated.

$$\text{Re}_v = \frac{(33{,}000\ \text{lb/hr-ft}^2)(2/12\ \text{ft})}{(0.89 \times 10^{-5}\ \text{ft}^2/\text{sec})(3600\ \text{sec/hr})} = 172{,}000$$

To read f from Figure 5.51, the quantity $(\sigma_\omega/\sigma)(\Gamma/\rho)$ must be known. The relative surface tension $\sigma_\omega/\sigma = 1$, and $\Gamma_{L\ \text{avg}}$ is

$$\Gamma_{L\ \text{avg}} = \frac{\left(\dfrac{0 + 30}{2}\ \dfrac{\text{lb}}{\text{min}}\right)\left(\dfrac{60\ \text{min}}{\text{hr}}\right)}{\pi(2/12\ \text{ft})} = 1720\ \frac{\text{lb}}{\text{hr-ft}}$$

thus

$$\frac{\sigma_\omega}{\sigma}\ \frac{\Gamma_{L\ \text{avg}}}{\rho} = \frac{1720\ \text{lb/hr-ft}}{59.9\ \text{lb/ft}^3} = 28.7\ \text{ft}^2/\text{hr}$$

The corresponding value of f is 0.013.

At this point we may also determine the exiting film Reynolds number as

$$\text{Re}_{L\ \text{exit}} = \frac{4\Gamma_{L\ \text{exit}}}{\mu} = \frac{4(3440\ \text{lb/hr-ft})}{0.195 \times 10^{-3}\ \text{lb/sec-ft}\left(3600\ \dfrac{\text{sec}}{\text{hr}}\right)}$$

$$= 19{,}600$$

and the condensate film is seen to be in turbulent flow.

With f determined above, we may evaluate τ_v and $\tau_v{}^*$ as follows:

$$\tau_v = 0.013\ \frac{(33{,}000\ \text{lb}_m/\text{hr-ft}^2)^2}{2(0.0723\ \text{lb}_m/\text{ft}^3)\left(32.2\ \dfrac{\text{lb}_m\text{-ft}}{\text{sec}^2\text{-lb}_f}\right)\left(3600\ \dfrac{\text{sec}}{\text{hr}}\right)^2}$$

$$= 0.235\ \text{lb}_f/\text{ft}^2$$

and

$$\tau_v{}^* = \frac{0.235\ \text{lb}_f/\text{ft}^2(32.2\ \text{lb}_m\text{ft}/\text{sec}^2\ \text{lb}_f)}{(32.2\ \text{ft}/\text{sec}^2)\left(59.9\ \dfrac{\text{lb}_m}{\text{ft}^3}\right)\left[\dfrac{(0.395 \times 10^{-5}\ \text{ft}^2/\text{sec})^2}{32.2\ \text{ft}/\text{sec}^2}\right]^{1/3}}$$

$$= 49.9$$

Using Figure 5.49, we obtain values of $(hm/k)(v^2/g)^{1/3}$ at $Pr = 1$ and 10, respectively, to be 0.58 and 1. For $Pr = 1.83$, an interpolated value of $(hm/k)(v^3/g)^{1/3}$ is 0.63. The quantities necessary to complete the solution are now as follows:

$$h = \frac{0.63k}{(v^2/g)^{1/3}} = \frac{0.63(0.383 \text{ Btu/hr-ft-}^\circ\text{F})}{(\text{ft}/12730)}$$

$$= 3150 \text{ Btu/hr-ft}^2\text{-}^\circ\text{F}$$

$$A = \frac{q}{h\,\Delta T} = \frac{(30 \text{ lb}_m/\text{min})(946 \text{ Btu/lb}_m)\left(\dfrac{60 \text{ min}}{\text{hr}}\right)}{(3150 \text{ Btu/hr-ft}^2\text{-}^\circ\text{F})(100^\circ\text{F})}$$

$$= 5.40 \text{ ft}^2$$

$$L = A/\pi D = 5.40 \text{ ft}^2/\pi(2/12 \text{ ft}) = 10.32 \text{ ft}$$

5.4-2.3   Film Condensation; Single Horizontal Cylinders

Nusselt[69] proposed the following expression for the mean heat transfer coefficient in the case of film condensation on a horizontal cylinder of diameter D:

$$h_{\text{avg}} = 0.725\left\{\frac{\rho_L g(\rho_L - \rho_v)k^3[h_{fg} + (3/8)c_{pL}(T_{\text{sat}} - T_o)]}{\mu D(T_{\text{sat}} - T_o)}\right\}^{1/4} \quad (5\text{-}227)$$

The reader may observe the similarity between equation (5-227) for a horizontal cylinder and equation (5-221) for a vertical cylinder. These expressions may be combined to yield the relation

$$\frac{h_{\text{vert}}}{h_{\text{horiz}}} = \frac{0.943}{0.725}\left(\frac{D}{L}\right)^{1/4} = 1.3\left(\frac{D}{L}\right)^{1/4} \quad (5\text{-}228)$$

With $h_{\text{vert}} = h_{\text{horiz}}$, equation (5-228) may be solved to yield the result that

$$\frac{L}{D} = 2.86 \quad (5\text{-}229)$$

This result indicates that, for a tube with length-to-diameter ratio of 2.86, equal amounts of heat transfer will occur for both horizontal and vertical orientation. As the ratio L/D increases, the greater heat transfer is possible with a horizontal tube.

5.4-2.4   Film Condensation; Banks of Horizontal Tubes

In the case of several horizontal tubes stacked in a vertical bank, vapor condensing on the uppermost tube will flow down and increase the amount

[69] Nusselt, 569.

of condensate on the tube below it. The total amount of condensate formed on a vertical bank is the cumulative effect of the heat transfer to all of the tubes in the bank. Obviously, the heat transfer and rate of condensation for a given tube are affected by the orientation of the tube in a bank and by its location relative to the others. Nusselt[70] analyzed the case of film condensation on a vertical bank of n horizontal tubes and suggested the following equation for the average heat transfer coefficient:

$$h_{avg} = 0.725 \left\{ \frac{\rho_L g (\rho_L - \rho_v) k^3 [h_{fg} + (3/8)(T_{sat} - T_o)]}{n D \mu (T_{sat} - T_o)} \right\} \qquad (5\text{-}230)$$

Chen[71] observed that values of h_{avg} from equation (5-230) were below experimental results. He included the effect of condensation on the liquid layer between tubes and obtained the expression

$$h_{avg} = 0.725 \left[1 + 0.02 \frac{c_{pL}(T_{sat} - T_o)}{h_{fg}} (n - 1) \right]$$

$$\times \left[\frac{\rho_L g (\rho_L - \rho_v) k^3 [h_{fg} + (3/8)(T_{sat} - T_o)]}{n D \mu (T_{sat} - T_o)} \right]^{1/4} \qquad (5\text{-}231)$$

which agrees reasonably well with experimental results. Equation (5-231) is valid for values of $[c_{pL}(T_{sat} - T_o)(n - 1)/h_{fg}] > 2$.

5.5 CLOSURE

Energy exchange processes occurring between a solid surface and an adjacent fluid have been considered in considerable detail in this chapter.

In our examination of convection fundamentals, we first considered the nature of fluid-solid interaction in a flow situation. The concept of a boundary layer was introduced, and boundary layer analysis was performed for both internal and external flow situations. Important parameters introduced included the coefficients of drag and skin friction, the Fanning friction factor, eddy diffusivity, and the Prandtl mixing length. Techniques employed for boundary layer analysis were exact solution of the governing differential equations, and approximate integral solutions. The nature of both laminar and turbulent flow was examined.

Fundamental aspects of convective heat transfer were examined in a similar fashion to those of fluid flow. The concept of a thermal boundary layer was introduced and analyzed by exact, integral, and analogical means.

---

[70] Ibid.
[71] M. M. Chen, *A.S.M.E. Trans., Series C* **83** (1961): 48.

Important nondimensional parameters for heat transfer by convection are the Nusselt number, Stanton number, Prandtl number, Grashof number, and Reynolds number.

Natural and forced convection situations were examined and analyzed by both theoretical and empirical means. Equations suitable for equipment and system design were presented for natural convection with various geometries and for forced convection with both internal and external flow. A detailed consideration of constant surface temperature and constant surface heat flux boundary conditions was made.

The final topic considered in the chapter was that of heat transfer accompanying a change in phase between liquids and vapors. Boiling and condensation processes were examined from the standpoints of both the fundamental mechanism and the means of describing and predicting heat transfer rates, the describing equations being largely empirical.

CHAPTER 6

RADIATION HEAT TRANSFER

Thermal radiation is that part of the electromagnetic emission of a substance that is characterized by light or by heat transfer. Emitted radiant energy can have extremely short wavelengths, as in the case of cosmic rays ($\lambda \sim 10^{-10}$ cm), or the wavelengths can be on the order of miles, as for some radio waves. The thermal band lies in the intermediate range and is generally taken to be between 0.1 and 100 microns (1 micron, μ, $= 10^{-6}$ meters). The electromagnetic spectrum is depicted in Figure 6.1.

In our considerations of conduction and convection in the preceding sections, we noted that the heat transfer rates vary as the temperature difference to approximately the first power. Radiant heat exchange between two bodies depends on the differences between their temperatures raised to approximately the *fourth* power. Another very significant difference between radiant heat exchange and the other types is that no medium is required for the propagation of radiant energy; radiant exchange between two surfaces is, in fact, a maximum when the space between the surfaces is evacuated.

The reader's attention is again drawn to Figure 6.1. The thermal radiation band is seen to include the visible light range extending between about 0.38 and 0.76 micron. The visible range extends from the violet to the red color band. It is well known that a heated object has a characteristic color, depending upon its temperature; the term "red hot" refers to the reddish-colored emission from a body at high temperature.

Our task in this chapter is to acquaint the reader with some of the basic concepts in radiant heat transfer and to introduce some techniques in the

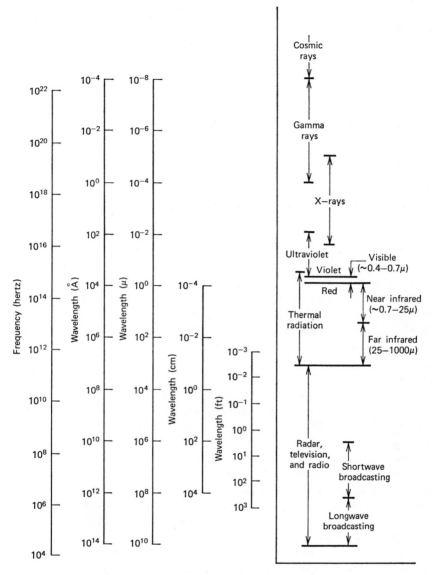

Figure 6.1 The electromagnetic radiation spectrum.

quantitative description of radiant heat transfer phenomena. A first step is to consider a body which radiates "ideally" and to use the behavior of such a surface as a standard against which other nonideal surfaces may be compared. In the next section the behavior of ideally radiating "black" surfaces is considered.

6.1   RADIATION FROM IDEAL (BLACK) SURFACES

A "black" surface is an idealization based upon the response of a surface to incident thermal radiation. It is necessary, in this regard, to consider what happens as radiant energy impinges on a body.

When radiation is incident on a homogeneous body, some of the energy penetrates into it and the remainder is reflected. Of that portion which penetrates into the surface, some may be absorbed and some may be transmitted through the body with little change in its nature. The word "opaque" describes a body which transmits none of the energy which is incident upon it. Absorbed energy is converted into internal energy of the body in question.

In complete fashion we may say that energy which is incident on a body is either reflected, absorbed, or transmitted. If we designate ρ, α, and τ as those portions of incident radiant energy which are reflected, absorbed, and transmitted, then we may write

$$\rho + \alpha + \tau = 1 \qquad (6\text{-}1)$$

where ρ is designated the *reflectivity*, α the *absorptivity*, and τ the *transmissivity*. According to our discussions thus far, it is apparent that for a perfectly absorbing (black) body, $\alpha = 1$, and for an *opaque* body, $\tau = 0$.

Reflections are of two types, regular and diffuse. Regular, or *specular*, reflection is that for which the angles of incidence and reflection are equal. This mirror-like behavior is reasonably familiar but is encountered less often in actual situations than the diffuse type. *Diffuse* reflection is that in which incoming radiation is reflected in all directions; this is sometimes likened to the situation in which incident energy is absorbed near the surface of a body and then re-emitted.

A body which absorbs all of the energy incident upon it is designated "black," as has been mentioned already. The term "black body" denotes the fact that a body which reflects no energy appears black to the eye. The closest approach, in actual practice, to a true black body is a cavity with only a small opening in it. Energy which enters the cavity through a small hole has very little chance of being reflected back out, thus is essentially all absorbed. The cavity itself may be made of a shiny material but will appear black when one looks in through the opening.

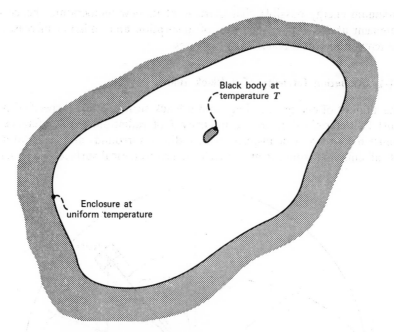

Figure 6.2 A black body inside an isothermal enclosure.

A black body has certain characteristics which are important from both a conceptual and a quantitative point of view. These will now be discussed briefly.

A black body is a perfect emitter. This fact is easily verified by considering the situation depicted in Figure 6.2. The black body and the enclosure which surrounds it will reach the same uniform temperature after a time due to heat transfer. When this state of thermal equilibrium is reached, the black body will absorb and emit at the same rate; if this were not true, its temperature would change—a direct violation of the second law of thermodynamics. Since the black body will, by definition, absorb the maximum possible energy regardless of direction or wavelength, it follows that its emission is likewise a maximum. A similar argument will indicate that maximum possible emission at every wavelength and in every direction is also characteristic of a black body.

The total radiation emitted by a black body is a function of temperature only. This is seen to be true by considering what happens as the enclosure temperature is changed to a different uniform value. The black body will adjust until its temperature is the same as that of the enclosure. Once this equilibration has occurred, the black body is again absorbing and emitting the maximum energy possible characteristic of its new temperature, hence the statement above is true. The rate of absorption and emission increases as the temperature of a black body increases.

6.1-1 Radiation Intensity of a Black Body

The amount of energy traveling from a black surface along a specified path must be determined from the intensity I of radiation. With reference to Figure 6.3, where a hemisphere with radius r surrounds an elemental area dA, all emission from dA must reach the hemispherical surface, which has a

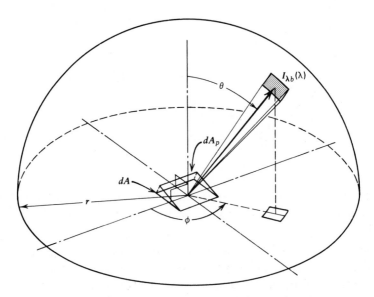

Figure 6.3 Emission from a plane black surface.

total area of $2\pi r^2$. Each small area on the hemispherical surface subtends a *solid angle* at dA. The unit of measure of a solid angle is the steradian (sr), which may be thought of as a unit of space. A solid angle in the case of a hemisphere is the area on the hemispherical surface divided by r^2, hence there are 2π sr in the hemisphere surrounding dA.

The *intensity of radiation* refers to the amount of energy emitted from a surface, such as dA, per unit time per unit area projected normal to a given direction per unit solid angle subtended at the emitter surface. One must also consider the *spectral intensity*, which includes the radiation in a small interval $d\lambda$ around a single wavelength. The *total intensity* includes all radiation over the entire wavelength spectrum. Total black body intensity I_b and spectral intensity $I_{b\lambda}$ are related as

$$I_b = \int_0^\infty I_{b\lambda} \, d\lambda \tag{6-2}$$

For a black body, the intensity of radiation is independent of the direction of emission.

6.1-2 Emissive Power of a Black Body

The *emissive power* is defined as the energy emitted by a surface per unit time per unit *unprojected* area. The emissive power of a black surface also exists in spectral form $E_{b\lambda}$ and in total form E_b. Referring again to Figure 6.3, the energy emitted from dA in a given direction at a given wavelength may be written as

$$dq_{b\lambda}(\lambda, \theta, \phi) = I_{b\lambda}(\lambda) \, dA \cos \theta \, d\omega \, d\lambda$$

$$= E_{b\lambda}(\lambda, \theta, \phi) \, dA \, d\omega \, d\lambda \tag{6-3}$$

A conclusion that may be reached from equation (6-3) is that spectral intensity and emissive power of a black body are related according to

$$E_{b\lambda}(\lambda, \theta, \phi) = I_{b\lambda}(\lambda) \cos \theta = E_{b\lambda}(\lambda, \theta) \tag{6-4}$$

or, in words, the black body emissive power is a function of λ and θ but is not a function of ϕ. The quantity $E_{b\lambda}(\lambda, \theta)$ is termed the *directional spectral emissive power* for a black surface. For some nonblack surfaces, the spectral emissive power E_λ will also depend on ϕ.

Equation (6-4) is known as *Lambert's cosine law*, for obvious reasons. Surfaces behaving in this manner are termed *diffuse*, or *Lambert* surfaces. Real surfaces, in general, depart from cosine law behavior.

6.1-3 Hemispherical Spectral Emissive Power of Black Surfaces

The *hemispherical spectral emissive power* of a black surface is the energy emitted from the surface per unit time per unit area at a given wavelength. It is determined by integrating over all solid angles subtended by a hemisphere centered at dA.

Consider the small area dA at the center of a hemisphere of radius r, as shown in Figure 6.4. The solid angle $d\omega$, subtended by the portion of surface shown, is

$$d\omega = \frac{(r \sin \theta \, d\phi)(r \, d\theta)}{r^2} = \sin \theta \, d\theta \, d\phi \tag{6-5}$$

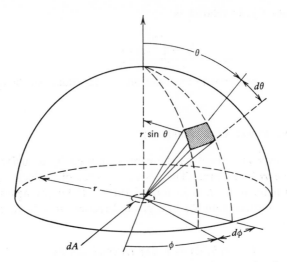

Figure 6.4   Area dA at the center of a hemisphere of radius r.

The spectral emission from dA may thus be written as

$$E_{b\lambda}(\lambda, \theta) \, d\omega = E_{b\lambda}(\lambda, \theta) \sin \theta \, d\theta \, d\phi \qquad (6\text{-}6)$$

which may be combined with the cosine law, equation (6-4), to yield

$$E_{b\lambda}(\lambda, \theta) \, d\omega = I_{b\lambda}(\lambda) \cos \theta \sin \theta \, d\theta \, d\phi \qquad (6\text{-}7)$$

This expression may be integrated over the entire hemispherical space, giving

$$E_{b\lambda}(\lambda) = I_{b\lambda}(\lambda) \int_0^{2\pi} \int_0^{\pi/2} \cos \theta \sin \theta \, d\theta \, d\phi$$

$$= \pi I_{b\lambda}(\lambda) \qquad (6\text{-}8)$$

The result given by equation (6-8) is an important one in the quantitative analysis of radiant energy exchange.

6.1-4   Spectral Distribution of Black Body Emissive Power; Planck's Law

Max Planck, in 1900, as part of his quantum theory, expressed the temperature and wavelength distribution of black body emissive power in a vacuum as

$$E_{b\lambda}(\lambda) = \pi I_{b\lambda}(\lambda) = \frac{2\pi C_1 \lambda^{-5}}{e^{C_2/\lambda T} - 1}$$

where the constants are $C_1 = hC_2$ and $C_2 = hc/\kappa$, h and κ being Planck's and Boltzmann's constants, respectively.

Equation (6-9) is plotted in Figure 6.5 as $E_{b\lambda}$ versus λ, with lines of constant absolute temperature shown. Certain general statements can be made regarding $E_{b\lambda}$ from the behavior depicted in this figure. One characteristic observed is that the total energy emitted increases with temperature. At any temperature the total energy emitted is the area under the appropriate curve in the figure. This increase in emission is true in total and is likewise true at every wavelength.

A second observation that can be made is that peak emission occurs at shorter wavelengths as the temperature increases. This characteristic is expressed quantitatively by *Wien's displacement law* as

$$\lambda_{max}T = 5215.6 \ \mu°R \tag{6-10}$$

where the units of λ and T are shown to be microns and degrees Rankine, respectively.

We further note that solar emission, at a temperature of about 10,000°R, includes a large amount of energy in the visible region. It has been suggested that the "visible" portion of the emission spectrum is due to the evolution of the eye to be sensitive to solar emission.

The separate curves for each temperature shown in Figure 6.5 can be consolidated, as shown in Figure 6.6, to a single curve by a simple manipulation of equation (6-9).

We divide this expression by T^5 and get

$$\frac{E_{b\lambda}(\lambda, T)}{T^5} = \frac{\pi I_{b\lambda}(\lambda, T)}{T^5} = \frac{2\pi C_1}{(\lambda T)^5 (e^{C_2/\lambda T} - 1)} \tag{6-11}$$

which expresses $E_{b\lambda}/T^5$ as a function of the combined variables λT. These are the quantities plotted in Figure 6.6.

6.1-5 Total Intensity and Emissive Power of a Black Body

The determination of total black body intensity from spectral intensity was already expressed by equation (6-2). The spectral and total emissive powers are similarly related, according to

$$E_b(T) = \int_0^\infty E_{b\lambda}(\lambda, T) \ d\lambda \tag{6-12}$$

The distribution of emissive power with wavelength given by Planck's law, equation (6-9), may be substituted into the above expression. Integration will

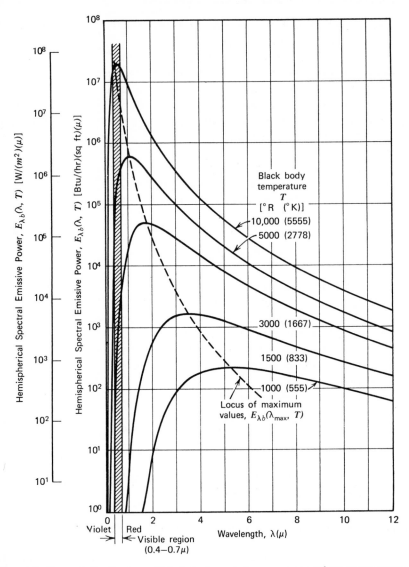

Figure 6.5 Spectral distribution of black body emissive power at various temperatures.

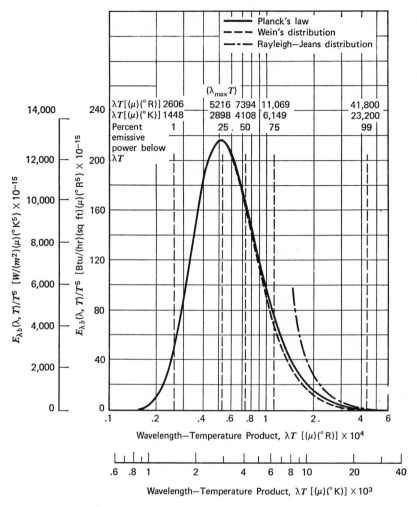

Figure 6.6   Monochromatic black body emissive power as a function of λT.

then yield the total hemispherical emissive power of a black surface to be

$$E_b(T) = \int_0^\infty E_{b\lambda}(\lambda, T)\, d\lambda = \int_0^\infty \pi I_{b\lambda}(\lambda, T) = \sigma T^4 \qquad (6\text{-}13)$$

where σ is the Stefan Boltzmann constant and T is absolute temperature. The integration which is indicated will yield $\sigma = 0.1714 \times 10^{-8}$ Btu/hr-ft²-°R⁴, while the accepted experimental value is slightly higher at 0.173×10^{-8} Btu/hr-ft²-°R⁴. We shall use the experimental value in this text.

6.1-6   Black Body Emissive Power within a Wavelength Band

It is often desirable to know how much emission occurs in a specific portion of the total wavelength spectrum. This is expressed most conveniently as a fraction of the total emissive power. The fraction between wavelengths λ_1 and λ_2 is designated $F_{\lambda_1-\lambda_2}$ and is given by

$$F_{\lambda_1-\lambda_2} = \frac{\displaystyle\int_{\lambda_1}^{\lambda_2} E_{b\lambda}(\lambda) \, d\lambda}{\displaystyle\int_{0}^{\infty} E_{b\lambda}(\lambda) \, d\lambda}$$

$$= \frac{1}{\sigma T^4} \int_{\lambda_1}^{\lambda_2} E_{b\lambda}(\lambda) \, d\lambda \qquad (6\text{-}14)$$

Equation (6-14) is conveniently broken down into two integrals, as follows:

$$F_{\lambda_1-\lambda_2} = \frac{1}{\sigma T^4}\left[\int_{0}^{\lambda_2} E_{b\lambda}(\lambda) \, d\lambda - \int_{0}^{\lambda_1} E_{b\lambda}(\lambda) \, d\lambda \right]$$

$$= F_{0-\lambda_2} - F_{0-\lambda_1} \qquad (6\text{-}15)$$

These values may thus be calculated at a given T and, between any two wavelengths, the fraction of total emission may be determined by subtraction.

Because values of $F_{0-\lambda}$ as expressed above will vary with temperature on additional manipulation, eliminating T as a separate variable is helpful. We may define F in terms of the product λT and modify equation (6-15) to read

$$F_{\lambda_1 T-\lambda_2 T} = \frac{1}{\sigma}\left[\int_{0}^{\lambda_2 T} \frac{E_{b\lambda}(\lambda)}{T^5} \, d(\lambda T) - \int_{0}^{\lambda_1 T} \frac{E_{b\lambda}(\lambda)}{T^5} \, d(\lambda T) \right]$$

$$= F_{0-\lambda_2 T} - F_{0-\lambda_1 T} \qquad (6\text{-}16)$$

It has already been shown that $E_{b\lambda}/T^5$ is a function of λT; thus $F_{0-\lambda T}$ may be evaluated and tabulated, as in Appendix E, or plotted, as in Figure 6.7. The use of this concept is illustrated in Example 6.1.

Example 6.1

A radiation measuring instrument detects all emission occurring between 0.65 and 4.5 μ but is unaffected by frequencies outside this range. What fraction of the total emission from a black surface will be detected for emitting surface temperatures of 1000°R, 5000°R, and 10,000°R? The quantities pertinent to the solution of this problem are listed in the table opposite.

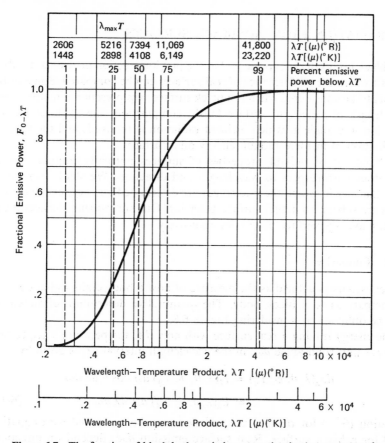

Figure 6.7 The fraction of black body emission occurring in the range 0 to λT.

Temperature (°R)	$\lambda_1 T$	$\lambda_2 T$	$F_{0-\lambda_1 T}$	$F_{0-\lambda_2 T}$	$F_{\lambda_1 T - \lambda_2 T}$
1,000	6,500	45,000	0.4061	0.9916	0.5585
5,000	32,500	225,000	0.9795	0.9999 −	0.0204
10,000	65,000	450,000	0.9971	0.9999 +	—

The results indicate that, at 1,000°R, approximately 56% of the total emission will be detected; at 5,000°R this is reduced to around 2%; and at 10,000°R, the temperature of the sun, the amount detected will be a negligible fraction of the total. Over 99.7% of the sun's emission occurs at wavelengths below 0.65μ.

6.2 RADIATION FROM NONBLACK SURFACES

The black body concepts presented in the previous section provide a standard against which the performance of real surfaces can be compared. Some of the factors to be considered when nonblack surfaces are present are temperature, wavelength of radiation, surface finish, composition, the angle of radiant emission, the angle at which incident radiation is received, and the spectral distribution of incident radiant energy on a surface.

Tabulations of radiant properties of surfaces normally contain average properties with respect to wavelength and direction. We shall briefly consider the way in which these averages are determined and how average properties are related to actual surface behavior. Our subsequent considerations will use average surface properties exclusively.

6.2-1 Emissivity

The emissivity is defined as the measure of how a body emits radiant energy in comparison to a black body. The quantitative description of emissivity may be obtained by considering Figure 6.8.

The energy leaving a real surface with area dA, temperature T_A, per unit time per unit solid angle $d\omega$, in the wavelength interval $d\lambda$, is given by

$$dq_\lambda(\lambda, \theta, \phi, T_A) = I_\lambda(\lambda, \theta, \phi, T_A)\, dA \cos\theta\, d\lambda\, d\omega$$

$$= E_\lambda(\lambda, \theta, \phi, T_A)\, dA\, d\lambda\, d\omega \qquad (6\text{-}17)$$

Equation (6-17) for real surface emission may be compared with equation (6-4) for a black surface. Real surface emission *does* depend on direction, this was shown not to be true for black body emission.

The energy leaving a black surface at T_A per unit area per unit time within $d\lambda$ and $d\omega$ is

$$dq_{b\lambda}(\lambda, \phi, T_A) = I_{b\lambda}(\lambda, T_A)\, dA \cos\theta\, d\lambda\, d\omega$$

$$= E_{b\lambda}(\lambda, \theta, T_A)\, dA\, d\lambda\, d\omega \qquad (6\text{-}18)$$

thus by the definition of emissivity, we have, for the directional spectral emissivity ε_λ,

$$\varepsilon_\lambda(\lambda, \theta, \phi, T_A) = \frac{dq_\lambda(\lambda, \theta, \phi, T_A)}{dq_{b\lambda}(\lambda, \theta, T_A)} = \frac{I_\lambda(\lambda, \theta, \phi, T_A)}{I_{b\lambda}(\lambda, T_A)}$$

$$= \frac{E_\lambda(\lambda, \theta, \phi, T_A)}{E_{b\lambda}(\lambda, \theta, T_A)} \qquad (6\text{-}19)$$

Averages may now be obtained with respect to wavelength and direction.

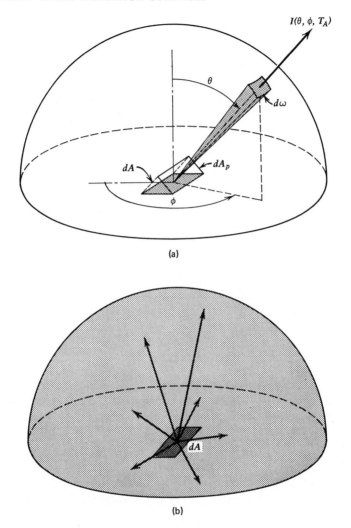

Figure 6.8 Geometric representation of directional and hemispherical radiation properties. (a) Directional emissivity $\varepsilon(\theta, \phi\, T_A)$. (b) Hemispherical emissivity $\varepsilon(T_A)$.

6.2-1.1 Directional Total Emissivity

The *directional total emissivity* is an average of $\varepsilon_\lambda(\lambda, \theta, \phi, T_A)$ over all wavelengths and is determined from

$$\varepsilon(\theta, \phi, T_A) = \frac{E(\theta, \phi, T_A)}{E_b(\theta, T_A)} \tag{6-20}$$

The directional total emissive power of a real surface and of a black surface, both at T_A, are given by

$$E(\theta, \phi, T_A) = \int_0^\infty E_\lambda(\lambda, \theta, \phi, T_A) \, d\lambda \qquad (6\text{-}21)$$

and

$$E_b(\theta, T_A) = \int_0^\infty E_{b\lambda}(\lambda, \theta, T_A) \, d\lambda = \frac{\sigma T_A^4}{\pi} \cos \theta \qquad (6\text{-}22)$$

Substitution into equation (6-20) now yields

$$\varepsilon(\theta, \phi, T_A) = \frac{\displaystyle\int_0^\infty E_\lambda(\lambda, \theta, \phi, T_A) \, d\lambda}{\dfrac{\sigma T_A^4}{\pi} \cos \theta} \qquad (6\text{-}23)$$

In terms of the directional spectral emissivity, the directional total emissivity is

$$\varepsilon(\theta, \phi, T_A) = \frac{\displaystyle\int_0^\infty \varepsilon_\lambda(\lambda, \theta, \phi, T_A) E_{b\lambda}(\lambda, \theta, T_A) \, d\lambda}{\dfrac{\sigma T_A^4}{\pi} \cos \theta} \qquad (6\text{-}24)$$

6.2-1.2 Hemispherical Spectral Emissivity

The *hemispherical spectral emissive power* $E_\lambda(\lambda, T_A)$ is obtained by integrating the spectral emissive power over all solid angles in the hemisphere enclosing the surface in question. This operation is specified as

$$E_\lambda(\lambda, T_A) = \int_H I_\lambda(\lambda, \theta, \phi, T_A) \cos \theta \, d\omega \qquad (6\text{-}25)$$

where $\int_H$ implies integration over the hemispherical solid angle. Equation (6-19) may be used to write equation (6-25) in simpler form

$$E_\lambda(\lambda, T_A) = I_{b\lambda}(\lambda, T_A) \int_H \varepsilon_\lambda(\lambda, \theta, \phi, T_A) \cos \theta \, d\omega \qquad (6\text{-}26)$$

The hemispherical spectral emissive power of a black body was previously expressed as

$$E_{b\lambda}(\lambda, T_A) = \pi I_{b\lambda}(\lambda, T_A) \qquad (6\text{-}8)$$

By the definition of emissivity, the *hemispherical spectral emissivity* can now be written as

$$\varepsilon_\lambda(\lambda, T_A) = \frac{E_\lambda(\lambda, T_A)}{E_{b\lambda}(\lambda, T_A)} = \frac{1}{\pi} \int_H \varepsilon_\lambda(\lambda, \theta, \phi, T_A) \cos \theta \, d\omega \qquad (6\text{-}27)$$

6.2-1.3 Hemispherical Total Emissivity

If integration of directional spectral emissive power is carried out over all wavelengths and over all solid angles comprising the hemisphere, the result is the hemispherical total emissive power. This quantity, divided by $\sigma T_A{}^4$, the black body hemispherical total emissive power, gives the *hemispherical total emissivity* $\varepsilon(T_A)$.

The hemispherical total emissivity may be expressed in terms of directional spectral emissivity as

$$\varepsilon(T_A) = \frac{E(T_A)}{E_b(T_A)} = \frac{\displaystyle\int_H \int_0^\infty E_\lambda(\lambda, \theta, \phi, T_A)\, d\lambda\, d\omega}{\sigma T_A{}^4}$$

$$= \frac{\displaystyle\int_H \int_0^\infty \varepsilon_\lambda(\lambda, \theta, \phi, T_A) I_{b\lambda}(\lambda, T_A)\, d\lambda \cos\theta\, d\omega}{\sigma T_A{}^4}$$

$$= \frac{\displaystyle\int_H \int_0^\infty \varepsilon_\lambda(\lambda, \theta, \phi, T_A) E_{b\lambda}(\lambda, \theta, T_A)\, d\lambda\, d\omega}{\sigma T_A{}^4} \tag{6-28}$$

In terms of directional total emissivity, $\varepsilon(T_A)$ is written

$$\varepsilon(T_A) = \frac{1}{\pi} \int_H \varepsilon(\theta, \phi, T_A) \cos\theta\, d\omega \tag{6-29}$$

and in terms of hemispherical spectral emissivity,

$$\varepsilon(T_A) = \frac{\displaystyle\int_0^\infty \varepsilon_\lambda(\lambda, T_A) I_{b\lambda}(\lambda, T_A)\, d\lambda}{\sigma T_A{}^4 / \pi}$$

$$= \frac{\displaystyle\int_0^\infty \varepsilon_\lambda(\lambda, T_A) E_{b\lambda}(\lambda, T_A)\, d\lambda}{\sigma T_A{}^4} \tag{6-30}$$

The variation in hemispherical spectral emissivity with wavelength may be represented, in general, as depicted in Figure 6.9. This information may be used to obtain a spectral distribution of emissive power. In Figure 6.10 the solid line depicts the black body spectral emissive power of a surface at T_A. The area under this curve is $\sigma T_A{}^4$.

At any wavelength the corresponding values of $\varepsilon_\lambda(\lambda, T_A)$ and $E_{b\lambda}(\lambda, T_A)$ may be multiplied to yield the hemispherical spectral emissive power for the real surface. Such a variation in $E_\lambda(\lambda, T_A)$ is given by the dashed line in

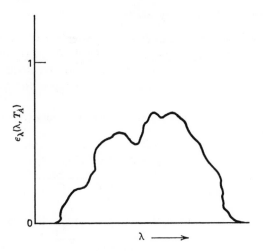

Figure 6.9   Wavelength variation of $\varepsilon_\lambda(\lambda, T_A)$ for a real surface.

Figure 6.10. The area under this curve is the integral in the numerator of equation (6-30). At each wavelength the hemispherical spectral emissivity is the ratio of the ordinates of the dashed and solid lines in Figure 6.10.

6.2-2   Absorptivity

The absorptivity has been defined previously as that fraction of the energy incident upon a surface which is absorbed. Incident radiation has characteristics which depend on the source, and it is these directional and spectral characteristics which make the description of absorptivity more complex than for emissivity.

6.2-2.1   Directional Spectral Absorptivity

The radiant energy which reaches elemental area dA from the θ, ϕ direction is shown in Figure 6.11.

The fraction of energy passing through dA_s on the hemispherical surface which is absorbed at dA is termed the *directional spectral absorptivity*, $\alpha_\lambda(\lambda, \theta, \phi, T_A)$, where the wavelength and directional properties are characteristic of the incoming radiant energy, while temperature dependence is that of the absorbing surface at T_A.

The energy reaching dA per unit time from the θ, ϕ direction in the wavelength interval $d\lambda$ is

$$dq_{\lambda,i}(\lambda, \theta, \phi) = I_{\lambda,i}(\lambda, \theta, \phi)\, dA_s \frac{dA \cos \theta}{r^2}\, d\lambda \qquad (6\text{-}31)$$

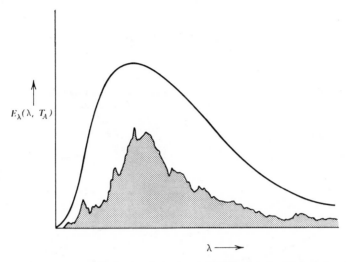

Figure 6.10   Hemispherical spectral emissive power distribution for black and real surfaces at T_A.

where the subscript i refers to incident, or incoming, radiation. The quantity $dA \cos \theta / r^2$ is the solid angle subtended by dA at dA_s. Equation (6-31) may be written in equivalent form as

$$dq_{\lambda,i}(\lambda, \theta, \phi) = I_{\lambda,i}(\lambda, \theta, \phi) \, d\omega \cos \theta \, dA \, d\lambda \qquad (6\text{-}32)$$

where $d\omega = dA_s/r^2$ is the solid angle subtended by dA_s at dA. The amount of incident energy absorbed at dA is given by $dq_{\lambda,i}(\lambda, \theta, \phi, T_A)|_{\text{abs}}$, and the directional spectral absorptivity is the ratio

$$\alpha_\lambda(\lambda, \theta, \phi, T_A) \equiv \frac{dq_{\lambda,i}(\lambda, \theta, \phi, T_A)|_{\text{abs}}}{dq_{\lambda,i}(\lambda, \theta, \phi)}$$

$$= \frac{dq_{\lambda,i}(\lambda, \theta, \phi, T_A)|_{\text{abs}}}{I_{\lambda,i}(\lambda, \theta, \phi) \, dA \cos \theta \, d\omega \, d\lambda} \qquad (6\text{-}33)$$

6.2-2.2   Directional Total Absorptivity

The averaging and/or summing techniques used in Section 6.2 with emissivity have equivalent counterparts when working with absorptivities. These quantities will be expressed without the detailed explanations of the section on emissivity.

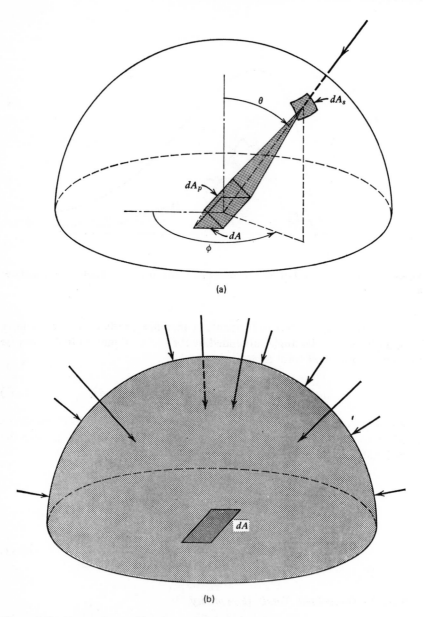

(a)

(b)

Figure 6.11 Geometric considerations of absorbed radiant energy. (a) Directional absorptivity $\alpha(\theta, \phi, T_A)$. (b) Hemispherical absorptivity $\alpha(T_A)$.

The *directional total absorptivity* is

$$\alpha(\theta, \phi, T_A) = \frac{dq_i(\theta, \phi, T_A)\big|_{\text{abs}}}{dq_i(\theta, \phi)}$$

$$= \frac{\displaystyle\int_0^\infty \alpha_{\lambda,i}(\lambda, \theta, \phi, T_A)I_{\lambda,i}(\lambda, \theta, \phi)\, d\lambda}{\displaystyle\int_0^\infty I_{\lambda,i}(\lambda, \theta, \phi)\, d\lambda} \tag{6-34}$$

6.2-2.3 Hemispherical Spectral Absorptivity

The fraction absorbed of the spectral energy which is incident on a surface from all directions over a surrounding hemisphere is termed the *hemispherical spectral absorptivity*. This quantity is expressed as

$$\alpha_\lambda(\lambda, T_A) = \frac{dq_{\lambda,i}(\lambda)\big|_{\text{abs}}}{dq_{\lambda,i}(\lambda)}$$

$$= \frac{\displaystyle\int_H \alpha_\lambda(\lambda, \theta, \phi, T_A)I_{\lambda,i}(\lambda, \theta, \phi)\cos\theta\, d\omega}{\displaystyle\int_H I_{\lambda,i}(\lambda, \theta, \phi)\cos\theta\, d\omega} \tag{6-35}$$

6.2-2.3 Hemispherical Total Absorptivity

When the fraction of incident energy absorbed by a surface over all wavelengths in all directions is determined, the result is the hemispherical total absorptivity $\alpha(T_A)$. This quantity is represented by the relation

$$\alpha(T_A) = \frac{dq_i(T_A)\big|_{\text{abs}}}{dq_i(T_A)}$$

$$= \frac{\displaystyle\int_H \int_0^\infty \alpha_\lambda(\lambda, \theta, \phi, T_A)I_{\lambda,i}(\lambda, \theta, \phi)\, d\lambda \cos\theta\, d\omega}{\displaystyle\int_H \int_0^\infty I_{\lambda,i}(\lambda, \theta, \phi)\, d\lambda \cos\theta\, d\omega} \tag{6-36}$$

In terms of directional total absorptivity, $\alpha(T_A)$ is written

$$\alpha(T_A) = \frac{\displaystyle\int_H \int_0^\infty I_{\lambda,i}(\lambda, \theta, \phi)\, d\lambda\, \alpha(\theta, \phi, T_A)\cos\theta\, d\omega}{\displaystyle\int_H \int_0^\infty I_{\lambda,i}(\lambda, \theta, \phi)\, d\lambda \cos\theta\, d\omega} \tag{6-37}$$

and, in terms of hemispherical spectral absorptivity, the relationship is

$$\alpha(T_A) = \frac{\int_0^\infty \int_H \alpha_\lambda(\lambda, T_A) I_{\lambda,i}(\lambda, \theta, \phi) \cos\theta \, d\omega \, d\lambda}{\int_0^\infty \int_H I_{\lambda,i}(\lambda, \theta, \phi) \cos\theta \, d\omega \, d\lambda} \qquad (6\text{-}38)$$

6.2-2.4   Kirchhoff's Law

Kirchhoff's radiation law relates the ability of a body to emit radiant energy to its absorption capability. This relationship is obtained by considering a nonblack body of surface area dA situated in a black enclosure. When thermal equilibrium is reached, the body and enclosure are at the same temperature T_A. The energy leaving dA per unit time per unit solid angle $d\omega$ in wavelength interval $d\lambda$ is given by equations (6-17) and (6-19) as

$$dq_\lambda(\lambda, \theta, \phi, T_A) = I_\lambda(\lambda, \theta, \phi, T_A) \, dA \cos\theta \, d\lambda \, d\omega$$

$$= \varepsilon_\lambda(\lambda, \theta, \phi, T_A) I_{\lambda b}(\lambda, T_A) \, dA \cos\theta \, d\omega \, d\lambda \qquad (6\text{-}39)$$

The absorbed energy at dA is given by equation (6-33)

$$dq_{\lambda,i}(\lambda, \theta, \phi, T_A)|_{\text{abs}} = \alpha_\lambda(\lambda, \theta, \phi, T_A) I_{\lambda,i}(\lambda, \theta, \phi) \, dA \cos\theta \, d\omega \, d\lambda \qquad (6\text{-}33)$$

Since dA and its enclosure are in thermal equilibrium, the emission and absorption from dA must be equal; thus, from the preceding two expressions,

$$\varepsilon_\lambda(\lambda, \theta, \phi, T_A) I_{\lambda b}(\lambda, T_A) = \alpha_\lambda(\lambda, \theta, \phi, T_A) I_{\lambda,i}(\lambda, \theta, \phi) \qquad (6\text{-}40)$$

and, since the energy incident on dA is emitted from black surroundings, $I_{\lambda,i}(\lambda, \theta, \phi) = I_{\lambda,b}(\lambda, T_A)$; thus the emissivity and absorptivity are equal, or

$$\varepsilon_\lambda(\lambda, \theta, \phi, T_A) = \alpha_\lambda(\lambda, \theta, \phi, T_A) \qquad (6\text{-}41)$$

This equation is true without restriction; thus the directional spectral emissivity and directional spectral absorptivity are equal for a body in thermal equilibrium with its surroundings.

The directional total emissivity and absorptivity are given by equations (6-24) and (6-34), respectively. Repeating these expressions for clarity, we have

$$\varepsilon(\theta, \phi, T_A) = \frac{\int_0^\infty \varepsilon_\lambda(\lambda, \theta, \phi, T_A) I_{b\lambda}(\lambda, T_A) \, d\lambda}{\dfrac{\sigma T_A^4}{\pi}} \qquad (6\text{-}24)$$

and

$$\alpha(\theta, \phi, T_A) = \frac{\displaystyle\int_0^\infty \alpha_{\lambda,i}(\lambda, \theta, \phi, T_A) I_{\lambda,i}(\lambda, \theta, \phi)\, d\lambda}{\displaystyle\int_0^\infty I_{\lambda,i}(\lambda, \theta, \phi)\, dA} \qquad (6\text{-}34)$$

If the incident ratiation $I_{\lambda,i}(\lambda, \theta, \phi)$ has a distribution which is proportional to that from a black body, $I_{\lambda,i}(\lambda, \theta, \phi) = K(\theta, \phi) I_{\lambda,b}(\lambda, T_A)$, then the proportionality constant $K(\theta, \phi)$ will cancel in equation (6-34), and these two expressions are equal, yielding the sought-for result

$$\varepsilon(\theta, \phi, T_A) = \alpha(\theta, \phi, T_A) \qquad (6\text{-}42)$$

We stress again that this expression is true only if $I_{\lambda,i}(\lambda, \theta, \phi, T_A)$ has the same wavelength distributions as a black body.

The hemispherical spectral emissivity and absorptivity are expressed by equations (6-27) and (6-35). If the incident spectral intensity $I_{\lambda,i}(\lambda, \theta, \phi, T_A)$ is *independent of direction*, i.e., perfectly diffuse, then it is written as $I(\lambda, T_A)$, and the emissivity and absorptivity are related as

$$\alpha(\lambda, T_A) = \frac{\displaystyle\int_H \alpha_\lambda(\lambda, \theta, \phi, T_A) I_{\lambda,i}(\lambda, \theta, \phi) \cos\theta\, d\omega}{\displaystyle\int_H I_{\lambda,i}(\lambda, \theta, \phi) \cos\theta\, d\omega}$$

$$= \int_H \alpha_\lambda(\lambda, \theta, \phi, T_A)\, d\omega = \varepsilon(\lambda, T_A) \qquad (6\text{-}43)$$

The hemispherical total emissivity and absorptivity are given by equations (6-28) and (6-36), respectively. These quantities are equal, i.e.,

$$\varepsilon(T_A) = \alpha(T_A) \qquad (6\text{-}44)$$

if the incident radiation is independent of direction with a spectral distribution proportional to that of a black body. There are some other special kinds of directional and spectral distributions of incident radiant energy for which equation (6-44) applies.

When radiant energy leaving a surface is independent of direction, it is termed *diffuse*. When the emissivity and absorptivity of a surface do not vary with wavelength, the surface is termed *gray*. The *gray body* approximation to real surface behavior is a tremendous simplification in radiant heat transfer analysis. We shall employ this approximation extensively in the sections to follow.

6.2-3   Reflectivity

When incident radiant energy is reflected from a surface, the amount reflected depends on the angle at which the incident energy strikes the surface and the directional characteristics of the reflected energy. For this reason the reflectivity is a more complicated property to specify than either the emissivity or the absorptivity.

The ratio of the reflected intensity in the θ_r, ϕ_r direction, to that which is incident on a surface in the θ, ϕ direction at wavelength λ, is termed the *bidirectional spectral reflectivity*. This quantity is defined as

$$\rho_\lambda(\lambda, \theta_r, \phi_r, \theta, \phi) = \frac{I_{\lambda,r}(\lambda, \theta_r, \phi_r, \theta, \phi)}{I_{\lambda,i}(\lambda, \theta, \phi) \cos \theta \, d\omega} \tag{6-45}$$

The bidirectional spectral reflectivity can be summed over the wavelength spectrum to give total reflectivity; it can be summed over all reflected solid angles, over all incident solid angles, and over both incident and reflected solid angles to give various combinations of directional and hemispherical reflectivities. These operations are summarized in complete fashion by Siegel and Howell.[1]

Our primary concern, with regard to reflected radiation, is that all energy incident on a surface is accounted for. Thus, for an opaque body, the energy per unit time, incident on dA per unit solid angle in the wavelength interval $d\lambda$, is either absorbed or reflected. Accordingly, we may write, in terms of directional spectral properties,

$$\alpha_\lambda(\lambda, \theta, \phi, T_A) + \rho_\lambda(\lambda, \theta, \phi, T_A) = 1 \tag{6-46}$$

Applying Kirchhoff's law, we may also write this expression as

$$\varepsilon_\lambda(\lambda, \theta, \phi, T_A) + \rho_\lambda(\lambda, \theta, \phi, T_A) = 1 \tag{6-47}$$

Similarly, if the fate of all energy striking dA from a given direction is considered, we have

$$\alpha(\theta, \phi, T_A) + \rho(\theta, \phi, T_A) = 1 \tag{6-48}$$

or, for a gray surface, this may also be written

$$\varepsilon(\theta, \phi, T_A) + \rho(\theta, \phi, T_A) = 1 \tag{6-49}$$

For incident spectral energy reaching dA from all directions,

$$\alpha(\lambda, T_A) + \rho(\lambda, T_A) = 1 \tag{6-50}$$

---

[1] R. Siegel and J. R. Howell, *Thermal Radiation Heat Transfer* (New York: McGraw-Hill, 1972).

and, when incident radiant intensity is independent of direction, Kirchhoff's law may be applied so that this expression takes the alternate form

$$\varepsilon(\lambda, T_A) + \rho(\lambda, T_A) = 1 \tag{6-51}$$

Finally, when incident energy is summed over all directions and wavelengths, the hemispherical total properties are related according to

$$\alpha(T_A) + \rho(T_A) = 1 \tag{6-52}$$

and with appropriate restrictions, the application of Kirchhoff's law will enable us to write

$$\varepsilon(T_A) + \rho(T_A) = 1 \tag{6-53}$$

6.3 RADIATIVE PROPERTIES FOR REAL SURFACES

The discussion in this section will be limited to opaque surfaces. Emphasis will be placed on real surface behavior and how this behavior can be expressed so that engineering calculations can be made.

Radiative properties of real surfaces vary with the direction of emission, wavelength, surface temperature, surface roughness, and surface impurities. These factors will be considered separately in the sections to follow.

6.3-1 Directional Variation of Radiant Emission

For a diffusely emitting surface, the emission varies as $\cos \theta$, where θ is the angle measured from the normal to the surface. This is the cosine variation that has been considered for black surfaces thus far.

For real surfaces, the cosine law is, at best, an approximation. The total directional emissivity for real surfaces is not a constant in all directions; this quantity varies with the angle θ, differently for different kinds of materials. Polar diagrams of directional total emissivity are helpful in portraying directional variation.

Figure 6.12 shows a polar plot for $\varepsilon(\theta)$ for several metallic conductors. The behavior shown is representative of conductors; the emissivity is relatively constant for small values of θ and then increases for larger angles. This type of angular variation is characteristic of directional spectral emissivity also, except at quite short wavelengths.

For nonconductors, the emissivity variation with direction is quite different than for conductors. Figure 6.13 shows how ε_θ varies with θ for several nonconducting materials. In this case, the maximum emissivity is experienced normal to the surface, with little change for values of θ up to $45°$ or more. At still larger θ values the emissivity decreases, approaching zero at $\theta = 90°$.

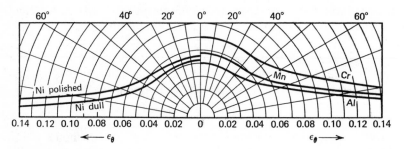

Figure 6.12 Emissivity variation with direction for several conducting materials.

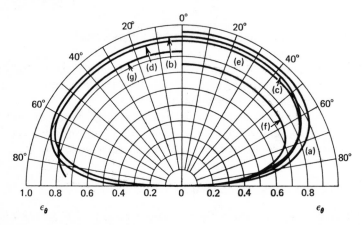

Figure 6.13 Emissivity variation with direction for several nonconducting materials. (a) Wet ice. (b) Wood. (c) Glass. (d) Paper. (e) Clay. (f) Copper oxide. (g) Aluminum oxide.

The hemispherical total emissivity can be determined from plots as shown and from data by using the expression

$$\varepsilon = \int_0^{\pi/2} \varepsilon_\theta \sin 2\theta \, d\theta \tag{6-54}$$

The value of ε so determined will differ from ε_0, the emissivity in the normal ($\theta = 0$) direction. For most bright metallic surfaces, the ratio $\varepsilon/\varepsilon_0$ has been found to be approximately 1.2. This ratio is slightly less than unity for nonconductors. Table 6.1 lists the ratio of ε to ε_0 for several bright metallic surfaces. This same ratio is listed in Table 6.2 for nonmetallic surfaces.

In most calculations the normal emissivity ε_0 is used without correction. The scarcity of data and uncertainty in actual surface conditions make such a practice acceptable.

Table 6.1   The Ratio $\varepsilon/\varepsilon_0$ for Several Bright Metallic Surfaces

Aluminum, bright rolled (338°F)	$\dfrac{0.049}{0.039} = 1.25$
Bismuth, bright (176°F)	$\dfrac{0.340}{0.336} = 1.08$
Chromium, polished (302°F)	$\dfrac{0.071}{0.058} = 1.22$
Iron, bright etched (302°F)	$\dfrac{0.158}{0.128} = 1.23$
Manganin, bright rolled (245°F)	$\dfrac{0.057}{0.048} = 1.19$
Nickel, bright matte (212°F)	$\dfrac{0.046}{0.041} = 1.12$
Nickel, polished (212°F)	$\dfrac{0.053}{0.045} = 1.18$

Table 6.2   The Ratio $\varepsilon/\varepsilon_0$ for Several Nonmetallic Surfaces

Copper oxide (300°F)	0.96
Fire clay (183°F)	0.99
Glass (200°F)	0.93
Ice (32°F)	0.95
Paper (200°F)	0.97
Plywood (158°F)	0.97

6.3-2   Wavelength Variation of Surface Emission

The effect of wavelength on radiant emission is less well established than that of direction. The general trend for metals is that the emissivity decreases with increased values of λ. This behavior is shown in Figure 6.14, which is taken from some data of Seban.[2] For most metals, the peak emissivity is exhibited near the visible wavelength region, with a decrease at values of λ higher and lower than this.

For nonmetals, the wavelength dependence of emissivity is quite weak. No generalizations of ε_λ versus λ can be made in this case.

[2] R. A. Seban, "Thermal Radiation Properties of Materials, Part III," *WADD TR*-60-370, PT III. Univ. of Calif. (Berkeley), 1963.

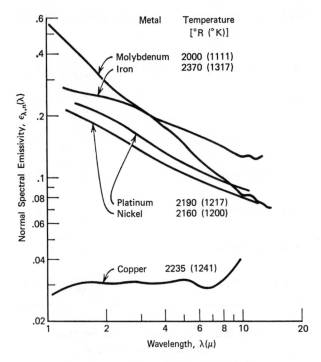

Figure 6.14 Normal spectral emissivity versus wavelength for several polished metals.

6.3-3 Emission Variation with Surface Temperature

From electromagnetic theory it can be shown that the hemispherical spectral emissivity of metallic surfaces varies as the one-half power of the electrical resistivity. Since resistivity is an increasing function of temperature, the spectral emissivity of metals will, in general, increase with temperature. This is true for wavelengths greater than approximately 5μ; for shorter wavelengths, the emissivity variation with temperature is reversed.

For nonmetals, the spectral emissivity generally decreases as the temperature is increased. For any material, one should use those values of emissivity listed at temperatures as close as possible to those actually experienced.

6.3-4 Surface Roughness Effects

An effect that substantially alters the radiant characteristics of a surface is the relationship between the wavelength of the radiant energy being considered and the size of the surface imperfections. An *optically smooth* surface is one for which the surface imperfections are much smaller than the wavelengths. A surface can, obviously, be optically smooth for the longer wavelength portion of the spectrum and quite rough at lower wavelengths. Properties of optically smooth surfaces can be predicted, using electromagnetic theory.

The subject of surface roughness effects on radiation characteristics of surfaces is quite complex. No good means exists for defining surface roughness. Tabulated properties are normally listed for various kinds of surface preparation, viz., lapping, polishing, etching, etc.

This subject will not be pursued here in any depth. We shall say, simply, that surface finish is a most important consideration in establishing the appropriate surface properties. We should recognize the need for tabulated properties which pertain to the conditions of a given analysis.

6.3-5 The Effects of Surface Impurities

A surface which has been carefully prepared to behave in an optically smooth manner can deviate from such behavior by a multitude of surface effects. Examples of these effects are thin layers of oxides formed or other products of chemical reaction at the surface, and surface adsorption, as in the case of water droplets.

Surface chemical reactions are of interest in the case of metals; obviously, most nonmetallic surfaces will have little or no such surface activity. The presence of an oxide layer, even if extremely thin, will generally increase the emissivity over the value for a metallic surface. The effect of oxide coatings on the emissivity of copper is shown in Figure 6.15. These curves are from Gubareff et al.[3]

6.3-6 Behavior of Real Surfaces in General

A few general comments can be made at this point to summarize, in part, the discussion in Section 6.3. Briefly, the pertinent comments are

1. Emissivity is a strong function of surface characteristics.
2. The emissivity of highly polished metallic surfaces is quite low.
3. The emissivity of all metallic surfaces increases with temperature.
4. Emissivity is increased appreciably by roughening the surface and by the formation of an oxide layer.
5. The ratio of $\varepsilon/\varepsilon_0$ is greater than 1 for bright metallic surfaces.

[3] G. C. Gubareff, J. F. Janssen, and R. A. Torberg, *Thermal Reduction Properties Survey*, 2nd ed. (Honeywell Research Center, 1960).

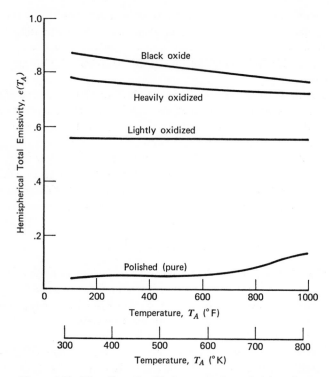

Figure 6.15 The effect of oxide layers on the emissivity of copper.

6. For nonmetallic surfaces, the emissivities are much higher than for metallic surfaces; these emissivities generally decrease as the temperature is increased.
7. For metals with colored oxides (Fe, Zn, Cr), the emissivities are much higher than for those with white oxides (Ca, Al, Mg).

Appendix D is an extensive listing of emissivities.

6.4 RADIANT ENERGY EXCHANGE BETWEEN BLACK ISOTHERMAL SURFACES

The exchange of radiant energy between two surfaces is, potentially, a very complex problem when all of the considerations of surface properties dealt with in the previous section are considered. We shall consider the simplest case at the outset, that which occurs between black surfaces. The exchange between black surfaces is simplified since, by definition, they are perfect absorbers, hence, no reflected energy need be considered; and they emit energy diffusely, thus the intensity of emitted radiation is independent of direction. We shall observe that the major complication when radiant exchange between black surfaces is considered is that of geometry or the orientation of the surfaces relative to one another. Additional complications when nonblack surfaces are present will be dealt with in the next section.

6.4-1 Radiant Exchange between Area Elements of Differential Extent

As an initial consideration of radiant exchange, we consider the two black surfaces shown in Figure 6.16. The surfaces designated A_1 and A_2 are at the uniform constant temperatures T_1 and T_2, respectively; the small elemental areas dA_1 and dA_2 represent differential portions of A_1 and A_2.

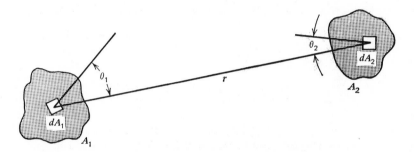

Figure 6.16 Radiant exchange between two black surfaces.

Using the notation and developments of the previous section, we have, for the monochromatic energy leaving dA_1 which is incident on dA_2,

$$dq_{b\lambda,dA_1 \to dA_2}(\lambda, \theta, T_1) = I_{b\lambda_1}(\lambda, T_1) \, dA_1 \cos \theta_1 \, d\omega_1 \, d\lambda \qquad (6\text{-}3)$$

where θ_1 is the angle between the normal to A_1 at dA_1 and the straight line drawn between dA_1 and dA_2, $d\omega_1$ is the solid angle subtended by dA_2 at dA_1, and dA is the wavelength interval of the given emission. The total radiation from dA_1 incident on dA_2 is obtained by integrating equation (6-3) over all wavelengths; the resulting expression is

$$dq_{dA_1 \to dA_2}(\theta, T_1) = I_{b1}(T_1) \, dA_1 \cos \theta_1 \, d\omega_1 \qquad (6\text{-}55)$$

The solid angle $d\omega_1$ may be written in terms of the projection of dA_2, as viewed from dA_1, as

$$d\omega_1 = \frac{\cos \theta_2 \, dA_2}{r^2} \qquad (6\text{-}56)$$

We may also recall that the radiant intensity and total emissive power of a black surface are related, according to

$$E_b(T) = \pi I_b(T) = \sigma T^4 \qquad (6\text{-}13)$$

These two expressions, substituted into equation (6-55), yield

$$dq_{dA_1 \rightarrow dA_2}(\theta) = E_{b1}(T_1) \frac{\cos \theta_1 \cos \theta_2 \, dA_1 \, dA_2}{\pi r^2} \qquad (6\text{-}57)$$

An exactly analogous expression can be developed for the total emission from dA_2 which is incident on dA_1; the result is

$$dq_{dA_2 \rightarrow dA_1}(\theta) = E_{b2}(T_2) \frac{\cos \theta_1 \cos \theta_2 \, dA_1 \, dA_2}{\pi r^2} \qquad (6\text{-}58)$$

The net energy exchange between these differential areas is the difference between the expressions given by equations (6-57) and (6-58). The result is

$$dq_{dA_1 \rightleftarrows dA_2} = [E_{b1}(T_1) - E_{b2}(T_2)] \frac{\cos \theta_1 \cos \theta_2 \, dA_1 \, dA_2}{\pi r^2} \qquad (6\text{-}59)$$

or

$$dq_{dA_1 \rightleftarrows dA_2} = \sigma(T_1^4 - T_2^4) \frac{\cos \theta_1 \cos \theta_2 \, dA_1 \, dA_2}{\pi r^2} \qquad (6\text{-}60)$$

Equations (6-59) and (6-60) are basic to all considerations of energy exchange between black surfaces. We will normally use equation (6-59) for developments and we will, henceforth, drop the function notation for purposes of brevity.

Examples 6.2 and 6.3 illustrate the use of equation (6-59).

Example 6.2

The sun and the earth are separated by a distance of 92.9×10^6 miles on the average. The diameter of the sun is approximately 860,000 miles and that of the earth is approximately 8000 miles. On a clear day solar irradiation has been measured at the earth's surface to be 360 Btu/hr-ft^2, with an additional 90 Btu/hr-ft^2 absorbed by the earth's atmosphere. Assuming the sun emits as a black body, estimate its surface temperature from this information.

The sun, emitting diffusely, will appear as a disk of area $\pi D_{\text{sun}}^2/4$. The energy interchange $dq_{\text{sun} \rightarrow \text{earth}}$, divided by the disk area of the earth, is the solar irradiation. Writing equation (6-59) in appropriate form, we have

$$\frac{dq_{\text{sun-earth}}}{dA_e} = (E_{bs} - E_{be}) \frac{\cos \theta_1 \cos \theta_2 \, dA_s}{\pi r^2}$$

For this problem, $\theta_1 = \theta_2 = 90°$; hence $\cos \theta_1 = \cos \theta_2 = 1$.

The emissive power of the earth may be neglected in comparison to E_{bs}. Solving for $E_{bs} = \sigma T_s^4$, we obtain

$$\sigma T_s^4 = \frac{dq_{s-e}}{dA_e} \frac{\pi r^2}{dA_s}$$

$$= 450 \text{ Btu/hr-ft}^2 \, \frac{\pi(93 \times 10^6)^2 \text{ mi}^2}{\pi(0.43 \times 10^6)^2 \text{ mi}^2}$$

$$T_s^4 = \frac{450 \text{ Btu/hr-ft}^2}{0.173 \times 10^{-8} \text{ Btu/hr-ft}^2\text{-}^\circ\text{R}} \frac{93^2}{0.43^2}$$

$$= 1.2167 \times 10^{16}$$

$$T_s \cong 10;500^\circ\text{R}$$

The sun's temperature is generally accepted to be of this approximate magnitude.

Example 6.3

An enclosure, with surfaces emitting as black bodies, has an opening which also interacts as though it were a black surface. Determine the radiant interchange between an element of enclosure surface and the opening with sizes and orientations as shown in Figure 6.17. The enclosure surface is uniform at $1,000^\circ\text{F}$, and the opening acts as a black surface at 250°F.

Equation (6-59) applies once more; the areas in question are small enough, compared to other dimensions, so that they may be represented as $dA_1 = dA_2 = 0.1 \text{ in.}^2$ We may write

$$dq_{dA_1 \rightleftarrows dA_2} = (E_{b1} - E_{b2}) \frac{\cos \theta_1 \cos \theta_2}{\pi r^2} dA_1 \, dA_2$$

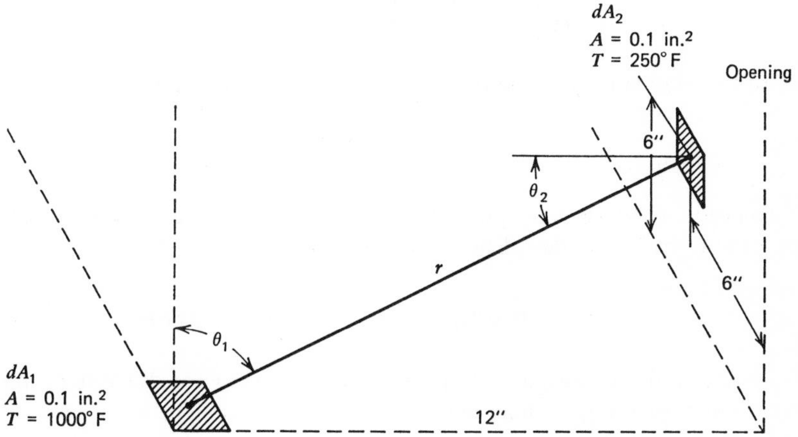

Figure 6.17 Radiant exchange between two small black surfaces.

The value of r is found from the given dimensions as

$$r = (12^2 + 6^2 + 6^2)^{1/2} = 14.7 \text{ in.}$$

thus $\cos \theta_1 = 6/14.7$ and $\cos \theta_2 = 12/14.7$. The net energy exchange may now be calculated; making appropriate substitutions, we have

$$dq_{1-2} = \left(0.173 \times 10^{-8} \frac{\text{Btu}}{\text{hr-ft}^2\text{-}^\circ\text{R}^4}\right)(1460^4 - 710^4)^\circ\text{R}^4$$

$$\times \frac{(6)(12)}{\pi(14.7)^4 \text{ in.}^2} \frac{(0.1 \text{ in.}^2)(0.1 \text{ in.}^2)}{144 \dfrac{\text{in.}^2}{\text{ft}^2}}$$

$$= \frac{(0.173)(45,400 - 2,540)(6)(12)(0.1)(0.1)}{\pi(14.7)^4(144)} \frac{\text{Btu}}{\text{hr}}$$

$$= 1.16 \times 10^{-5} \text{ Btu/hr}$$

Equations (6-57), (6-58), and (6-59) represent radiant energy transfer in terms of a driving force E_b, and a term accounting for geometry, $\cos \theta_1 \cos \theta_2 \, dA_1 \, dA_2 / \pi r^2$. A convenient designation can be made if one defines a *geometric configuration factor*, or *view factor*, which represents the fraction of energy leaving black surface element dA_1 that is incident on black surface element dA_2; this quantity is designated F_{d1-d2}. The view factor for differential areas dA_1 and dA_2 is defined according to

$$F_{d1-d2} \equiv \frac{dq_{dA_1 \rightarrow dA_2}}{E_{b1} \, dA_1} = \frac{E_{b1} \dfrac{\cos \theta_1 \cos \theta_2}{\pi r^2} dA_1 \, dA_2}{E_{b1} \, dA_1}$$

$$\equiv \frac{\cos \theta_1 \cos \theta_2 \, dA_2}{\pi r^2} \tag{6-61}$$

In an exactly analogous manner, one can represent F_{d2-d1} as

$$F_{d2-d1} \equiv \frac{\cos \theta_2 \cos \theta_1 \, dA_1}{\pi r^2} \tag{6-62}$$

If equation (6-61) is multiplied by dA_1 and equation (6-62) multiplied by dA_2, the right-hand sides of both will be equal; thus

$$dA_1 F_{d1-d2} = dA_2 F_{d2-d1} = \frac{\cos \theta_1 \cos \theta_2}{\pi r^2} dA_1 \, dA_2 \tag{6-63}$$

Equation (6-63) is designated the *reciprocity relation* for differential areas. With this expression we may rewrite equation (6-59) in the form

$$dq_{dA_1-dA_2} = (E_{b1} - E_{b2}) \, dA_1 F_{d1-d2} = (E_{b1} - E_{b2}) \, dA_2 F_{d2-d1} \tag{6-64}$$

The defining relationships for F_{d1-d2} and F_{d2-d1} are used in calculating numerical values for $dq_{dA_1-dA_2}$; Examples 6.2 and 6.3 include such calculations.

6.4-2   Radiant Exchange between Area Elements of Finite Extent

In most problems of engineering interest, surface areas involved in radiant energy exchange are finite rather than of differential size. The expressions developed in the previous section may be extended, by the process of integration, to allow finite areas to be evaluated properly.

Consider two areas, as shown in Figure 6.18. The two differential areas

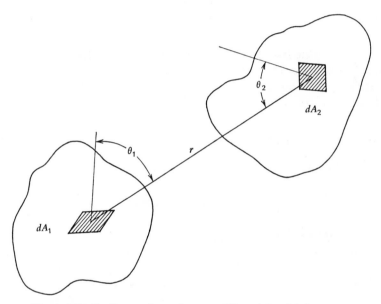

Figure 6.18   Radiant exchange between differential and finite areas.

dA_1 and dA_2 are portions of finite black surfaces. As before, we may write, for the energy leaving dA_1 which reaches dA_2,

$$dq_{dA1 \to dA2} = E_{b1} \frac{\cos \theta_1 \cos \theta_2 \, dA_1 \, dA_2}{\pi r^2} \qquad (6\text{-}59)$$

If we now consider the energy transmitted from dA_1 to the finite area A_2, we integrate over the surface of A_2, according to

$$dq_{dA1 \to A2} = E_{b1} \int_{A_2} \frac{\cos \theta_1 \cos \theta_2 \, dA_1 \, dA_2}{\pi r^2} \qquad (6\text{-}65)$$

The corresponding view factor is the radiant transfer $q_{dA_1 \to A_2}$, divided by the energy leaving dA_1

$$F_{d1-2} \equiv \frac{q_{dA1 \to A2}}{E_{b1}\, dA_1} = \frac{E_{b1} \int_{A2} \dfrac{\cos\theta_1 \cos\theta_2 \, dA_1 \, dA_2}{\pi r^2}}{E_{b1}\, dA_1}$$

$$\equiv \int_{A2} \frac{\cos\theta_1 \cos\theta_2 \, dA_2}{\pi r^2} \qquad (6\text{-}66)$$

The actual limits of integration on A_2 are those which define the portion of A_2 seen by dA_1. The integrand of this expression can be seen to be F_{d1-d2}, according to equation (6-61). We may thus write

$$F_{d1-2} = \int_{A2} F_{d1-d2} \qquad (6\text{-}67)$$

or, in words, the fraction of energy leaving dA_1 which strikes A_2 is the sum of that which is incident on every differential portion of A_2.

Turning the above analysis around, we now consider the energy leaving finite area A_2 which arrives at dA_1. We may write

$$dq_{A2 \to dA1} = E_{b2}\, dA_1 \int_{A2} \frac{\cos\theta_1 \cos\theta_2 \, dA_2}{\pi r^2} \qquad (6\text{-}68)$$

Since the total energy leaving black surface A_2 is

$$q_{2\,\text{total}} = \int_{A2} E_{b2}\, dA_2$$

the view factor is the ratio $q_{A2-dA1}/q_{2\,\text{total}}$, or

$$F_{2-d1} \equiv \frac{E_{b2}\, dA_1 \int_{A2} \dfrac{\cos\theta_1 \cos\theta_2 \, dA_2}{\pi r^2}}{E_{b2} \int_{A2} dA_2}$$

$$\equiv \frac{dA_1 \int_{A2} (\cos\theta_1 \cos\theta_2 \, dA_2)/\pi r^2}{A_2} \qquad (6\text{-}69)$$

Again using equation (6-61), the integrand in the above expression can be rewritten, and an alternate form of equation (6-69) is

$$F_{2-d1} = \frac{dA_1 \int_{A2} F_{d1-d2}}{A_2} \qquad (6\text{-}70)$$

The reciprocity theorem for exchange between a differential surface area and a finite surface area may now be obtained by relating equations (6-67) and (6-70); the expression is

$$dA_1 F_{d1-2} = A_2 F_{2-d1} \qquad (6\text{-}71)$$

Radiant energy exchange between a differential and a finite surface may be evaluated by subtracting $q_{A_2-dA_1}$ from $q_{dA_1-dA_2}$, as given by equations (6-65) and (6-68). The expression which is obtained is

$$q_{dA1-A2} = (E_{b1} - E_{b2})\, dA_1 F_{d1-2}$$

$$= (E_{b2} - E_{b1}) A_2 F_{2-d1} \qquad (6\text{-}72)$$

We now go the final step of evaluating radiant exchange between two finite black surfaces. The energy leaving A_1 and arriving at A_2 is obtained from equation (6-59) to be

$$q_{A1-A2} = E_{b1} \int_{A_1} \int_{A_2} \frac{\cos\theta_1 \cos\theta_2\, dA_1\, dA_2}{\pi r^2} \qquad (6\text{-}73)$$

The view factor F_{1-2} is the ratio of $q_{A_1-A_2}$ to the total energy $E_{b1}A_1$ leaving A_1. The expression for F_{1-2} is

$$F_{1-2} \equiv \frac{E_{b1} \int_{A_1} \int_{A_2} \dfrac{\cos\theta_1 \cos\theta_2\, dA_1\, dA_2}{\pi r^2}}{E_{b1} A_1}$$

$$\equiv \frac{1}{A_1} \int_{A_1} \int_{A_2} \frac{\cos\theta_1 \cos\theta_2\, dA_1\, dA_2}{\pi r^2} \qquad (6\text{-}74)$$

In terms of the previously defined configuration factors, F_{1-2} may also be written as

$$F_{1-2} = \frac{1}{A_1} \int_{A_1} \int_{A_2} F_{d1-d2}\, dA_1 = \frac{1}{A_1} \int_{A_1} F_{d1-2}\, dA_1 \qquad (6\text{-}75)$$

In an exactly analogous development as that for F_{1-2}, we may obtain, for F_{2-1}, the expression

$$F_{2-1} \equiv \frac{1}{A_2} \int_{A_1} \int_{A_2} \frac{\cos\theta_1 \cos\theta_2\, dA_1\, dA_2}{\pi r^2} \qquad (6\text{-}76)$$

The reciprocity theorem for radiant exchange between finite areas is obtained from equations (6-74) and (6-76) as

$$A_1 F_{1-2} = A_2 F_{2-1} \qquad (6\text{-}77)$$

Energy exchange between finite black surfaces may now be expressed as

$$q_{1-2} = (E_{b1} - E_{b2}) A_1 F_{1-2} = (E_{b1} - E_{b2}) A_2 F_{2-1} \qquad (6\text{-}78)$$

6.4-3   View-Factor Algebra

Thus far the expressions for view factors between combinations of differential and finite areas have been given in equation form. We may generalize the expressions for finite areas which comprise an enclosure in the following manner.

All energy which leaves one surface, designated i, in an enclosure must reach all surfaces in the enclosure which it can "see." The enclosure will be considered to have n surfaces, with any surface that receives energy from i designated j. (Note that, in the case of a concave surface, $F_{ii} \neq 0$). This concept may be expressed formally as

$$\sum_{j=1}^{n} F_{ij} = 1 \qquad (6\text{-}79)$$

The reciprocity relation, repeated here for clarity, is, in general,

$$A_i F_{ij} = A_j F_{ji} \qquad (6\text{-}77)$$

These two expressions comprise the basis for view-factor algebra, which is a most useful technique for evaluating view factors whose evaluation might, at first glance, appear quite difficult.

Simplified notation will be used by introducing the symbol G_{ij}, defined as

$$G_{ij} \equiv A_i F_{ij} \qquad (6\text{-}80)$$

This definition permits equations (6-79) and (6-77) to be written as

$$\sum_{j=1}^{n} G_{ij} = A_i \qquad (6\text{-}81)$$

and

$$G_{ij} = G_{ji} \qquad (6\text{-}82)$$

The quantity symbolized by G_{ij} is designated the *geometric flux*. Relations involving geometric fluxes are provided by energy conservation requirements.

Some special symbolism must now be explained. Between surface 1 and two other surfaces "seen" by 1, designated 2 and 3, we may write

$$G_{1-(2+3)} = G_{1-2} + G_{1-3} \qquad (6\text{-}83)$$

Equation (6-83) is reduced quite easily as follows:

$$A_1 F_{1-(2+3)} = A_1 F_{1-2} + A_1 F_{1-3}$$

or

$$F_{1-(2+3)} = F_{1-2} + F_{1-3}$$

which says, simply, that the energy leaving surface 1 and striking both surfaces 2 and 3 is the total of that striking each separately.

A second expression, involving four surfaces, is written $F_{(1+2)-(3+4)}$, which is interpreted as

$$G_{(1+2)-(3+4)} = G_{1-(3+4)} + G_{2-(3+4)} \tag{6-84}$$

The reciprocity relation for equation (6-84) may be obtained easily; it is

$$G_{(3+4)-(1+2)} = G_{(3+4)-1} + G_{(3+4)-2} \tag{6-85}$$

The third relation to be given here is a decomposition of equation (6-84)

$$G_{(1+2)-(3+4)} = G_{1-3} + G_{1-4} + G_{2-3} + G_{2-4} \tag{6-86}$$

which is a direct result of equations (6-84) and (6-83).

The example problems which follow illustrate the use of the preceding equations in the evaluation of view factors for the determination of radiant energy exchange between surfaces.

Example 6.4

Evaluate the view factor F_{d1-d2} between differential areas dA_1 and dA_2 oriented on plane surfaces which are perpendicular to one another, as shown in Figure 6.19. The expression which applies is equation (6-59). The terms to be evaluated are

$$\cos\theta_1 = \frac{y}{r} \qquad \cos\theta_2 = \frac{z}{r}$$

$$r = [x^2 + y^2 + z^2]^{1/2}$$

Substitution of these terms into equation (6-61) yields

$$F_{d1-d2} = \frac{yz}{\pi r^4}\,dA_2$$

$$= \frac{yz\,dx_2\,dy_2}{\pi[x^2 + y^2 + z^2]^2} \tag{6-87}$$

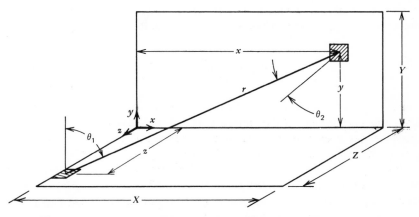

Figure 6.19 View-factor determination for perpendicular plane surfaces.

Example 6.5

Extend the result from Example 6.4 to obtain the view factor between a differential plane area and a finite perpendicular plane area.

The extension of equation (6-87) for F_{d1-d2} to evaluate F_{d1-2}, A_2 being perpendicular to the plane of $d1$, is accomplished by using equation (6-67). The procedure is

$$F_{d1-2} = \int_{A_2} F_{d1-d2}$$

$$= \int_{A_2} \frac{yz \, dx_2 \, dy_2}{\pi [x^2 + y^2 + z^2]^2}$$

$$= \int_0^Y \int_0^X \frac{yz \, dx_2 \, dy_2}{\pi [x^2 + y^2 + z^2]^2} \qquad (6\text{-}88)$$

The integration which is indicated is reasonably formidable; the result will be written without the detailed steps in the solution. The expression which results is

$$F_{d1-2} = \frac{1}{2\pi}\left[\tan^{-1}\frac{1}{B} - BC \tan^{-1} AC \right] \qquad (6\text{-}89)$$

where

$$A = Y/X, \qquad B = Z/X, \quad \text{and} \quad C = (A^2 + B^2)^{-1/2}$$

The next step in the sequence is the extension of equation (6-89) to express the view factor between two finite rectangular areas which are perpendicular to each other and share the line of intersection as a common side. The expression for F_{1-2} in this case is, according to equation (6-75),

$$F_{1-2} = \frac{1}{A_1} \int_{A1} F_{d1-2} \, dA_1 \qquad (6\text{-}75)$$

which, even without the substitution and definition of integration limits, is complex and the analytical evaluation most formidable. It is fortunate that this case has been solved; the results are available in both equation and graphical form. The graphical solution for F_{1-2} is presented in Figure 6.20.

Having a plot for the case represented by Figure 6.20, we can obtain view factors for several other related geometries. The next example illustrates how this plot, reciprocity, and view-factor algebra may be used to obtain view factors for other finite area configurations.

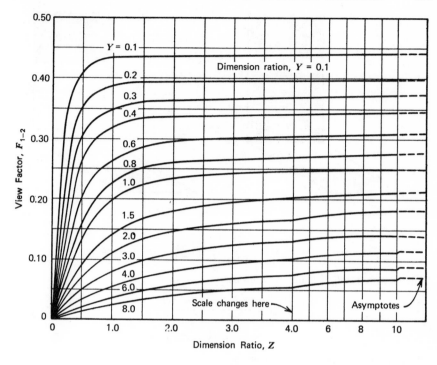

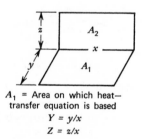

A_1 = Area on which heat—
transfer equation is based
$Y = y/x$
$Z = z/x$

Figure 6.20 View factors for perpendicular rectangles with a common side. [From H. C. Hottel, "Radiant Heat Transmission," *Mechanical Engineering* 52 (1930). By permission of the publishers.]

Example 6.6

Determine the view factors F_{1-2} for the finite areas depicted in the figure on p. 332 Inspection indicates that the view factors F_{2-A} and $F_{2-(1+A)}$ can be read directly from Figure 6.20. Flux algebra will be used to relate F_{1-2} to these. Using flux algebra, we have

$$G_{2-(1+A)} = G_{2-1} + G_{2-A}$$

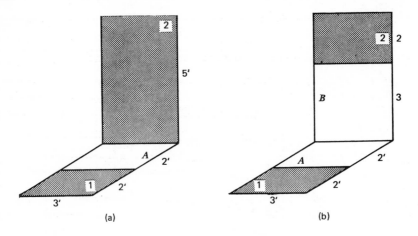

(a)   (b)

or

$$G_{2-1} = G_{2-(1+A)} - G_{2-A}$$

The view factor F_{2-1} is now written as

$$F_{2-1} = \frac{A_2 F_{2-(1+A)} - A_2 F_{2-A}}{A_2}$$

$$= F_{2-(1+A)} - F_{2-A}$$

Finally, by reciprocity, we may write

$$F_{1-2} = \frac{A_2}{A_1} F_{2-1} = \frac{A_2}{A_1} (F_{2-(1+A)} - F_{2-A})$$

From Figure 6.20 we read

$$F_{2-(1+A)} = 0.15 \qquad F_{2-A} = 0.10$$

and the answer for configuration (a) is

$$F_{1-2} = \frac{5}{2} (0.15 - 0.10) = 0.125$$

For (b), the sequence of solution is

$$G_{1-2} = G_{1-(2+B)} - G_{1-B}$$

or

$$F_{1-2} = F_{1-(2+B)} - F_{1-B}$$

We may use the results of (a) to find

$$F_{1-(2+B)} = \left(\frac{A_2 + A_B}{A_1}\right) [F_{(2+B)-(1+A)} - F_{(2+B)-A}]$$

$$F_{1-B} = \frac{A_B}{A_1} [F_{B-(1+A)} - F_{B-A}]$$

Each of the view factors on the right-hand side of these expressions may be read from Figure 6.20; they are

$$F_{(2+B)-(1+A)} = 0.15 \qquad F_{B-(1+A)} = 0.22$$

$$F_{(2+B)-A} = 0.10 \qquad\quad F_{B-A} = 0.165$$

The problem is now completed according to

$$F_{1-(2+B)} = \frac{(5)(3)}{(2)(3)}(0.15 - 0.10) = 0.125$$

$$F_{1-B} = \frac{(3)(3)}{(2)(3)}(0.22 - 0.165) = 0.0825$$

yielding, for (b), the result

$$F_{1-2} = 0.125 - 0.0825 = 0.0425$$

The preceding examples have all dealt with plane areas or elements thereof which are at right angles to one another. Obviously, this is a very special situation, yet numerous cases of practical importance can be solved.

View factors for other geometries of interest are given in the figures which follow. This compilation of view factors is by no means exhaustive. For more

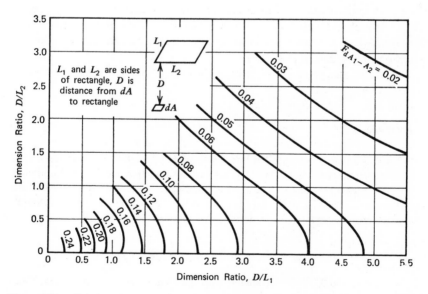

Figure 6.21   View factor for a differential plane element oriented below a corner of a parallel finite rectangle. [From H. C. Hottel, "Radiant Heat Transmission," *Mechanical Engineering* 52 (1930). By permission of the publishers.]

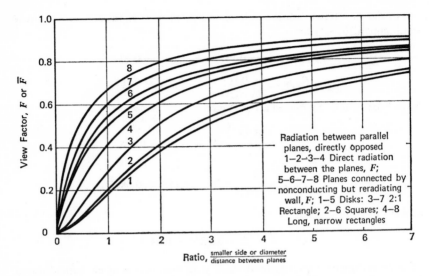

Figure 6.22 View factors for parallel squares, rectangles, and disks. [From H. C. Hottel, "Radiant Heat Transmission," *Mechanical Engineering* 52 (1930). By permission of the publishers.]

complete compilations, the reader is referred to Wiebelt,[4] Sparrow and Cess,[5] Love,[6] Hottel and Sarofim,[7] and other advanced treatises on radiant heat transfer. All view factors available in any form are based on the integral equations developed in this section.

The use of the preceding figures is illustrated in the example problems which follow.

Example 6.7

A 3/8-inch-diameter hole is drilled into a slab of metal as shown. If the metal is at a uniform temperature and behaves as a black surface, how much of the emission from the cavity surface escapes to the surroundings? If the walls in the cavity are designated as surface 1 and the hole is considered to be a circular surface, 3/8-inch in diameter and designated 2, we wish to find the view factor F_{1-2}.

[4] J. A. Wiebelt, *Engineering Radiation Heat Transfer* (New York: Holt, Rinehart and Winston, 1966).
[5] E. M. Sparrow and R. D. Cess, *Radiation Heat Transfer*, (Belmont, Calif.: Brooks/Cole, 1966).
[6] T. J. Love, *Radiative Heat Transfer*, (Columbus, Ohio: Merrill, 1968).
[7] H. C. Hottel and A. F. Sarofim, *Radiative Transfer*, (New York: McGraw-Hill, 1967).

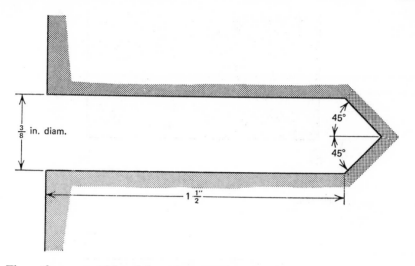

The surface areas of 1 and 2 are determined as follows:

$$A_1 = \pi \left(\frac{3}{8}\right)(1.5)\ \text{in.}^2 + \frac{1}{2}\ \frac{(\pi)\left(\frac{3}{8}\right)\left(\frac{3}{16}\right)\text{in.}^2}{\sin 45^\circ} = 1.916\ \text{in.}^2$$

$$A_2 = \frac{\pi}{4}\left(\frac{3}{8}\ \text{in.}\right)^2 = 0.110\ \text{in.}^2$$

The view factor between 2 and 1 is

$$F_{2-1} = 1$$

By reciprocity, we obtain

$$F_{1-2} = \frac{A_2 F_{2-1}}{A_1} = \frac{(0.110\ \text{in.}^2)(1)}{1.916\ \text{in.}^2} = 0.057$$

or, approximately 6% of the emission from the surface of the cavity is lost through the opening.

Example 6.8

If a 1-inch hole is drilled completely through a 2-inch-thick metal plate which is maintained at a uniform temperature of 350°F, how much energy will be lost per hour to the surroundings at 60°F? Both the metallic surface and surroundings may be considered black.

The cavity surface will be designated 1, and the two ends of the 1-inch-diameter hole will be treated as black disks and designated surfaces 2 and 3, as shown.

From Figure 6.22 the view factors F_{2-3} and F_{3-2} were read as

$$F_{2-3} = F_{3-2} = 0.07$$

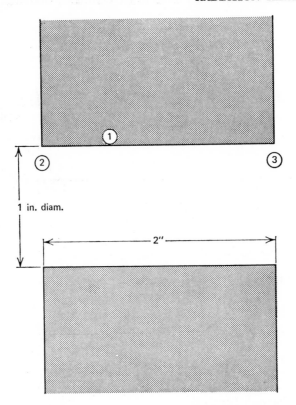

We may now determine F_{2-1} and F_{3-1} as

$$F_{2-1} = F_{3-1} = 1 - F_{2-3} = 0.93$$

and, using reciprocity, we obtain

$$F_{1-2} = F_{1-3} = \frac{A_2 F_{2-1}}{A_1} = \frac{\dfrac{\pi (1 \text{ in.})^2}{4} (0.93)}{\pi (1 \text{ in.})(2 \text{ in.})}$$

$$= 0.116$$

The total energy loss is the sum of the amounts leaving through each end of the hole. The solution is completed as follows:

$$q = (A_1 F_{1-2} + A_1 F_{1-3})(E_{b1} - E_{b \text{ surroundings}})$$

$$= A_1 (F_{1-2} + F_{1-3}) \sigma (T_1^4 - T_{\text{sur}}^4)$$

$$= \frac{\pi (1 \text{ in.})(2 \text{ in.})}{114 \text{ in.}^2/\text{ft}^2} (0.116 + 0.116) \left(0.173 \frac{\text{Btu}}{\text{hr-ft}^2\text{-}^\circ\text{R}} \right)$$

$$\left[\left(\frac{810}{100} \,^\circ\text{R} \right)^4 - \left(\frac{520}{100} \,^\circ\text{R} \right)^4 \right]$$

$$= 6.26 \text{ Btu/hr}$$

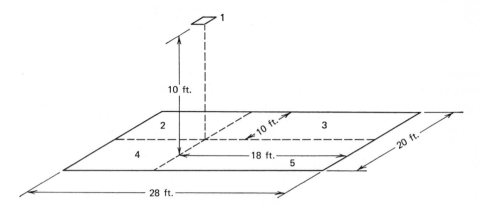

Example 6.9

Determine the view factor between a 1-foot-square skylight oriented as shown above a floor which measures 20 ft by 28 ft. The relative areas are such that the skylight, designated 1, may be approximated as a differential area when compared with the floor. The total view factor $F_{1-\text{floor}}$ may thus be expressed as

$$G_{1-(2+3+4+5)} = G_{1-2} + G_{1-3} + G_{1-4} + G_{1-5}$$

or

$$F_{1-\text{floor}} = F_{1-2} + F_{1-3} + F_{1-4} + F_{1-5}$$

Using Figure 6.21, we obtain

$$F_{1-2} = F_{1-4} = 0.14$$
$$F_{1-3} = F_{1-5} = 0.164$$

giving, finally,

$$F_{1-\text{floor}} = 0.14 + 0.164 + 0.14 + 0.164 = 0.608$$

6.4-4 Special Reciprocity

The reciprocity expression, developed earlier, has been shown to be extremely useful. Another *special reciprocity* theorem is of sufficient importance to warrant examination at this time.

The development of special reciprocity will be referenced to Figure 6.23. Of interest is the direct exchange between areas A_1 and A_4, oriented as shown. The view factor between these areas may be expressed, according to equation (6-74), in the form

$$A_1 F_{1-4} = A_4 F_{4-1} = \int_{A_1} \int_{A_4} \frac{\cos \theta_1 \cos \theta_4 \, dA_4 \, dA_1}{\pi r^2}$$

$$= \int_0^Z \int_0^a \int_0^Y \int_a^X \frac{\cos \theta_1 \cos \theta_4 \, dx_4 \, dy \, dx_1 \, dz}{\pi r^2} \qquad (6\text{-}90)$$

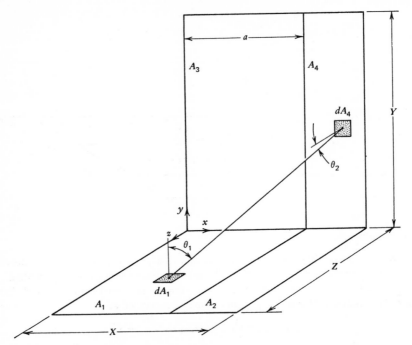

Figure 6.23 Surface orientation for the development of special reciprocity.

A similar expression for the exchange between A_2 and A_3 may be written as

$$A_2 F_{2-3} = A_3 F_{3-2} = \int_{A_3} \int_{A_2} \frac{\cos \theta_2 \cos \theta_3 \, dA_2 \, dA_3}{\pi r^2}$$

$$= \int_0^Y \int_0^a \int_0^Z \int_a^X \frac{\cos \theta_2 \cos \theta_3 \, dx_2 \, dz \, dx_3 \, dy}{\pi r^2} \qquad (6\text{-}91)$$

We may observe that the integrands and the limits of integration in equations (6-90) and (6-91) are symmetric; thus the *special reciprocity* relation

$$A_1 F_{1-4} = A_2 F_{2-3} \qquad (6\text{-}92)$$

is seen to be valid.

The preceding development was related specifically to areas in perpendicular rectangles that were defined by lines running parallel to two sides of the larger rectangular areas. In an exactly analogous manner, the areas shown in Figure 6.24 are related by special reciprocity, according to equation (6-92). These areas are designated such that the subscript notation in equation (6-92) applies to both orientations (a) and (b) in this figure.

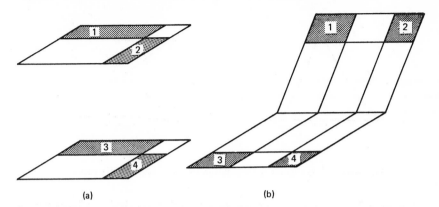

(a)   (b)

Figure 6.24   Additional surface orientation for special reciprocity expressions.

The view factor F_{1-4} may be determined, for the situation depicted in Figure 6.23, as follows. By a total energy consideration, we have

$$G_{(1+2)-(3+4)} = G_{1-3} + G_{1-4} + G_{2-3} + G_{2-4}$$

Using special reciprocity, we may write

$$G_{1-4} = G_{2-3}$$

thus

$$2G_{1-4} = G_{(1+2)-(3+4)} - G_{1-3} - G_{2-4}$$

and it follows, finally, that

$$F_{1-4} = \frac{(A_1 + A_2)F_{(1+2)-(3+4)} - A_1 F_{1-3} - A_2 F_{2-4}}{2A_1}$$

The view factor F_{1-4} is written in terms of the three view factors $F_{(1+2)-(3+4)}$, F_{1-3}, and F_{2-4}, each of which may be read directly from Figure 6.20.

6.4-5   View-Factor Determination for Elongated Surfaces: Hottel's Crossed-String Method

In the case of surfaces which extend sufficiently far in one direction as to be considered infinite, Hottel[8] has presented an extremely simple and useful technique for determining view factors.

---

[8] H. C. Hottel, "Radiant-Heat Transmission," in W. H. McAdams, *Heat Transmission,* 3rd ed., (New York: McGraw-Hill, 1954).

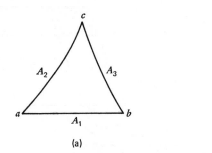 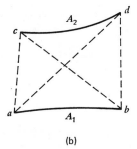

(a)   (b)

Figure 6.25 View-factor determination by the crossed-strings method.

Figure 6.25(a) shows an enclosure consisting of three surfaces whose dimension normal to the paper is considered infinite. By view-factor algebra, we have

$$\sum_{1}^{3} G_{1-j} = A_1 F_{1-2} + A_1 F_{1-3} = A_1$$

$$\sum_{1}^{3} G_{2-j} = A_2 F_{2-1} + A_2 F_{2-3} = A_2$$

$$\sum_{1}^{3} G_{3-j} = A_3 F_{3-1} + A_3 F_{3-2} = A_3$$

The flux $G_{1-2} = A_1 F_{1-2}$ may be solved for as

$$G_{1-2} = A_1 F_{1-2} = \frac{A_1 + A_2 - A_3}{2} \tag{6-93}$$

We next consider surfaces A_1 and A_2, which are parts of a multisurfaced enclosure, as shown in Figure 6.25(b). The procedure is to imagine strings to be tightly stretched between the end points of A_1 and A_2; two of these strings, $\overline{ac}$ and $\overline{bd}$, may be thought of as surfaces which form an enclosure with A_1 and A_2.

Again, from view-factor algebra, we may write

$$\sum_{j=1}^{n} G_{1j} = A_1 F_{1-2} + A_1 F_{1-\overline{ac}} + A_1 F_{1-\overline{bd}} = A_1 \tag{6-94}$$

The crossed strings $\overline{ad}$ and $\overline{bc}$ are now similarly considered. The three-sided enclosures $\overline{acb}$ and $\overline{adb}$ are formed and, using equation (6-92), we may write

$$G_{1-\overline{ac}} = A_1 F_{1-\overline{ac}} = \frac{A_1 + A_{\overline{ac}} - A_{\overline{cb}}}{2}$$

and

$$G_{1-\overline{bd}} = A_1 F_{1-\overline{bd}} = \frac{A_1 + A_{\overline{bd}} - A_{\overline{ad}}}{2}$$

These expressions are substituted into equation (6-94) to yield

$$A_1 F_{1-2} = A_1 - A_1 F_{1-\overline{ac}} - A_1 F_{1-\overline{bd}}$$

$$= \frac{A_{\overline{cb}} + A_{\overline{ad}}}{2} - \frac{A_{\overline{ac}} + A_{\overline{bd}}}{2}$$

or

$$F_{1-2} = \frac{A_{\overline{cb}} + A_{\overline{ad}}}{2A_1} - \frac{A_{\overline{ac}} + A_{\overline{bd}}}{2A_1}$$

Since the areas extend infinitely normal to the paper, each may be represented by the lengths of the tightly stretched strings, yielding

$$F_{1-2} = \frac{(\overline{cb} + \overline{ad}) - (\overline{ac} + \overline{bd})}{2\overline{ab}} \tag{6-95}$$

This expression, in general terms, may be written as

$$F_{1-2} = \frac{(\text{sum of crossed strings}) - (\text{sum of uncrossed strings})}{2 \times \text{length of 1}} \tag{6-96}$$

Example 6.10 illustrates how the crossed-strings technique is used as a very simple device for view-factor determination.

Example 6.10

Evaluate the view factor between the rectangular areas shown in Figure 6.26, whose length, normal to the plane of the paper, is very long compared to other dimensions.

Employing the crossed-strings technique, we may envision four strings stretched between the extremities of A_1 and A_2. The lengths of the uncrossed strings are Y, and those of the crossed strings are $(X^2 + Y^2)^{1/2}$. According to equation (6-95), the view factor F_{1-2} is given by

$$F_{1-2} = \frac{2(X^2 + Y^2)^{1/2} - 2Y}{2X}$$

$$= \frac{(1 + Z^2)^{1/2} - 1}{Z} \tag{6-97}$$

where $Z = X/Y$ is the ratio of the small side of the opposed rectangles to the distance between the planes.

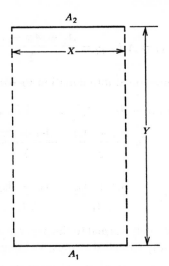

Figure 6.26 View-factor determination between long, narrow rectangles by the crossed-strings method.

Equation (6-97) is the analytical form of line 4 in Figure 6.22. The direct integration approach to obtain the view factor in this case is much more involved than the development used here.

6.4-6 Radiation between Black Surfaces; Electrical Network Analogy

The radiant exchange between two black surfaces was expressed in an earlier section as

$$q_{1-2} = A_1 F_{1-2}(E_{b1} - E_{b2}) \qquad (6\text{-}78)$$

This expression may be generalized to represent the exchange between surface A_i, in an enclosure consisting of several black surfaces, and the n other surfaces which it "sees." The net rate of energy flow from A_i is

$$q_{i\,\text{net}} = \sum_{j=1}^{n} A_i F_{i-j}[E_{bi} - E_{bj}] \qquad (6\text{-}98)$$

Equation (6-98), along with equation (6-79)

$$\sum_{j=1}^{n} F_{i-j} = 1 \qquad (6\text{-}79)$$

provides the basis for an analogy between black body radiant exchange and the flow of electric current.

In the analogy the quantity being transferred in the heat transfer case, $q_{i\,\text{net}}$, is analogous to electric current; the driving force $E_{bi} - E_{bj}$ is analogous to electrical potential difference ΔV; and the conductance in the heat transfer

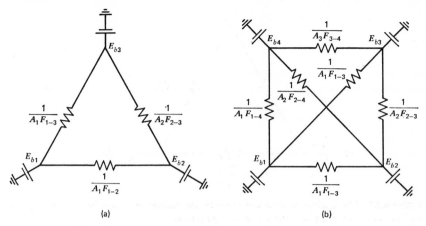

Figure 6.27 Analogous electrical networks for radiant exchange in black enclosures with (a) three surfaces and (b) four surfaces.

case is $A_i F_{i-j}$. In Figure 6.27 the equivalent networks for three- and four-surface enclosures are depicted.

In all cases, a sketch of the equivalent electrical network is helpful for problem solution. Analytical solutions become very tedious for cases involving four or more surfaces; in such cases, it is usually easier to construct an equivalent circuit for a radiation problem and measure the desired quantity.

6.4-7 Radiation between Black Surfaces with Nonconducting Reradiating Surfaces Present

In Figure 6.27 all surfaces depicted have an emissive power E_{bi}, characterized by a battery; and all are absorbing surfaces, that is, there is a path to ground. In practice, certain surfaces behave as nonabsorbers of radiant energy, or they emit and absorb radiant energy at the same rate. The equivalent network with three surfaces, one a reradiating, or *refractory*, surface, is diagrammed in Figure 6.28.

In this figure the reradiating surface has emissive power E_3, which "floats" between the emissive powers E_{b1} and E_{b2} at a level which depends on the magnitudes of E_{b1} and E_{b2}, as well as the resistances involved in the equivalent circuit. Energy exchange between A_1 and A_2 includes that which passes directly through the resistance $1/A_1 F_{1-2}$, and that which takes the parallel path to A_3, then to A_2 through the equivalent resistance $1/A_1 F_{1-3} + 1/A_2 F_{2-3}$. The net radiant exchange may thus be expressed as

$$q_{1-2} = \frac{E_{b1} - E_{b2}}{R_{\text{equiv}}}$$

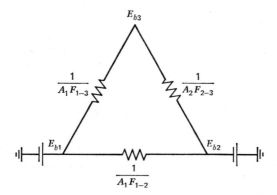

Figure 6.28 The analogous electrical network for radiant exchange between two black surfaces in the presence of a third reradiating surface.

where the equivalent resistance for the parallel situation is given by

$$\frac{1}{R_{\text{equiv}}} = A_1 F_{1-2} + \frac{1}{1/A_1 F_{1-3} + 1/A_2 F_{2-3}}$$

The net exchange between A_1 and A_2 may now be expressed as

$$q_{1-2} = A_1 \left[F_{1-2} + \frac{A_2 F_{2-3} F_{1-3}}{A_1 F_{1-3} + A_2 F_{2-3}} \right] (E_{b1} - E_{b2}) \qquad (6\text{-}99)$$

or, more simply, as

$$q_{1-2} = A_1 \bar{F}_{1-2} (E_{b1} - E_{b2}) \qquad (6\text{-}100)$$

In equation (6-100) the symbol $\bar{F}_{1-2}$ has replaced the square-bracketed term in the previous equation. This term is a modified view factor which includes the effect of a reradiating surface between the two black surfaces which are exchanging radiant energy. Values of $\bar{F}_{1-2}$ for four cases of opposed parallel plane surfaces are given by the curves designated 5, 6, 7, and 8 in Figure 6.22.

6.5　RADIANT ENERGY EXCHANGE BETWEEN GRAY ISOTHERMAL SURFACES

It is now our task to describe the radiant interchange between nonblack surfaces. The gray body concept, as introduced earlier, will be employed in the present analysis. In the case of a gray body, all emissions and reflections are diffuse, and the surface properties are considered independent of wavelength.

6.5-1 Radiosity and Irradiation

The added consideration of reflections in the case of nonblack surfaces introduces some considerable complexity into calculations of energy exchange. Two quantities which are helpful in describing such exchanges are the *radiosity* and the *irradiation*.

Radiosity, symbolized J, is the total radiant energy per unit time per unit area leaving a surface. Irradiation, symbolized G, is the rate at which radiant energy reaches a surface per unit time per unit area. Both of these quantities may be thought of as fluxes, usually expressed in units of Btu per hour per square foot.

Figure 6.29 shows a surface element which is exchanging radiant energy with its surroundings. In considering a hypothetical boundary a, just outside the surface, the net rate of energy gain by the surface is given by

$$\frac{q}{A}\bigg|_{net} = G(T_s) - J(T_A, T_s) \qquad (6\text{-}101)$$

where the functional notation indicates that the incident energy is characteristic of T_s, the temperature of the surroundings, and the leaving energy is characteristic of both the surface temperature T_A and that of the surroundings, T_s.

A second point of view is represented by the hypothetical boundary, designated b in the figure, which is immediately below the surface in question. With this reference boundary the net energy *gain* by this surface is represented as

$$\frac{q}{A}\bigg|_{net} = \alpha(T_A, T_s)G(T_s) - E(T_A) \qquad (6\text{-}102)$$

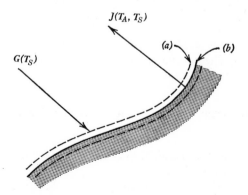

Figure 6.29   Surface element exchanging radiant energy with its surroundings.

where the functional notation involving T_A and T_s is as before; $\alpha(T_A, T_s)$ is the hemispherical spectral absorptivity which was discussed in Section 6.2–2.3. Henceforth all properties expressed in this section will be the hemispherical, spectral ones.

Either equation (6-101) or (6-102) may be used to evaluate net radiant flux at a surface. These expressions may be equated to represent radiosity in terms of irradiation, and vice versa. Such an exercise for an opaque surface where $\alpha + \rho = 1$, yields

$$J(T_A, T_s) = \rho(T_A, T_s)G(T_s) + \varepsilon(T_A)E_b(T_A) \qquad (6\text{-}103)$$

or

$$G(T_s) = \frac{J(T_A, T_s) - \varepsilon(T_A)E_b(T_A)}{\rho(T_A, T_s)} \qquad (6\text{-}104)$$

Equation (6-104) may now be used with equation (6-101) to express $q/A|_{net}$ in terms of J and E_b, according to

$$\left.\frac{q}{A}\right|_{net} = \frac{J(T_A, T_s) - \varepsilon(T_A)E_b(T_A)}{\rho(T_A, T_s)} - J(T_A, T_s)$$

$$= \frac{[1 - \rho(T_A, T_s)]}{\rho(T_A, T_s)} J(T_A, T_s) - \frac{\varepsilon(T_A)}{\rho(T_A, T_s)} E_b(T_A) \qquad (6\text{-}105)$$

In the case of an opaque surface, $\alpha + \rho = 1$, and this expression becomes

$$\left.\frac{q}{A}\right|_{net} = \frac{\alpha(T_A, T_s)}{\rho(T_A, T_s)} J(T_A, T_s) - \frac{\varepsilon(T_A)}{\rho(T_A, T_s)} E_b(T_A) \qquad (6\text{-}106)$$

and, finally, using Kirchhoff's law, we obtain

$$\left.\frac{q}{A}\right|_{net} = \frac{\varepsilon(T_A, T_s)J(T_A, T_s) - \varepsilon(T_A)E_b(T_A)}{\rho(T_A, T_s)} \qquad (6\text{-}107)$$

The functional notation used in these expressions may be somewhat confusing at this point. The consideration of a specific case may be helpful in clarifying the nature of all terms.

6.5-2 Radiant Exchange between Infinite, Parallel, Isothermal Gray Surfaces

The system of two infinite, parallel, gray surfaces, each at a constant temperature, is represented in Figure 6.30.

The radiosity J_1 is the sum of all energy terms shown as arrows leaving surface 1 in (a) and (b) of Figure 6.30. The total energy leaving surface 1 due

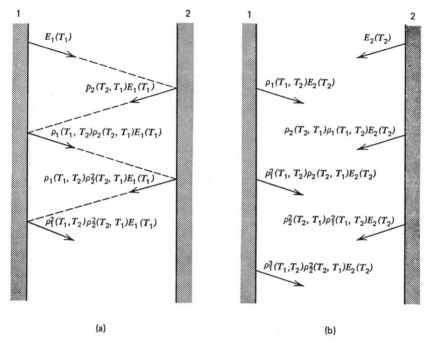

Figure 6.30 Radiant exchange between infinite, parallel, isothermal, gray surfaces. (a) Energy originating at surface 1. (b) Energy originating at surface 2.

to that which originates at 1, is given in (a) of the figure. This sum is an infinite series of the form

$$E_1(T_1)[1 + \rho_1(T_1, T_2)\rho_2(T_2, T_1) + \rho_1{}^2(T_1, T_2)\rho_2{}^2(T_2, T_1) + \cdots]$$

which may be expressed as

$$\frac{E_1(T_1)}{1 - \rho_1(T_1, T_2)\rho_2(T_2, T_1)}$$

Similarly, the energy which leaves surface 1, resulting from $E_2(T_2)$, that which originates at surface 2, is found by summing the quantities in (b) of the figure, which yields

$$\frac{\rho_1(T_1, T_2)E_2(T_2)}{1 - \rho_1(T_1, T_2)\rho_2(T_2, T_1)}$$

The sum of these two terms is the radiosity from 1; thus we may write

$$J_1(T_1, T_2) = \frac{E_1(T_1) + \rho_1(T_1, T_2)E_2(T_2)}{1 - \rho_1(T_1, T_2)\rho_2(T_2, T_1)} \tag{6-108}$$

Substitution of equation (6-108) into (6-107) produces, for the net energy gain of surface 1, the expression

$$\frac{q}{A}\bigg|_{net,1} = \frac{\varepsilon_1(T_1, T_2)}{\rho_1(T_1, T_2)}\left[\frac{E_1(T_1) + \rho_1(T_1, T_2)E_2(T_2)}{1 - \rho_1(T_1, T_2)\rho_2(T_2, T_1)}\right] - \frac{\varepsilon_1(T_1)E_{b1}(T_1)}{\rho_1(T_1, T_2)} \quad (6\text{-}109)$$

Certain assumptions, compatible with the gray body concept, will permit simplification of equation (6-109). These are

$$\alpha_1(T_1, T_2) = \varepsilon_1(T_1, T_2) = \varepsilon_1(T_1) \quad (6\text{-}110a)$$

$$\alpha_2(T_2, T_1) = \varepsilon_2(T_2, T_1) = \varepsilon_2(T_2) \quad (6\text{-}110b)$$

$$1 - \rho_1(T_1, T_2) = \varepsilon_1(T_1) \quad (6\text{-}110c)$$

$$1 - \rho_2(T_2, T_1) = \varepsilon_2(T_2) \quad (6\text{-}110d)$$

The first two of the above expressions state that absorption and emission properties of a surface are functions of the temperature of that surface only. The latter two expressions relate this concept to the reflective property which applies to opaque surfaces. This assumption and the inaccuracies involved are discussed by Hottel[9] and Wiebelt.[10] With the inclusion of these simplifying assumptions, the expression for net radiant exchange between infinite, parallel, gray, isothermal surfaces becomes

$$\frac{q}{A}\bigg|_{net,1} = \frac{E_{b1}(T_1) - E_{b2}(T_2)}{\dfrac{1}{\varepsilon_1(T_1)} + \dfrac{1}{\varepsilon_2(T_2)} - 1} \quad (6\text{-}111)$$

6.5-3 Radiant Exchange between Finite Isothermal Gray Surfaces

The results of the previous section may be extended to include geometric factors when the surfaces considered are finite in extent. The terms that must be considered in a similar approach to this problem are shown in Figure 6.31. Only a few of the terms are shown; the obvious difference between this and the infinite parallel case is the inclusion of areas and view factors between the surfaces considered. The complete analysis would obviously be quite messy.

An alternate approach which is much simpler, more direct, and more general is presented in the following section.

[9] Hottel, op. cit.
[10] Wiebelt, op. cit.

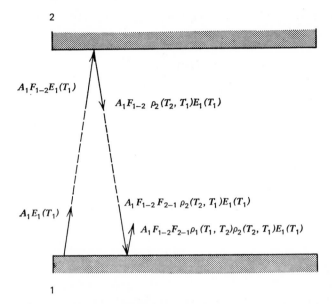

2

$A_1 F_{1-2} E_1(T_1)$

$A_1 F_{1-2}\ \rho_2(T_2,\ T_1)E_1(T_1)$

$A_1 E_1(T_1)$

$A_1 F_{1-2} F_{2-1}\ \rho_2(T_2,\ T_1)E_1(T_1)$

$A_1 F_{1-2} F_{2-1}\rho_1(T_1,\ T_2)\rho_2(T_2,\ T_1)E_1(T_1)$

1

Figure 6.31   Radiant exchange between isothermal, gray, finite surfaces.

6.5-4   Radiant Exchange between Gray Isothermal Surfaces: the Electrical Analogy Approach

Electrical analogs were considered earlier for exchange between surfaces in a black enclosure. The same concept is possible in the case of gray surfaces, with the added consideration of surface emissivity.

With the simplifying assumptions given by equations (6-110), the net heat flux gain of a surface, expressed by equation (6-107), becomes

$$\frac{q}{A}\bigg|_{\text{net}} = \frac{\varepsilon}{\rho}\,[J - E_b] \qquad (6\text{-}112)$$

This expression, along with equations (6-79) and (6-98), allows many problems to be solved with relative ease.

Consider, as an initial example, the case of two infinite, parallel, gray isothermal planes. Assuming that $T_1 > T_2$, so that net energy transfer is from surface 1 to surface 2, we have

Rate of energy loss from surface 1: $\dfrac{q}{A}\bigg|_1 = \dfrac{\varepsilon_1}{\rho_1}\,(E_{b1} - J_1)$

Rate of energy exchange between 1 and 2: $\dfrac{q}{A}\bigg|_{1-2} = F_{1-2}(J_1 - J_2)$

Rate of energy gain by surface 2: $\dfrac{q}{A}\bigg|_2 = \dfrac{\varepsilon_2}{\rho_2}\,(J_2 - E_{b2})$

Since $(q/A)|_1 = (q/A)|_{1-2} = (q/A)_2 = (q/A)_{net}$, we may solve for the respective driving forces

$$E_{b1} - J_1 = \frac{q}{A}\bigg|_{net}\bigg/\frac{\rho_1}{\varepsilon_1}$$

$$J_1 - J_2 = \frac{q}{A}\bigg|_{net}\bigg/\frac{1}{F_{1-2}}$$

$$J_2 - E_{b2} = \frac{q}{A}\bigg|_{net}\bigg/\frac{\rho_2}{\varepsilon_2}$$

and, adding, we obtain

$$E_{b1} - E_{b2} = \frac{q}{A}\bigg|_{net}\left[\frac{\rho_1}{\varepsilon_1} + \frac{1}{F_{1-2}} + \frac{\rho}{\varepsilon_2}\right]$$

Finally,

$$\frac{q}{A}\bigg|_{net} = \frac{E_{b1} - E_{b2}}{\dfrac{\rho_1}{\varepsilon_1} + \dfrac{1}{F_{1-2}} + \dfrac{\rho_2}{\varepsilon_2}}$$

In the preceding expressions all functional notation has been deleted; the usual dependence on T_1 and T_2 prevails.

In the case of infinite parallel planes, $F_{1-2} = 1$, and this expression reduces to equation (6-111).

$$\frac{q}{A}\bigg|_{net} = \frac{E_{b1} - E_{b2}}{\dfrac{1}{\varepsilon_1} + \dfrac{1}{\varepsilon_2} - 1} \qquad (6\text{-}111)$$

A generalization of equations (6-96) and (6-112) leads to 'an analogy between thermal and electrical quantities, with heat flow rate and current being common quantities, electrical potential (voltage) difference being analogous to $E_b - J$ or ΔJ, and equivalent thermal resistance being $\rho/A\varepsilon$ for a surface and $1/A_1F_{1-2}$ for exchange between surfaces. The equivalent electrical network for the infinite parallel plane problem is shown in Figure 6.32.

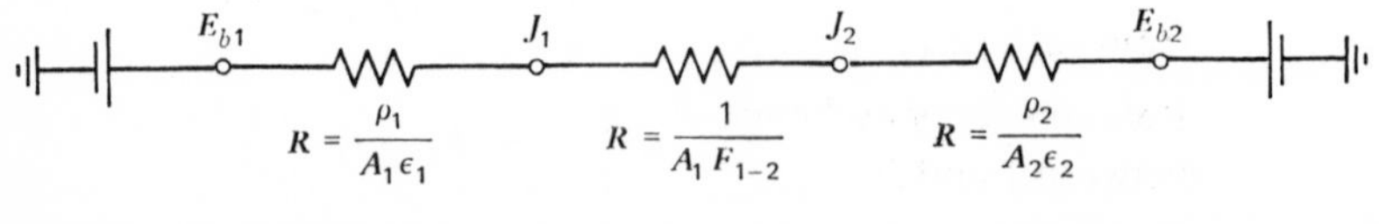

Figure 6.32 Equivalent electrical circuit for radiant energy exchange between parallel, isothermal, gray planes of infinite extent.

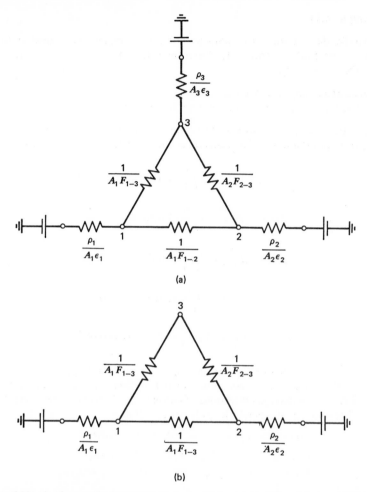

Figure 6.33 Analogous electrical circuits for radiant energy exchange in an enclosure consisting of three gray isothermal surfaces.

In the case of an enclosure consisting of 3 gray surfaces, the equivalent electrical network is shown in Figure 6.33 for (a) all three surfaces being isothermal radiating-absorbing surfaces and (b) one surface being a nonconducting, reradiating surface.

A comparison between Figure 6.33 and Figures 6.27(a) and 6.28 indicates that the earlier considerations were merely special cases of the condition shown above. For black surfaces, $\varepsilon_1 = \varepsilon_2 = \varepsilon_3 = 1$ and $\rho_1 = \rho_2 = \rho_3 = 0$, and the circuits shown in Figure 6.33 become those considered previously.

A quantitative example, using these concepts, will now be considered.

Example 6.11

Two parallel disks, both 5 ft in diameter and 5 ft apart, are maintained uniformly at 1040°F and 540°F, respectively. Determine the net rate of heat loss from the disk at the higher temperature if

(a) the surroundings are black at 0°R.
(b) a reradiating surface extends between the two plates.

The emissivities of the two disk surfaces are 0.6 and 0.8, respectively. Figure 6.34 applies to both (a) and (b). The resistances shown have values as follows:

$$R_1 = 1/A_1F_{1-3} = 1 \left/ \frac{\pi}{4} \right. (5 \text{ ft})^2 (0.82) = 0.062$$

$$R_2 = 1/A_2F_{2-3} = 1 \left/ \frac{\pi}{4} \right. (5 \text{ ft})^2 (0.82) = 0.062$$

$$R_3 = 1/A_1F_{1-2} = 1 \left/ \frac{\pi}{4} \right. (5 \text{ ft})^2 (0.18) = 0.283$$

$$R_4 = \rho_1/A_1\varepsilon_1 = 0.4 \left/ \frac{\pi}{4} \right. (5 \text{ ft})^2 (0.6) = 0.034$$

$$R_5 = \rho_2/A_2\varepsilon_2 = 0.2 \left/ \frac{\pi}{4} \right. (5 \text{ ft})^2 (0.8) = 0.013$$

For the case of heat loss to the surroundings, Figure 6.34(a) applies. Analysis will require writing three "loop" equations for the portions of the circuit designated ①, ②, and ③. The assumed directions of current (i.e., heat) flow are shown by the arrows. The loop equations which apply are as follows:

Loop ①: $E_{b1} - E_{b3} = I_1R_4 + (I_1 + I_3)R_1$

Loop ②: $E_{b2} - E_{b3} = I_2R_5 + (I_2 - I_3)R_2$

Loop ③: $0 = I_3R_3 + (I_3 + I_1)R_1 + (I_3 - I_2)R_2$

Inserting numerical values for E_{b1}, E_{b2}, E_{b3}, and the resistances, we have

$$8690 = 0.034I_1 + 0.062(I_1 + I_3)$$

$$1714 = 0.013I_2 + 0.062(I_2 - I_3)$$

$$0 = 0.283I_3 + 0.062(I_1 + I_3) + 0.062(I_2 - I_3)$$

The solution to these three simultaneous algebraic equations yields

$$I_1 = 10.59 \times 10^4 \text{ Btu/hr}$$

$$I_2 = 0.317 \times 10^4 \text{ Btu/hr}$$

$$I_3 = -2.38 \times 10^4 \text{ Btu/hr}$$

The answer sought, the net heat loss from the higher-temperature disk, is I_1, the value being 10.59×10^4 Btu/hr.

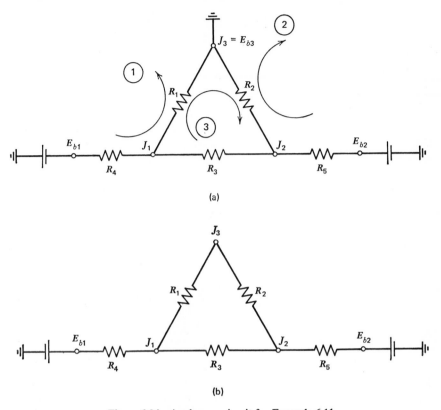

Figure 6.34 Analogous circuit for Example 6.11.

In the case where disks 1 and 2 are separated by a nonabsorbing, reradiating surface 3, the mathematics of the solution become easier. The equivalent resistance of the parallel portion of the circuit shown in Figure 6.34(b) is evaluated as

$$\frac{1}{R_{\text{equiv}}} = \frac{1}{R_3} + \frac{1}{R_1 + R_2}$$

$$= \frac{1}{0.283} + \frac{1}{0.062 + 0.062} = 11.60$$

or

$$R_{\text{equiv}} = 1/11.60 = 0.086$$

The total resistance between surfaces 1 and 2 is thus

$$\sum R = R_4 + R_{\text{equiv}} + R_5$$

$$= 0.034 + 0.086 + 0.013 = 0.133$$

The net heat loss from the higher-temperature surface is now determined as

$$q_{net} = \frac{E_{b1} - E_{b2}}{\sum R}$$

$$= \frac{8690 - 1714}{0.133} = 52,500 \ \text{Btu/hr}$$

which is approximately 50% of the loss calculated in (a).

It should be noted that in both cases there is net energy transfer to the cooler disk; however, the rate with a reradiating surface present is over 17 times as great as when energy is radiated to black surroundings at $0°R$.

Example 6.12

A cylindrical rod is used as a heater; its diameter is 2 in., its effective emissivity is 0.7, and it is maintained at $1640°F$ by electrical resistance heating. The room in

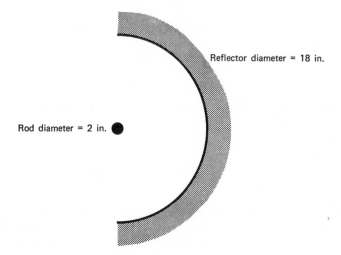

Reflector diameter = 18 in.

Rod diameter = 2 in.

Figure 6.35   Radiant heat loss from a circular heating rod.

which the heating rod is placed has walls at $60°F$ whose effective emissivity is 0.6. Determine the energy which must be supplied per foot of rod length to the rod (a) under the conditions stated and (b) if an insulated, half-circular reflector with 18 in. diameter is placed as shown relative to the rod. For the case of the rod without the reflector the equivalent circuit is

$$R_1 = \frac{\rho_R}{A_R \epsilon_R} \qquad R_2 = \frac{1}{A_R F_{R-W}} \qquad R_3 = \frac{\rho_W}{A_W \epsilon_W}$$

$E_{b \, \text{Rod}}$ $E_{b \, \text{Walls}}$

The emissive powers and thermal resistances have the following values:

$$E_{b\ \text{Rod}} = \sigma T_R^4 = 0.173\left(\frac{2100}{100}\right)^4 = 3.36 \times 10^4$$

$$E_{b\ \text{walls}} = \sigma T_w^4 = 0.173\left(\frac{520}{100}\right)^4 = 126$$

$$R_1 = \frac{\rho_R}{A_R \varepsilon_R} = \frac{0.3}{\pi\left(\frac{2}{12}\right)(0.7)} = 0.819$$

$$R_2 = \frac{1}{A_R F_{R-w}} = \frac{1}{\pi\left(\frac{2}{12}\right)(1)} = 1.91$$

$$R_3 = \frac{\rho_w}{A_w \varepsilon_w} \cong 0$$

The radiant heat loss from the bare rod now becomes

$$q = \frac{E_{bR} - E_{bw}}{\sum R} = \frac{3.36 \times 10^4 - 126}{2.729}$$

$$= 12{,}270 \frac{\text{Btu}}{\text{hr}} \text{ per foot}$$

With the reradiating reflector present, the equivalent circuit is

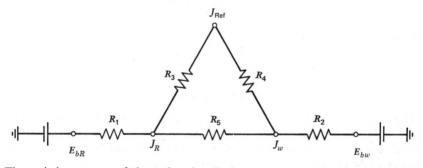

The emissive powers of the rod and walls have the same values as before. The thermal resistances are as follows:

$$R_1 = \frac{\rho_R}{A_R \varepsilon_R} = 0.819$$

$$R_2 = \frac{\rho_w}{A_w \varepsilon_w} \cong 0$$

$$R_3 = \frac{1}{A_R F_{R-\text{ref}}} = \frac{1}{\pi\left(\frac{2}{12}\right)(0.5)} = 3.82$$

$$R_4 = \frac{1}{A_\text{ref} F_{\text{ref}-w}}$$

For purposes of evaluating $F_{\text{ref}-w}$, the reciprocity theorem can be used, where the walls may be thought of as an equivalent plane covering the part of the opening to the surroundings exclusive of the heater rod. The terms which apply are

$$A_{\text{ref}}F_{\text{ref}-w} = A_w F_{w-\text{ref}}|_{\text{equiv}}$$

and, since $F_{w-\text{ref}}|_{\text{equiv}} = 1$, the product $A_{\text{ref}}F_{\text{ref}-w}$ may be replaced by $A_w|_{\text{equiv}}$, and the resistance R_4 becomes

$$R_4 = \frac{1}{A_{\text{ref}}F_{\text{ref}-w}} = \frac{1}{A_{w \text{ equiv}}} = \frac{1}{\left(\dfrac{18-2}{12}\right)} = 0.75$$

$$R_5 = \frac{1}{A_R F_{R-w}} = \frac{1}{\pi\left(\dfrac{2}{12}\right)(0.5)} = 3.82$$

The total thermal resistance is now evaluated as

$$R = R_1 + R_2 + \frac{1}{\dfrac{1}{R_5} + \dfrac{1}{R_3 + R_4}}$$

$$= 0.818 + 0 + \frac{1}{\dfrac{1}{3.82} + \dfrac{1}{4.57}} = 2.90$$

The net heat loss from the heating rod may now be determined as

$$q_{\text{net}} = \frac{E_{\text{br}} - E_{\text{bw}}}{\sum R} = \frac{3.36 \times 10^4 - 126}{2.90}$$

$$= 11,540 \text{ Btu/hr per foot}$$

The presence of the reflector in this case is seen to affect total heat loss by approximately 6%. The energy is concentrated, however, opposite the reflector in the second case so that a greater radiant flux will be experienced by one standing in the path of the radiant energy with the reflector present.

6.5-5 Radiant Energy Exchange between Gray Surfaces: Numerical Solution

The problem of energy exchange among a large number of surfaces becomes quite tedious even with some simplifying approaches such as the electrical analog.

If each of the surfaces involved can be considered isothermal with constant properties, an expression can be written for the net energy to each of the surfaces. The simultaneous solution to such a set of equations is amenable to the digital computer, using iterative techniques as considered earlier.

For any surface i in an enclosure of N total surfaces, the radiosity J_i is expressed, according to equation (6-103), as

$$J_i = \rho_i G_i + \varepsilon_i E_{bi} \qquad (6\text{-}103)$$

In this and subsequent expressions the functional notation will not be used; it is assumed that $\rho_i = \rho_i(T_i)$, $\alpha_i = \varepsilon_i$, and T_i is constant over i.

The irradiation of i is a function of the energy radiated from each surface k in the system which has a view of surface i. We may write, for G_i, the expression

$$G_i = \frac{1}{A_i} \sum_{k=1}^{N} J_k F_{ki} A_k \qquad (6\text{-}113)$$

or, using reciprocity, $A_k F_{ki} = A_i F_{ik}$, this may be rewritten in equivalent form as

$$G_i = \frac{1}{A_i} \sum_{k=1}^{N} J_k F_{ik} A_i \qquad (6\text{-}114)$$

Cancelling A_i, we have, finally, for G_i,

$$G_i = \sum_{k=1}^{N} J_k F_{ik} \qquad (6\text{-}115)$$

Equation (6-115) may now be substituted into equation (6-103) to yield, for J_i

$$J_i = \rho_i \sum_{\substack{k=1 \\ k \neq i}}^{N} J_k F_{ik} + \rho_i J_i F_{ii} + \varepsilon_i E_{bi} \qquad (6\text{-}116)$$

where the irradiation of surface i by itself, involving F_{ii}, has been separated from the rest of the summation.

We are now able to solve for J_i to get

$$J_i(1 - \rho_i F_{ii}) = \rho_i \sum_{\substack{k=1 \\ k \neq i}}^{N} J_k F_{ik} + \varepsilon_i E_{bi}$$

and, finally,

$$J_i = \frac{\rho_i}{1 - \rho_i F_{ii}} \sum_{\substack{k=1 \\ k \neq i}}^{N} J_k F_{ik} + \frac{\varepsilon_i}{1 - \rho_i F_{ii}} E_{bi} \qquad (6\text{-}117)$$

Equation (6-117) for J_i, equation (6-115) for G_i, along with equation (6-101) for $q/A|_{\text{net}}$ *received* by surface i,

$$\frac{q}{A}\bigg|_{\text{net}} = G_i - J_i \qquad (6\text{-}101)$$

comprise the set necessary for our iterative solution.

One should note before continuing what happens in the case of a reradiating surface. For a nonconducting, reradiating surface j, $q/A|_{\text{net},j} = 0$, and according to equation (6-101),

$$G_j = J_j$$

Equation (6-103) thus requires that

$$J_j = G_j = \frac{\varepsilon_j}{1 - \rho_j} E_{bj}$$

which, since $1 - \rho_j = \varepsilon_j$, reduces to

$$J_j = G_j = E_{bj} = \sigma T_j^4 \tag{6-118}$$

We will now set up the equations to be solved numerically. For convenience, the coefficients B_i and C_i are defined as

$$B_i \equiv \frac{\varepsilon_i}{1 - \rho_i F_{ii}} E_{bi} \tag{6-119}$$

$$C_i \equiv \frac{\rho_i}{1 - \rho_i F_{ii}} \tag{6-120}$$

The set of equations which pertains to a radiant heat transfer problem with N total surfaces involved includes the following:

$$J_1 = B_1 + C_1[J_2 F_{12} + J_3 F_{13} + \cdots + J_N F_{1N}]$$

$$J_2 = B_2 + C_2[J_1 F_{21} + J_3 F_{23} + \cdots + J_N F_{2N}]$$

$$J_i = B_i + C_i[J_1 F_{i1} + \cdots + J_{i-1} F_{i,i-1} + J_{i+1} F_{i,i+1} + \cdots + J_N F_{iN}]$$

$$J_N = B_N + C_N[J_1 F_{N1} + J_2 F_{N2} + \cdots + J_{N-1} F_{N,N-1}]$$

This set of N equations is solved by Gauss-Seidel iteration. The procedure is to assume values for J_i, $i \neq 1$, and solve for J_1. With the new value of J_1 and assumed values of J_i, the solution for J_2 is obtained. This procedure is repeated, using the most recent values for J_i, until some criterion for convergence is satisfied. In our case, the criterion will be for some error limit, where the error is defined as

$$\text{error} = QA/QE$$

The quantities QA and QE are defined as follows:

$$QE \equiv \sum_{i=1}^{N} |Q_i|$$

$$= \sum_{i=1}^{N} |(G_i - J_i)A_i|$$

$$QA \equiv \left| \sum_{i=1}^{N} Q_i \right|$$

$$= \left| \sum_{i=1}^{N} (G_i - J_i)A_i \right|$$

which will be zero for an exact solution.

The following simple example will illustrate the iterative solution technique.

Example 6.13

Consider a cubical enclosure with two opposite walls maintained at 1260°R and 540°R; the other side walls are reradiating surfaces. The emissivities of the high- and low-temperature walls are 0.57 and 0.67, respectively. Assuming no absorbing medium present within the enclosure, determine the net radiant energy flux at the source and sink surfaces. Determine the temperature of the side walls also.

For this very simple case, we may solve the problem by setting up a table to keep track of calculated quantities. Three zones are considered: A_1 will designate the 1260°R surface, A_2 will designate the reradiating walls, and A_3 will designate the 540°R sink. Values of J_2 and J_3 assumed initially will be 2000 Btu/hr-ft² and 1000 Btu/hr-ft², respectively. The table on p. 360 is used to keep track of the quantities as they are calculated.

After five iterations the values of J_i, G_i, and $q/A_{A|net}$ are given by

Surface	J_i	G_i	$q/A_{A|net}$
1	3224	1718	−1506
2	2056	2056	0
3	889	2395	+1506

The net heat leaving surface 1 and that absorbed at surface 3 are seen to be equal with a value of 1506 Btu/hr-ft².

The temperature of the reradiating walls is determined from equation (6-118) as

$$T_2^4 = \frac{J_2}{100} = \frac{2056}{0.173 \times 10^{-8}} = 11884 \times 10^8$$

yielding

$$T_2 = 1044°R = 584°F$$

In Example 6.13 the reradiating surfaces were lumped together into a single "zone" for analytical purposes. The single-zone approach, in which the entire reradiating zone is assumed to be at one temperature, is, obviously, a gross simplification. The temperature at the hot end of zone 2 will be near 1260°R, and it will approach 540°R at the cold end. While the single-zone approach may be sufficient in many cases, there are others for which a more accurate model is needed. The situation described in Example 6.13, for instance, could be worked with the side walls considered as two, three, or any desired number of zones, each at a constant temperature. Figure 6.36 indicates the nature of 3-, 4-, and 5-zone configurations. An iterative approach is necessary for any problem involving more than three zones.

The preceding example could have been worked, using the electrical analog technique, had we wished. The utility of the iterative technique

Table for Example 6.13

$\varepsilon_1 = 0.57 \qquad \rho_1 = 0.43$

$\varepsilon_3 = 0.67 \qquad \rho_3 = 0.33$

Nodes	F_{ik}	J_k	J_kF_{ik}	J_k	J_kF_{ik}	J_k	J_kF_{ik}	J_k	J_kF_{ik}	J_k	J_kF_{ik}
1–2	0.71	2000	1420	2113	1500	2073	1472	2062	1464	2056	1460
1–3	0.29	1000	290	904	262	895	260	891	258	889	258
$\sum J_kF_{1k}=$			1710		1762		1732		1722		1718
$\rho_1\sum J_kF_{1k}=$			735		758		745		740		739
$\varepsilon_1E_{b1}=$			2485		2485		2485		2485		2485
$J_1=$			3220		3243		3230		3225		3224
2–1	0.1775	3220	572	3245	576	3230	573	3225	572	3224	572
2–3	0.1775	1000	178	904	160	895	159	891	158	889	158
$\sum J_kF_{2k}=$			750		736		732		730		730
$J_2=\dfrac{1}{0.355}\sum J_kF_{2k}=$			2113		2073		2062		2056		2056
3–1	0.29	3220	939	3245	940	3230	937	3225	935	3224	935
3–2	0.71	2113	1500	2073	1472	2062	1464	2056	1460	2056	1460
$\sum\sum J_kF_{3k}=$			2439		2412		2401		2395		2395
$\rho_3\sum J_kF_{3k}=$			805		796		792		790		790
$\rho_3E_{b3}=$			99		99		99		99		99
$J_3=$			904		895		891		889		889

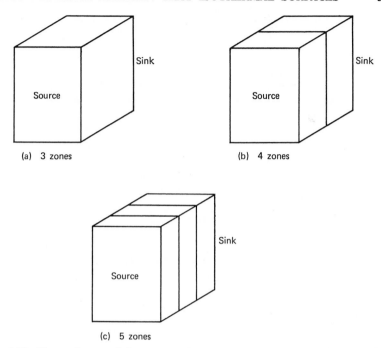

Figure 6.36 Three-, four-, and five-zone configurations for a rectangular enclosure with end walls radiating and side walls reradiating.

becomes obvious when many surfaces are involved. The digital computer becomes the only practical tool for such a problem, and the program to be used will likely follow the steps that have been outlined in this section.

Example 6.14

Consider once more the problem of the cubical enclosure with side walls reradiating and the ends at 1260°R and 540°R, respectively, as analyzed in Example 6.13. Work this problem again, using the digital computer, for the cases of

(a) the side walls assumed to be at a single constant temperature (a 3-zone problem).
(b) the side walls taken to be in two equal parts, each at a single constant temperature (a 4-zone problem).
(c) the side walls considered as three equal isothermal zones (a 5-zone problem).

The configurations pertaining to (a), (b), and (c) are depicted in Figure 6.36. Case (a) is identical to the one solved manually in the previous example. The other cases are more realistic, the actual situation being one in which the temperature of the side walls varies continuously from the 540°R surface to the 1260°R surface. This would be a situation with an infinite number of zones.

The flow diagram (Figure 6.37) and computer program listing for this problem follow.

```
      PROGRAM RAD
C        THIS PROGRAM COMPUTES THE RADIATION HEAT FLUX
C        FROM A CONFIGURATION OF N GREY SURFACES. THE
C        TEMPERATURE OF ANY RERADIATING SURFACE IS ALSO
C        CALCULATED.
      DIMENSION F(9,9),E(9),T(9),A(9),RJ(9),G(9),R(9)
      DIMENSION EB(9),Q(9)
      N=TTYIN(4HNUMB,4HER O,4HF SU,4HRFAC,4HES =)
      DELTA=TTYIN(4HCONV,4HERGE,4HNCE ,4HCRIT,3H. =)
      WRITE(61,100)
100 FORMAT(1H0,20X,12HVIEW FACTORS)
      DO 10 I=1,N
      WRITE(61,101)I,N
101 FORMAT(1H0,3H F(,I1,19H,K) — K FROM 1 TO ,I1)
      READ(41,102)(F(I,J),J=1,N)
      READ(41,105)E(I),A(I),T(I)
105 FORMAT(F6.4,F6.2,F7.2)
 10 WRITE(61,99)(F(I,J),J=1,N)
 99 FORMAT(1H0,5(F6.4,3X))
102 FORMAT(5F9.4)
      WRITE(61,103)
103 FORMAT(1H-,4HNODE,3X,10HEMISSIVITY,3X,9HAREA SQFT,
     13K,13HTEMPERATURE R)
      DO 20 I=1,N
      IF(E(I).EQ.0.0)GO TO 19
      WRITE(61,106)I,E(I),A(I),T(I)
106 FORMAT(1H0,2X,I1,6X,F6.4,7X,F6.2,7X,F7.2)
      GO TO 20
 19 WRITE(61,104)I,E(I),A(I)
104 FORMAT(1H0,2X,I1,6X,F6.4,7X,F6.2,5X,11HRERADIATING)
 20 CONTINUE
C        INITIAL ASSUMPTION FOR RADIOSITY
      DO 30 I=1,N
 30 RJ(I)=1000.0
C        CALCULATE REFLECTIVITY AND BLACK BODY EMISSIVE
C        POWER
      DO 40 I=1,N
      EB(I)=(1.73E-009)*T(I)**4.
 40 R(I)=1.0-E(I)
C        ITERATE FOR RADIOSITY
 54 L=0
 64 DO 50 I=1,N
```

```
      SUM =0.0
      DO 55 K=1,N
      IF(K.EQ.I)GO TO 55
      SUM =SUM +RJ(K)*F(I,K)
   55 CONTINUE
   50 RJ(I) =(E(I)*EB(I) +R(I)*SUM)/(1.0 −R(I)*F(I,I))
      L =L+1
      IF(L.LT.5)GO TO 64
C        CHECK FOR CONVERGENCE
      DO 80 I=1,N
      SUM1 =0.0
      DO 75 K=1,N
   75 SUM1 =SUM1 +RJ(K)*F(I,K)
   80 G(I) =SUM1
      QE =0.0
      QA =0.0
      DO 90 I=1,N
      D =G(I) −RJ(I)
      IF(E(I).EQ.0.0)GO TO 85
      QA =QA +D
      GO TO 90
   85 QE =QE +ABS(D)
   90 CONTINUE
      ERR =ABS(QA) +QE
      IF(ERR.GT.DELTA)GO TO 54
      DO 95 I=1,N
      Q(I) =G(I) −RJ(I)
      IF(E(I).NE.0.0)GO TO 95
      T(I) =((G(I)/.173)**.25)*100.0
   95 CONTINUE
      WRITE(61,108)
  103 FORMAT(//////,5H NODE,3X,12HNET HT. FLUX,3X,13HTEM-
      PERATURE R)
      DO 96 I=1,N
   96 WRITE(61,109)I,Q(I),T(I)
  109 FORMAT(1H0,1X,I1,7X,F9.2,6X,F7.2)
      END
```

The computer solution uses equations (6-117), (6-115), and (6-101) to determine J_i, G_i, and $q/A_{|net}$. Emissivity and reflectivity values were specified for the nonblack surfaces, and the view factors in each case were determined prior to starting the iterative solution.

The output for (a), (b), and (c) follow in order. Note that the results for (a) are identical to those obtained by hand in Example 6.13, except that more accuracy is obtained with the computer since more significant figures are carried.

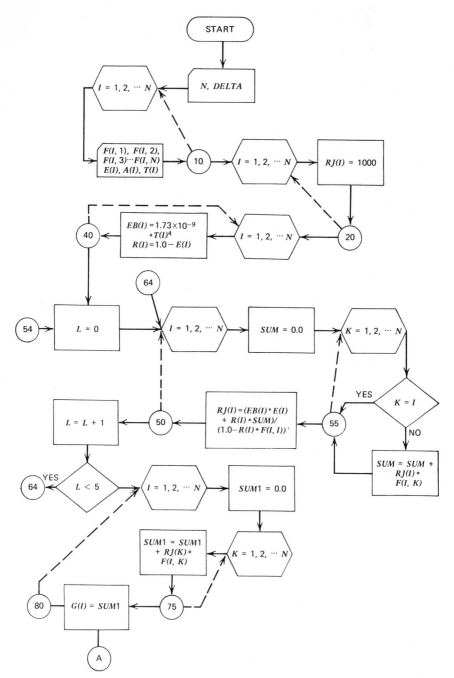

Figure 6.37 Flow chart for computer program in Example 6.14.

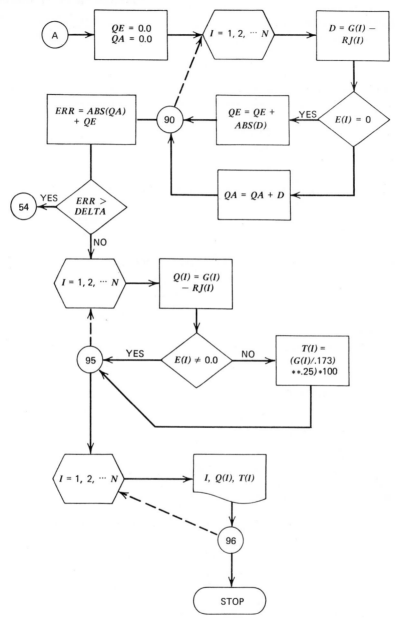

Figure 6.37 (*Continued*)

NUMBER OF SURFACES = 3
CONVERGENCE CRIT. = .005

VIEW FACTORS

F(1,K) — K FROM 1 TO 3
  0   .7100   .2900
F(2,K) — K FROM 1 TO 3
.1775   .6450   .1775
F(3,K) — K FROM 1 TO 3
.2900   .7100   0

NODE	EMISSIVITY	AREA SQFT	TEMPERATURE R
1	.5700	1.00	1260.00
2	0	4.00	RERADIATING
3	.6700	1.00	540.00

NODE	NET HT. FLUX	TEMPERATURE R
1	−1506.20	1260.00
2	.00	1044.18
3	1506.20	540.00

NUMBER OF SURFACES = 4
CONVERGENCE CRIT. = .005

VIEW FACTORS

F(1,K) — K FROM 1 TO 4
  0   .5200   .2900   .1900
F(2,K) — K FROM 1 TO 4
.2600   .4800   .0950   .1650
F(3,K) — K FROM 1 TO 4
.2900   .1900   0   .5200
F(4,K) — K FROM 1 TO 4
.0950   .1650   .2600   .4800

NODE	EMISSIVITY	AREA SQFT	TEMPERATURE R
1	.5700	1.00	1260.00
2	0	2.00	RERADIATING
3	.6700	1.00	540.00
4	0	2.00	RERADIATING

NODE	NET HT. FLUX	TEMPERATURE R
1	−1453.31	1260.00
2	.00	1079.89
3	1453.31	540.00
4	−0.00	1006.34

NUMBER OF SURFACES = 5
CONVERGENCE CRIT. = .005

VIEW FACTORS

F(1,K) — K FROM 1 TO 5
| 0 | .4600 | .2900 | .2320 | .0180 |

F(2,K) — K FROM 1 TO 5
| .3450 | .3100 | .0135 | .1710 | .1605 |

F(3,K) — K FROM 1 TO 5
| .2900 | .0180 | 0 | .2320 | .4600 |

F(4,K) — K FROM 1 TO 5
| .1740 | .1710 | .1740 | .3100 | .1710 |

F(5,K) — K FROM 1 TO 5
| .0135 | .1605 | .3450 | .1710 | .3100 |

NODE	EMISSIVITY	AREA SQFT	TEMPERATURE R
1	.5700	1.00	1260.00
2	0	1.33	RERADIATING
3	.6700	1.00	540.00
4	0	1.33	RERADIATING
5	0	1.33	RERADIATING

NODE	NET HT. FLUX	TEMPERATURE R
1	−1387.66	1260.00
2	.00	1102.50
3	1387.66	540.00
4	.00	1046.14
5	0	978.84

When the side walls are treated more realistically, as in (b) and (c), the net heat flux is seen to decrease. An exact solution for net heat flux would require that an infinite number of zones be used. Note that the consideration of the wall as three isothermal zones rather than one reduces the net heat flux by almost 8%.

6.6 RADIANT ENERGY EXCHANGE WITH ABSORBING AND RERADIATING GASES PRESENT

The consideration of radiant exchange thus far has neglected any participation of the gaseous media which may be present between surfaces. Gas molecules contain bands of energy which are associated with allowed energy levels due to rotational and vibrational motion. Some of the energy incident upon a gas molecule will be absorbed, provided the frequencies of the incident energy correspond to those of the allowed states within the molecule.

Similarly, any energy emitted by a gas will be at discrete frequencies as the energy levels change between allowed states.

It has been seen that energy emission from a solid comprises a continuous frequency spectrum. Gaseous absorption and emission, on the other hand, is restricted to bands of frequencies which are characteristic of the allowed energy levels for whatever kind of gas is present. Figure 6.38 shows the emission and absorption bands for water vapor and carbon dioxide at 2000°R.

Certain types of gaseous molecules may be considered transparent insofar as thermal radiation is concerned. These include the inert gases such as argon and neon, and those with symmetric diatomic molecules such as oxygen, nitrogen, and hydrogen. Air is seen to be essentially transparent.

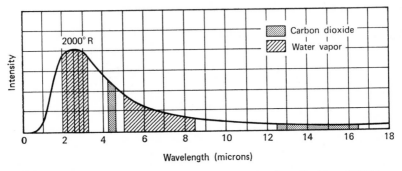

Figure 6.38   Emission bands for water vapor and carbon dioxide at 2000°R.

Gaseous molecules which are unsymmetrical, such as carbon monoxide, and polyatomic gases such as carbon dioxide, water vapor, and oxides of sulfur and nitrogen, as well as practically all organic vapors, are absorbers and emitters of thermal radiation. It is apparent that the products of combustion fall into the category of emitting-absorbing gases.

The description of gaseous emission and absorption involves temperature, composition, density, and geometry. A complete analysis of gas-surface radiation interaction is quite difficult, well beyond the scope of this text. For our purposes some idealized cases will be considered which will make a crude analysis possible; the idealizations are

1. The assumption will be made that gases are *gray;* this will allow absorption and emission characteristics to be described by a single parameter, i.e., $\alpha = \varepsilon$.
2. Gases will be assumed to be in thermal equilibrium; thus a single temperature will indicate the thermodynamic state.

6.6-1   Monochromatic Absorption and Transmission through a Gas Layer

Consider a gas layer of thickness L, with monochromatic incident radiation of intensity $I_{\lambda 0}$ passing through it. The intensity of radiation at some distance x through the gas layer will be $I_{\lambda x}$, and the difference $I_{\lambda 0} - I_{\lambda x}$ is the absorption by the gas. The amount of absorption by a gas layer with thickness dx is given by

$$dI_{\lambda x} = -k_\lambda I_{\lambda x}\, dx \tag{6-121}$$

where k_λ is the monochromatic absorption coefficient, a function of gas density (which is a function of pressure and temperature).

Separating variables and integrating between the boundaries of the gas layer, we have

$$\int_{I_{\lambda 0}}^{I_{\lambda L}} \frac{dI_{\lambda x}}{I_{\lambda x}} = -\int_0^L k_\lambda\, dx$$

which yields

$$I_{\lambda L} = I_{\lambda 0} e^{-k_\lambda L} \tag{6-122}$$

The energy absorbed by the gas layer is now given by

$$I_{\lambda 0} - I_{\lambda L} = I_{\lambda 0}(1 - e^{-k_\lambda L}) \tag{6-123}$$

The bracketed quantity is the monochromatic absorptivity of the layer. The summation over all wavelengths gives the effective absorptivity of the gas layer. One may note that a gas layer will approach black body behavior for large values of L.

6.6-2   Gray Gas Approximations for H_2O and CO_2

Hottel[11] has developed an approximate method of describing absorption and emission characteristics of gray gases which is relatively simple and sufficiently accurate for most engineering calculations. Hottel's calculations for gas emissivities for various temperatures and pressures are presented in graphical form. Figures 6.39 and 6.40 are for water vapor and carbon dioxide, respectively. These figures represent properties for a hemispherical gas mass at temperature T_G with partial pressures P_w and P_c of water vapor and carbon dioxide, respectively, exchanging energy with a black element of surface at temperature T_s located on the base of the hemisphere at its center.

The effective mean beam length for other than hemispherical enclosures may be found in Table 6.3. For those shapes not included in the table, the mean beam length may be approximated as 3.4(volume)/surface area.

[11] Hottel, op. cit.

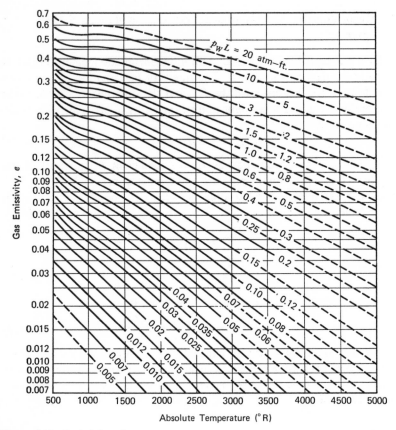

Figure 6.39 Emissivity of water vapor at one atmosphere total pressure and near zero partial pressure.

Table 6.3   Mean Beam Lengths L for Various Gas Shapes

Shape	L
Sphere	2/3 × diameter
Infinite cylinder	1 × diameter
Right circular cylinder, height = diameter	2/3 × diameter
Space between infinite parallel planes	2 × distance between planes
Cube	2/3 × side

In the figures for water vapor and carbon dioxide, the emissivity is represented as a function of absolute temperature for a range in values of the

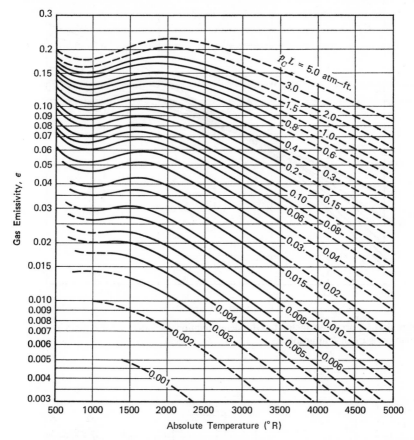

Figure 6.40 Emissivity of carbon dioxide at one atmosphere total pressure and near zero partial pressure.

product of partial pressure and mean beam length. In each case, the values are accurate if the absorbing constituent has a partial pressure near zero and the total system pressure is 1 atmosphere. A correction factor must be used if system pressure differs from 1 atmosphere. Figure 6.41 allows the appropriate correction factor to be obtained for water vapor, and Figure 6.42 applies to carbon dioxide. The emissivity at total pressure P is the product of the correction factor, obtained from one of these plots, and the emissivity value for a total pressure of 1 atmosphere.

One additional correction factor is necessary when both water vapor and carbon dioxide are present in a gaseous mixture.

As one can see from Figure 6.38, there is some overlap in the emission

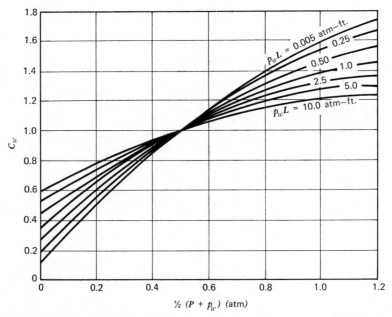

Figure 6.41 Correction factor for the determination of water vapor emissivity at a total system pressure of P atmospheres.

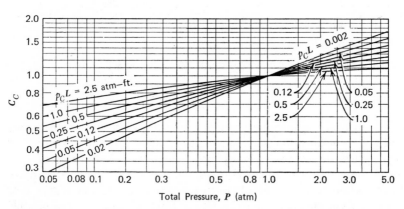

Figure 6.42 Correction factor for the determination of carbon dioxide emissivity at a total system pressure of P atmospheres.

bands of these two gases. When both are present, the total emissivity of the gas mixture is determined as

$$\varepsilon_{\text{total}} = \varepsilon_{\text{H}_2\text{O}} + \varepsilon_{\text{CO}_2} - \Delta\varepsilon \qquad (6\text{-}124)$$

where $\Delta\varepsilon$, the correction for overlap, may be obtained from the plot in Figure 6.43.

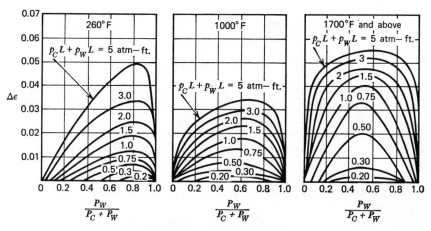

Figure 6.43 Emissivity correction for gas mixtures containing both water vapor and carbon dioxide.

6.6-3 Radiant Exchange between Absorbing Gases and a Black Surface

The rate of heat exchange between a black surface at temperature T_s and an absorbing gas at T_G involves using gaseous emissivities and/or absorptivities obtained from the plots in the previous section. The emission characteristics of the gas are embodied in the emissivity ε_G, evaluated at T_G. The absorption by the gas of radiant energy emitted by a black surface at T_s requires that the effective absorptivity α_s be evaluated at T_s. The net rate of heat transfer is the difference between emitted and absorbed energy and is expressed as

$$q_{\text{net}} = \sigma A(\varepsilon_G T_G^4 - \alpha_s T_s^4) \qquad (6\text{-}125)$$

The use of equation (6-114) and much of the preceding information is illustrated in Example 6.15.

Example 6.15

An exhaust system for a stationary internal combustion engine is to be designed. The engine will burn fuel which may be taken to be octane (C_8H_{18}) with 150% of stoichiometric air. The combustion products will pass through the exhaust system at a temperature of $2000°F$ while the walls of the exhaust pipe are at a temperature of $450°F$. Approximating the exhaust pipe as an infinite black cylinder 3 inches in diameter, estimate the amount of radiant exchange between the hot combustion gases and the pipe wall. The combustion gases may be assumed to be at a pressure of 1.2 atmospheres.

The partial pressures of H_2O and CO_2 will be determined from an analysis of the combustion reaction which is, per mole of C_8H_{18},

$$C_8H_{18} + 1.5\left(\frac{25}{2}\right)O_2 + 1.5\left(\frac{79}{21}\right)\left(\frac{25}{2}\right)N_2 = 9H_2O + 8CO_2$$

$$+ 0.5\left(\frac{25}{2}\right)O_2 + 1.5\left(\frac{79}{21}\right)\left(\frac{25}{2}\right)N_2$$

The total moles in the exhaust, per mole of fuel, include 9 moles of H_2O, 8 moles of CO_2, 6.25 moles of O_2, and 70.5 moles of N_2, making a total of 93.75 moles of gas. The mole fractions of H_2O and CO_2 are $9/93.75 = 0.096$ and $8/93.75 = 0.085$, respectively. The partial pressures, which are required for calculation, now become $1.2(0.096) = 0.115$ atmosphere for H_2O and $1.2(0.085) = 0.102$ atmosphere for CO_2.

The mean beam length L for the exhaust line, approximated as an infinite cylinder, is evaluated, using Table 6.3, to be 1/4 ft. The parameters required to use Figures 6.39 and 6.40 are

$$p_wL = 0.115(0.25) = 0.029 \text{ atm-ft}$$

$$p_cL = 0.102(0.25) = 0.026 \text{ atm-ft}$$

We now read the emissivities and absorptivities to be

for H_2O: $\varepsilon_G = 0.014$ Figure 6.39
                         $\alpha_s = 0.045$ Figure 6.39
                         $C_w = 1.2$ Figure 6.41
for CO_2: $\varepsilon_G = 0.036$ Figure 6.40
                         $\alpha_s = 0.043$
                         $C_c = 1.1$ Figure 6.42
                         $\Delta\varepsilon \simeq 0$ Figure 6.43

The total corrected emissivity and absorptivity values are

$$\varepsilon_{G \text{ total}} = 0.014(1.2) + 0.036(1.1) = 0.0564$$

$$\alpha_{s \text{ total}} = 0.045(1.2) + 0.043(1.1) = 0.1013$$

The heat flux from the gas to the pipe walls may now be obtained from appropriate substitution into equation (6-125)

$$\frac{q_{net}}{A} = \sigma(\varepsilon_G T_G^4 - \alpha_s T_s^4)$$

$$= 0.173\left[0.0564\left(\frac{2460}{100}\right)^4 - 0.1013\left(\frac{910}{100}\right)^4\right]$$

$$= 0.173[0.0564(36.6 \times 10^4) - 0.1013(0.686 \times 10^4)]$$

$$= 3450 \text{ Btu/hr per square foot}$$

In all discussion thus far, as well as in this example, no consideration has been given to exchange between a gas and a nonblack surface. The addition of multiple reflections to an already difficult situation involves complications much too involved for this treatment. The gray-gas approximation and plots apply only to exchange with black or near-black surfaces. For surfaces with emissivities less than 0.7, a more rigorous, and much more involved, analysis is necessary. With surface emissivities of 0.7 or greater, the radiant exchange given by equation (6-125) may be corrected by multiplying the result by $(\varepsilon_s + 1)/2$, ε_s being the surface emissivity. This correction, although crude, will yield satisfactory results in most cases.

6.7 THE RADIATION HEAT TRANSFER COEFFICIENT

Radiation and convection often occurs simultaneously in heat transfer situations. In these instances it is convenient to use the *radiation heat transfer coefficient*, which describes the radiation contribution in a manner analogous to the convective coefficient. The effect of such a treatment is the determination of a total or effective surface conductance defined as

$$h_{\text{total}} = h_{\text{convection}} + h_{\text{radiation}} \tag{6-126}$$

where

$$h_r = \frac{q/A|_{\text{radiation}}}{(T - T_{\text{ref}})}$$

$$= \frac{\mathscr{F}_{1-2}(E_{b1} - E_{b2})}{(T - T_{\text{ref}})} \tag{6-127}$$

In equation (6-127) the term $\mathscr{F}$ is the gray body view factor, which accounts for all geometric effects and all deviations from black body behavior of the surfaces involved.

6.8 CLOSURE

In this chapter the phenomenon of thermal radiation has been considered. Some of the complexities involved in a thorough analysis of radiant energy exchange between real surfaces both with and without an absorbing/emitting gas present have been alluded to. In all cases, the simplifications which make possible the solutions to problems of engineering interest have been presented and example problems worked.

An ideal, or black, surface has been defined and analyzed. Planck's law was presented as the relation between black body emissive power, wavelength, and absolute temperature.

The properties emissivity, absorptivity, transmissivity, and reflectivity were defined and related to direction and spectral distribution. For a surface in thermal equilibrium with its surroundings, Kirchhoff's law allowed the emissivity and absorptivity to be equated.

The assumption of *gray body behavior*, i.e., that surface emissivity and absorptivity do not vary with wavelength, was seen to afford great simplification in treating nonblack surface behavior without sacrificing appreciable accuracy.

In treating radiant exchange between surfaces, it was necessary to consider the orientation and configuration of the surfaces involved. Geometric considerations were included in the geometric view factor F; the determination of F for numerous situations consumed much of the effort of this chapter. Charts were presented for evaluating view factors for common configurations; numerous techniques were examined for relating chart values as well as evaluating view factors for additional cases.

Radiant exchange between two surfaces and between all surfaces forming an enclosure was examined. The treatment of radiant exchange by using an analogous electrical circuit was seen to be helpful. Electrical analogs were examined when all surfaces comprising an enclosure were black, when one or more surfaces were reradiating, and when one or more surfaces were gray.

The presence of an absorbing and emitting gas was seen to be a serious complication in radiation heat transfer analysis. The simple case of a gray gas was examined; more complicated analysis of gaseous behavior is well beyond the scope of this text.

Wherever possible, example problems were worked in detail. Techniques of problem formulation and solution have been presented in each applicable situation.

CHAPTER 7

HEAT TRANSFER EQUIPMENT

Not infrequently in large industrial processes is it necessary to transfer relatively large amounts of thermal energy between the system and its surroundings or between different parts of a given system. A device whose primary purpose is the transfer of heat between two fluids is called a *heat exchanger*. Three categories are normally used to classify heat exchangers:

(a) regenerators.
(b) open-type exchangers.
(c) closed-type exchangers, or recuperators.

Regenerators are exchangers in which a hot fluid, then a cold fluid, flow through the same space alternatively, with as little physical mixing as possible occurring between the two streams. The surface, which alternatively receives, then releases, thermal energy, is important in such a device. Surface material properties as well as flow and fluid properties of the fluid streams, along with system geometry, are quantities which must be known for analyzing or designing regenerators. Analysis of such a device is possible by means already discussed in Chapters 4 and 5.

Open-type heat exchangers are, as the name implies, devices wherein the entering fluid streams flow into an open chamber, and complete physical mixing of the two streams occurs. Hot and cold streams entering such an exchanger separately will leave as a single stream. Analysis of open-type exchangers involves the law of conservation of mass and the first law of thermodynamics; no rate equations are necessary for analysis or design of this type of exchanger.

Closed-type exchangers are those in which heat transfer occurs between two fluid streams which do not mix or physically contact each other. The fluid streams so involved are separated from one another by a pipe or tube wall, or by any other surface which may be involved in the heat transfer path. Heat transfer will thus occur by convection from the hotter fluid to the solid surface, by conduction through the solid, thence by convection from the solid surface to the cooler fluid. Each of these processes has been considered in detail in the preceding chapters. Our task in this chapter is to analyze those situations wherein these heat transfer processes occur in series and result in the continuous temperature change of at least one, but usually both, of the fluid streams involved.

The term *heat exchanger*, although applying to all three categories listed above, will be used in the remainder of this chapter to designate the closed-type exchanger. Our concern with these devices will be with regard to thermal analysis primarily. First law considerations, as for an open-type exchanger, may pose the limiting condition on closed-type exchangers, as we shall see directly. A complete design of a heat exchanger requires that structural, economic, and flow constraints be accounted for in addition to satisfying the basic heat exchange requirements. Such a complete design and analysis is beyond the scope of this text.

7.1 HEAT EXCHANGER CLASSIFICATION

A recuperator is classified according to its configuration. The information conveyed by such classification is the relative flow directions of the two fluid streams and the number of passes made by each fluid as it traverses the exchanger.

Relative directions of flow of the fluid streams are specified as *counterflow* or *countercurrent* flow when the fluid streams flow in opposite directions, *cocurrent* flow or *parallel* flow when the streams flow in the same direction, and *crossflow* if the fluid streams flow at right angles to one another. Illustrations of these simple single-pass configurations are shown in Figure 7.1.

It is possible to have variations on the crossflow configuration with one or the other fluid stream *mixed*. In Figure 7.1(c) the situation is one where both fluids are unmixed. If the baffles which are shown were removed in the case of one stream, it would be unseparated, or mixed. When the crossflow condition is as shown in Figure 7.1, the fluid stream at a given layer will have a temperature variation from one side to the other since each section contains an adjacent fluid stream at a different temperature. It is usually desirable to have one or both fluid streams unmixed.

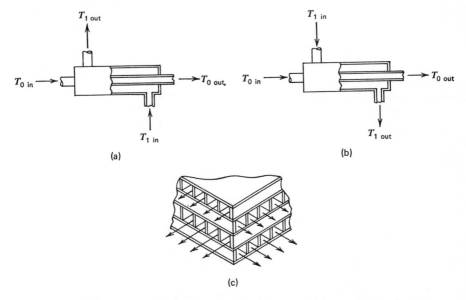

Figure 7.1   Simple single-pass heat exchanger configurations. (a) Counterflow. (b) Parallel flow. (c) Crossflow.

Design of heat exchangers generally begins with the determination of the required area to transfer the necessary heat between fluid streams which enter at specified temperatures with certain flow rates. Other quantities of interest are the exit temperatures of the two streams. Various configurations of heat exchangers have been developed to incorporate the required heat transfer area into as small a volume as possible. Some "compact" heat exchanger configurations are shown in Figure 7.2. An excellent reference on heat exchanger design for compact configuration is the book by Kays and London.[1]

A common type of compact exchanger configuration is that designated *shell-and-tube*, in which a large chamber—the shell—houses many tubes which may make one, two, or many passes within the shell. A schematic picture of a shell-and-tube exchanger is shown in Figure 7.3. In this diagram the shell-side fluid makes one pass and the tube-side fluid two passes. Note that the shell-side fluid is forced to flow back and forth across the tubes by the baffles shown. Without these baffles present, the shell-side fluid has a tendency to "channel" or proceed from inlet to exit by a relatively direct route. Were channeling allowed to occur, the shell-side fluid would remain

[1] W. M. Kays and A. L. London, *Compact Heat Exchangers*, 2nd ed. (New York: McGraw-Hill, 1964).

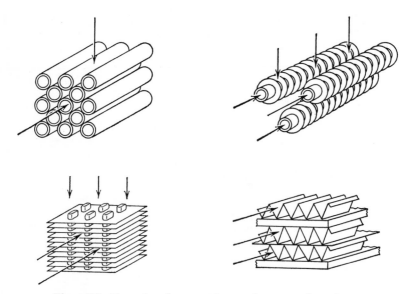

Figure 7.2   Examples of compact heat exchanger configurations.

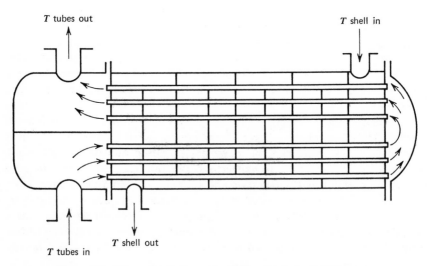

T tubes out

T shell in

T tubes in

T shell out

Figure 7.3   Schematic diagram of a shell-and-tube heat exchanger.

stagnant in certain locations and these stagnant or "dead" regions would result in less than optimum performance of the heat exchanger.

Shell-and-tube heat exchangers can be enormous, measuring many feet in diameter and length, and may include thousands of tubes. Several tube passes may be incorporated into a single shell; rarely are more than two shell passes used.

The determination as to which fluid should be used in the tubes and which in the shell should be mentioned briefly. One practical aspect is that of cleaning. If one fluid is very corrosive or causes a film or scum to build up on the solid surface, it is usually introduced on the tube side because the insides of the tubes can be cleaned with relative ease compared to the outsides and the rest of the shell. The other consideration of importance in this regard is that of pressure drop and/or pumping requirements for the two fluids. The larger pressure drop is usually encountered on the tube side; thus this will influence the selection of fluids for such an exchanger. Other considerations, specific to a certain application, may control the selection of tube-side and shell-side fluids.

The analysis of compact exchangers and shell-and-tube exchangers is relatively involved compared to the single-pass case. Each of these complex arrangements is, as a matter of fact, merely a combination of several single-pass effects. It is thus appropriate that we initially consider the single-pass case for beginning our heat exchanger analysis.

7.2 SINGLE-PASS HEAT EXCHANGER ANALYSIS

In Figure 7.4 temperature profiles are represented for the fluid streams in the four basic single-pass double-pipe configurations. Note that in cases (c) and (d), only one of the fluid streams experiences a temperature change. This is due, of course, to the fact that the other fluid undergoes a phase change as it releases or receives heat, such a phase change occurring at constant temperature.

In each of the cases shown, the temperature variation of the two fluid streams is intuitively correct. Note that the temperature of the hotter fluid T_h decreases as it releases heat except in case (c), where the hot fluid is condensing. Similarly, the cool fluid temperature T_c rises as the stream passes through the exchanger except in case (d), where the heat received causes the cool fluid to evaporate, or change from liquid to vapor. Directions of flow are shown for each fluid stream except in cases (c) and (d), where the direction of the condensing and evaporating fluid streams are not involved; the profiles will be the same regardless of whether the streams flow in parallel or counter flow.

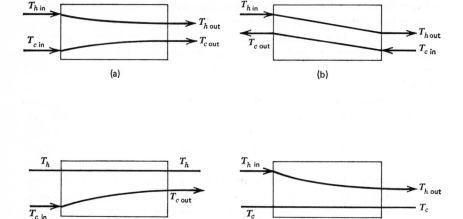

(a)   (b)

(c)   (d)

Figure 7.4   Temperature profiles for four single-pass, double-pipe heat exchanger configurations. (a) Parallel flow. (b) Counterflow. (c) Condenser. (d) Evaporator.

Flow direction in condensers and evaporators will be significant if the phase change is complete within the exchanger. Figure 7.5 depicts such a case for fluid streams flowing in opposite directions when the hot fluid condenses, then subcools from its saturated condition, as heat transfer continues to occur within the double-pipe exchanger. Such a case can be treated as if the exchanger were in two parts. One part would be the condenser, the other a counterflow exchanger with the hot fluid entering as saturated liquid. The superposition of these two parts to yield a complete temperature profile for such a condition is shown.

Referring to Figure 7.4, it is apparent that a basic difference exists in the heat transfer potential of the parallel flow and the counterflow configurations. The second law of thermodynamics requires that, at any location along a double-pipe heat exchanger, there can be no change in role between the streams entering as "hot" and "cold" fluids. If an exchanger were extended to provide infinite heat transfer area in the limit, the two streams in a parallel

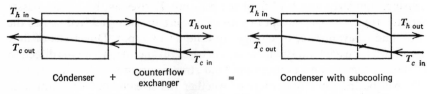

Condenser   +   Counterflow   =   Condenser with subcooling
                exchanger

Figure 7.5   Temperature profiles in a counterflow condenser with subcooling.

flow exchanger would leave at the same temperature. It is a simple exercise to show that this would be the same temperature as the single leaving fluid stream in the case of an open-type heat exchanger.

In the case of a counterflow exchanger with infinite area, on the other hand, one of the fluid streams would leave at the *entering* temperature of the other. We shall consider these two limits in greater detail presently. For the moment it is sufficient to observe that the heat transfer potential for a given pair of fluids in a given heat exchanger is greater for counterflow than for parallel flow. The single-pass counterflow configuration is, thus, the one to which we shall direct our attention for a detailed analysis.

7.2-1 Analysis of the Single-Pass, Double-Pipe Counterflow Heat Exchanger

The discussion to follow will refer to the terminology and case shown in Figure 7.6. The sketch is similar to Figure 7.4(b), except that more detail

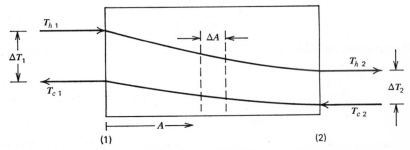

Figure 7.6 Temperature profiles in a double-pipe counterflow heat exchanger.

is shown. The hot fluid enters at (1), the left side of the exchanger shown, and the cold fluid enters at (2). The abscissa is A, heat exchanger area, which is directly related to the length of the double-pipe configuration being considered.

Two basic tools are necessary for the analysis of this case, these being the first law of thermodynamics and the applicable rate equation expressed in terms of an overall heat transfer coefficient, equation (1-15).

A first law consideration of each fluid with mass flow rate $\overset{\circ}{m}$, for an incremental area ΔA of the exchanger yields

$$\Delta q = (\overset{\circ}{m} c_p)_c \, \Delta T_c$$

and

$$\Delta q = (\overset{\circ}{m} c_p)_h \, \Delta T_h$$

As the incremental area approaches differential size, we may write

$$dq = (\mathring{m}c_p)_c \, dT_c = C_c \, dT_c \tag{7-1}$$

$$dq = (\mathring{m}c_p)_h \, dT_h = C_h \, dT_h \tag{7-2}$$

where the *capacity rate C* is introduced and used in place of the more cumbersome product $\mathring{m}c_p$.

Quite obviously, the heat transfer from the hot fluid and the heat transfer to the cool fluid will be equal as both streams traverse the exchanger. Thus equations (7-1) and (7-2) may be integrated from one end of the heat exchanger to the other, yielding

$$q = \int_1^2 dq_c = C_c \int_1^2 dT_c = C_c(T_{c2} - T_{c1}) \tag{7-3}$$

$$q = \int_1^2 dq_h = C_h \int_1^2 dT_h = C_h(T_{h2} - T_{h1}) \tag{7-4}$$

Equating the results, we obtain, for the ratio of capacity rates,

$$\frac{C_h}{C_c} = \frac{T_{c2} - T_{c1}}{T_{h2} - T_{h1}} \tag{7-5}$$

The development thus far involved the application of the first law of thermodynamics only.

Another expression for dq is achieved, using rate equations this time, by writing equation (1-15) for conditions at dA; the expression which is obtained is

$$dq = U \, dA(T_h - T_c) \tag{7-6}$$

The temperature difference $T_h - T_c$ will now be written as ΔT; it follows that

$$d(\Delta T) = dT_h - dT_c \tag{7-7}$$

Equations (7-1) and (7-2) will now be used to substitute for dT_h and dT_c. The result is

$$d(\Delta T) = dq\left(\frac{1}{C_h} - \frac{1}{C_c}\right) = \frac{dq}{C_h}\left(1 - \frac{C_h}{C_c}\right) \tag{7-8}$$

The ratio C_h/C_c, from equation (7-5), will now be substituted and, with some work, we obtain

$$
\begin{aligned}
d(\Delta T) &= \frac{dq}{C_h}\left(1 - \frac{T_{c2} - T_{c1}}{T_{h2} - T_{h1}}\right) \\
&= \frac{dq}{C_h}\left(\frac{T_{h2} - T_{h1} - T_{c2} + T_{c1}}{T_{h2} - T_{h1}}\right) \\
&= \frac{dq}{C_h}\frac{\Delta T_2 - \Delta T_1}{T_{h2} - T_{h1}}
\end{aligned}
\tag{7-9}
$$

We now utilize equations (7-4), (7-6), and (7-9); combining these expressions, we have

$$d(\Delta T) = \frac{U \, dA \, \Delta T}{q} (\Delta T_2 - \Delta T_1)$$

from which a separation of variables will yield

$$\frac{d(\Delta T)}{\Delta T} = \frac{\Delta T_2 - \Delta T_1}{q} U \, dA$$

Integrating this expression between limits at either end of the exchanger, we obtain

$$\int_{\Delta T_1}^{\Delta T_2} \frac{d \, \Delta T}{\Delta T} = \frac{\Delta T_2 - \Delta T_1}{q} \int_1^2 U \, dA$$

or

$$\ln \frac{\Delta T_2}{\Delta T_1} = \frac{\Delta T_2 - \Delta T_1}{q} \int_1^2 U \, dA \qquad (7\text{-}10)$$

Equation (7-10) is commonly written in the form

$$q = \frac{\Delta T_2 - \Delta T_1}{\ln (\Delta T_2/\Delta T_1)} \int_1^2 U \, dA \qquad (7\text{-}11)$$

which becomes, for constant U,

$$q = UA \frac{\Delta T_2 - \Delta T_1}{\ln (\Delta T_2/\Delta T_1)} \qquad (7\text{-}12)$$

The temperature driving force on the right-hand side of equation (7-12) is the *logarithmic mean temperature difference*,

$$\Delta T_{\mathrm{lm}} \equiv \frac{\Delta T_2 - \Delta T_1}{\ln (\Delta T_2/\Delta T_1)} \qquad (7\text{-}13)$$

thus equations (7-11) and (7-12) are frequently written

$$q = \left(\int_1^2 U \, dA \right) \Delta T_{\mathrm{lm}} \qquad (7\text{-}14)$$

and

$$q = UA \, \Delta T_{\mathrm{lm}} \qquad (7\text{-}15)$$

The results of the foregoing development, equations (7-14) and (7-15), were achieved assuming counterflow. The identical expressions would have resulted—and are, thus, equally valid—for any of the single-pass configurations depicted in Figure 7.4.

Special mention should be made of when one may use equation (7-15) rather than equation (7-14), i.e., when the overall heat transfer coefficient is properly considered constant. The work in Chapter 5 would indicate that, in general, U is not constant when fluids experience any significant change in temperature. Usually, calculations based on a value of U taken midway between the ends of an exchanger are sufficiently accurate. Should the temperature change be extreme, particularly for those fluids whose viscosities are strong functions of temperature, then a finite-difference solution of equation (7-14) would be appropriate. For such conditions, the exchanger would be divided into n increments and the total heat transfer expressed as

$$q = \sum_{i=1}^{n} \Delta q_i = \sum_{i=1}^{n} U_i \, \Delta A_i \, \Delta T_{\mathrm{lm}} \qquad (7\text{-}16)$$

The tedium of evaluating U_i several times suggests that a solution of the sort indicated by equation (7-17) would be a reasonably lengthy task.

A special case arises in evaluating ΔT_{lm} when ΔT has the same value at each end of a counterflow heat exchanger. In such a case,

$$\Delta T_{\mathrm{lm}} = \frac{\Delta T - \Delta T}{\ln (\Delta T / \Delta T)} = \frac{0}{0}$$

which is indeterminant. This case may be handled simply by applying L'Hospital's rule as follows:

$$\lim_{\Delta T_2 \to \Delta T_1} \frac{\Delta T_2 - \Delta T_1}{\ln (\Delta T_2 / \Delta T_1)} = \lim_{(\Delta T_2 / \Delta T_1) \to 1} \frac{\Delta T_1 \left\{ \left[\dfrac{\Delta T_2}{\Delta T_1} \right] - 1 \right\}}{\ln (\Delta T_2 / \Delta T_1)}$$

$$= \lim_{F \to 1} \frac{\Delta T(F - 1)}{\ln F}$$

where the ratio $\Delta T_2 / \Delta T_1$ is designated F. Differentiating numerator and denominator with respect to F, and taking the limit, we have

$$\lim_{F \to 1} \frac{\Delta T}{1/F} = \Delta T$$

When $\Delta T_1 = \Delta T_2$, equation (7-15) becomes simply

$$q = UA \, \Delta T \qquad (7\text{-}17)$$

When ΔT_1 and ΔT_2 have nearly equal values, equation (7-17) is sufficiently accurate. The rule of thumb in this case is that when $\Delta T_{\mathrm{max}} / \Delta T_{\mathrm{min}} <$ 1.5, a simple arithmetic mean value for ΔT will yield results which are accurate to within 1 %.

The use of single-pass, double-pipe results for evaluating heat exchanger performance or design criteria is illustrated in Example 7.1.

Example 7.1

Benzene is obtained from a fractionating column as saturated vapor at 176°F. Determine the heat transfer area required to condense and subcool 8000 lb_m/hr of benzene to 115°F if the coolant is water with a flow of 40,000 lb_m/hr available at 55°F. Compare the areas required for (a) counterflow and (b) parallel flow configurations. An overall heat transfer coefficient of 200 Btu/hr-ft²-°F may be used for this case.

Determining the total heat to be transferred, we have, for the condensing section,

$$q = 8000 \frac{lb_m}{hr} \left(169.6 \frac{Btu}{lb_m}\right) = 1.36 \times 10^6 \frac{Btu}{hr}$$

and in the subcooling section,

$$q = 8000 \frac{lb_m}{hr} (176 - 115)°F \left(0.42 \frac{Btu}{lb_m\text{-}°F}\right)$$

$$= 205,000 \text{ Btu/hr}$$

Analysis of the counterflow and parallel cases will refer to the sketches in Figure 7.7. For the counterflow case, in the subcooling section,

$$q = 205,000 \frac{Btu}{hr} = C_w(T_{wc} - 55°F)$$

$$T_{wc} = 55°F + \frac{205,000 \text{ Btu/hr}}{40,000 \text{ lb}_m/hr \left(1 \frac{Btu}{lb_m\text{-}°F}\right)}$$

$$= 55°F + 5.12°F = 60.1°F$$

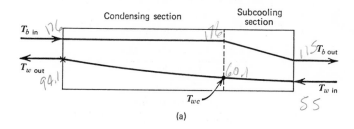
(a)

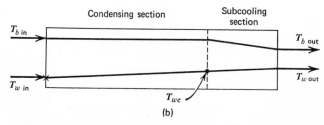
(b)

Figure 7.7 Temperature profiles for condensing benzene with subcooling—Example 7.1.

In the subcooling section for (a), then,

$$\Delta T_{\mathrm{lm}} = \frac{(176 - 60.1) - (115 - 55)}{\ln \dfrac{176 - 60.1}{115 - 55}}$$

$$= \frac{115.9 - 60}{\ln 115.9/60} = \frac{55.9}{0.658} = 84.9$$

$$A = \frac{q}{U\,\Delta T_{\mathrm{lm}}} = \frac{205{,}000 \text{ Btu/hr}}{(200 \text{ Btu/hr ft}^2\text{-}^\circ\text{F})(84.9^\circ\text{F})} = 12.1 \text{ ft}^2$$

In the condensing section,

$$q = 1.36 \times 10^6\,\frac{\text{Btu}}{\text{hr}} = C_w(T_{w\text{ out}} - 60.1^\circ\text{F})$$

$$T_{w\text{ out}} = 60.1^\circ\text{F} + \frac{1.36 \times 10^6 \text{ Btu/hr}}{(40{,}000 \text{ Btu/hr-}^\circ\text{F})} = 94.1^\circ\text{F}$$

$$\Delta T_{\mathrm{lm}} = \frac{(176 - 94.1) - (176 - 60.1)}{\ln \dfrac{176 - 94.1}{176 - 60.1}}$$

$$A = \frac{1.36 \times 10^6 \text{ Btu/hr}}{\left(200\,\dfrac{\text{Btu}}{\text{hr}}\text{ ft}^2\text{-}^\circ\text{F}\right)(97.9^\circ\text{F})} = 69.5 \text{ ft}^2$$

The required area for the counterflow case is the sum of the two parts, or 81.6 ft². For the parallel flow case, in the condensing section,

$$q = 1.36 \times 10^6 \text{ Btu/hr} = C_w(T_{wc} - 55^\circ\text{F})$$

$$T_{wc} = 55^\circ\text{F} + \frac{1.36 \times 10^6 \text{ Btu/hr}}{40{,}000 \text{ Btu/hr-}^\circ\text{F}} = 89^\circ\text{F}$$

$$\Delta T_{\mathrm{lm}} = \frac{(176 - 55) - (176 - 89)}{\ln \dfrac{176 - 55}{176 - 89}} = 103.1^\circ\text{F}$$

$$A = \frac{1.36 \times 10^6 \text{ Btu/hr}}{(200 \text{ Btu/hr ft}^2\text{-}^\circ\text{F})(103.1^\circ\text{F})} = 66.0 \text{ ft}^2$$

In the subcooling section,

$$q = 205{,}000 \text{ Btu/hr} = C_w(T_{w\text{ out}} - 89^\circ\text{F})$$

$$T_{w\text{ out}} = 89^\circ\text{F} + \frac{205{,}000 \text{ Btu/hr}}{40{,}000 \text{ Btu/hr-}^\circ\text{F}} = 94.1^\circ\text{F}$$

$$\Delta T_{\mathrm{lm}} = \frac{(176 - 89) - (115 - 94.1)}{\ln \dfrac{176 - 89}{115 - 94.1}} = 46.3^\circ\text{F}$$

$$A = \frac{205{,}000 \text{ Btu/hr}}{(200 \text{ Btu/hr-ft}^2\text{-}^\circ\text{F})(46.3^\circ\text{F})} = 22.1 \text{ ft}^2$$

The total area for the parallel flow case is approximately 88.1 ft².

The more efficient nature of counterflow as opposed to parallel flow is evident from these results. In this one case, the parallel flow configuration requires 8% more area.

7.3 SHELL-AND-TUBE AND CROSSFLOW HEAT EXCHANGER ANALYSIS

When a large amount of heat must be transferred, it is frequently undesirable or impossible to devote the space which would be necessary if a single-pass double-pipe exchanger were to be used. Space considerations generally dictate the use of more compact heat exchanger configurations such as those depicted in Figures 7.2 and 7.3.

These more complex configurations are much more difficult to treat analytically than the single-pass cases. In the case of shell-and-tube and counterflow exchangers, equation (7-15) may be used along with a correction factor. The correction factors are available in chart form in Figures 7.8 and 7.9. These charts are taken from the work of Bowman, Mueller, and Nagel;[2] others are available from the Tubular Exchanger Manufacturers Association.[3] Figure 7.8 presents correction factors F for three shell-and-tube configurations, and Figure 7.9 gives the correction factors for three crossflow arrangements.

The correction factor F is shown in each case as a function of two parameters, Y and Z, which are defined as shown with each figure.

The correction factor F, found from the appropriate plot, is used to modify equation (7-15) in the form

$$q = UAF \, \Delta T_{lm} \qquad (7\text{-}18)$$

with ΔT_{lm} *always* determined on the basis of counterflow.

The use of equation (7-18) and the correction factor F is illustrated in Example 7.2.

Example 7.2

Benzene, available as saturated liquid at 170°F, is to be cooled to 115°F for transporting. The benzene is supplied at a rate of 8000 lb_m/hr; the cooling water is available at 55°F and a flow of 5,000 lb_m/hr. Determine the required heat exchanger area for the following configurations!
(a) single-pass, counterflow.
(b) shell-and-tube, with water making one shell pass and four tube passes.
(c) crossflow single-pass, with the water mixed and benzene unmixed.

[2] R. A. Bowman, A. C. Mueller, and W. M. Nagle, *A.S.M.E. Transactions* **62** (1940): 283.
[3] Tubular Exchanger Manufacturers Association, *TEMA Standards*, 3rd ed. (New York, 1952).

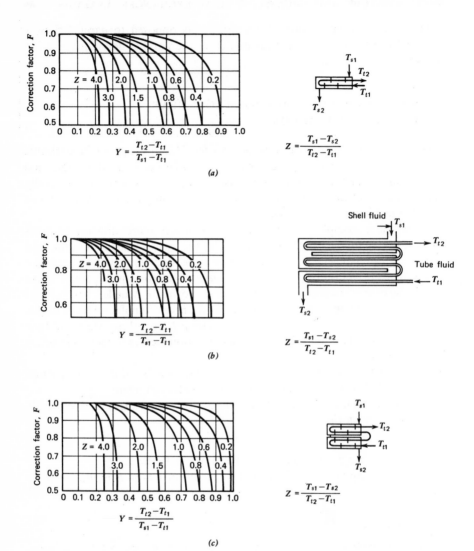

Figure 7.8 Correction factors for three shell-and-tube arrangements. (a) One shell pass and two, four, or any multiple of two tube passes. (b) One shell pass and three, or multiples of three tube passes. (c) Two shell passes and four, eight, or any multiple of four tube passes. [From R. A. Bowman, A. C. Mueller, and W. M. Nagle, *Trans. A.S.M.E.* 62 (1940): 283. By permission of the publishers.]

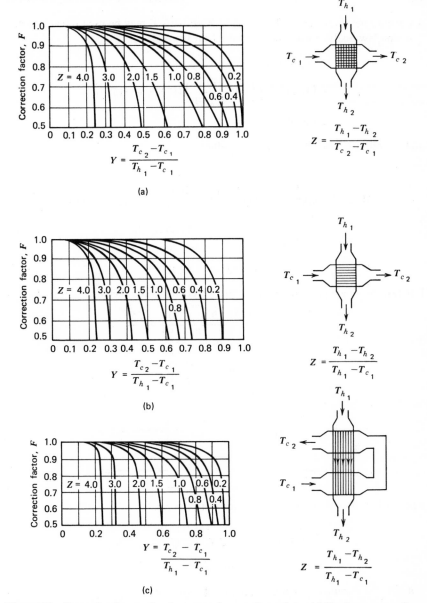

Figure 7.9 Correction factors for three crossflow configurations. (a) Single-pass crossflow, both fluids unmixed. (b) Single-pass crossflow, one fluid unmixed. (c) Two-pass crossflow, tube passes mixed; flows over first and second passes in series. [From R. A. Bowman, A. C. Mueller, and W. M. Nagle, *Trans. A.S.M.E.* 62 (1940): 283. By permission of the publishers.]

An overall heat transfer coefficient of 55 Btu/hr-ft²-°F may be used in each case. A first law analysis will yield the total heat transfer

$$q = C_{\text{Benz}} \, \Delta T_{\text{Benz}}$$

$$= \left(8000 \, \frac{\text{lb}_m}{\text{hr}}\right) \left(0.42 \, \frac{\text{Btu}}{\text{lb}_m\text{-}°\text{F}}\right) (170°\text{F} - 115°\text{F})$$

$$= 185,000 \text{ Btu/hr}$$

and the required exciting water temperature

$$q = C_{\text{H}_2\text{O}} \, \Delta T_{\text{H}_2\text{O}}$$

$$185,000 \text{ Btu/hr} = \left(5,000 \, \frac{\text{lb}_m}{\text{hr}}\right) \left(1 \, \frac{\text{Btu}}{\text{lb}_m\text{-}°\text{F}}\right) (T_{\text{H}_2\text{O exit}} - 55°\text{F})$$

$$T_{\text{H}_2\text{O exit}} = 55 + \frac{185,000 \text{ Btu/hr}}{5,000 \text{ Btu/hr-}°\text{F}}$$

$$= 55 + 37 = 92°\text{F}$$

For (a), the single-pass counterflow case (referring to the sketch),

$$\Delta T_{\text{lm}} = \frac{(170 - 92) - (115 - 55)}{\ln \dfrac{170 - 92}{115 - 55}}$$

$$= \frac{18}{\ln 78/60} = 68.6°\text{F}$$

Application of equation (7-15) will thus yield the result

$$A = \frac{q}{U \, \Delta T_{\text{lm}}} = \frac{185,000 \text{ Btu/hr}}{(55 \text{ Btu/hr ft}^2\text{-}°\text{F})(68.6°\text{F})}$$

$$= 49.0 \text{ ft}^2$$

For (b), the parameters Y and Z are first determined as

$$Y = \frac{115 - 170}{55 - 170} = \frac{55}{115} = 0.478$$

$$Z = \frac{55 - 92}{115 - 170} = \frac{37}{55} = 0.673$$

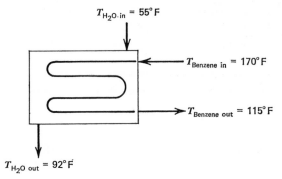

The correction factor is now read from Figure 7.8(a) as 0.93. The appropriate values may now be used with equation (7-18) to yield

$$A = \frac{q}{UF \, \Delta T_{lm}} = \frac{185,000 \text{ Btu/hr}}{(55 \text{ Btu/hr ft}^2\text{-}^\circ\text{F})(0.93)} \, (68.6^\circ\text{F})$$
$$= 52.7 \text{ ft}^2$$

For (c), Figure 7.9(b) must be used. The parameters Y and Z are

$$Y = \frac{115 - 170}{55 - 170} = \frac{55}{115} = 0.478$$

$$Z = \frac{55 - 92}{115 - 170} = \frac{37}{55} = 0.673$$

and, from the plot, $F = 0.93$. The area is determined from equation (7-18) to be

$$A = \frac{q}{UF \, \Delta T_{lm}} = \frac{185,000 \text{ Btu/hr}}{(55 \text{ Btu/hr ft}^2\text{-}^\circ\text{F})(0.93)} \, (68.6^\circ\text{F})$$
$$= 52.7^\circ\text{F}$$

In the preceding example it is seen that both of the heat exchanger configurations designated "compact" require more heat transfer area. Thus the effect of accomplishing the same amount of heat transfer in a smaller volume or a more convenient shape is accompanied by a decrease in the *effectiveness* of the heat transfer surface. This term will be discussed in detail directly. The fact remains, however, that long runs of a double-pipe, single-pass arrangement are usually undesirable, and a more convenient "compact" configuration is used even though more area is required to transfer the same amount of heat.

Each of the preceding examples has illustrated the use of the logarithmic mean temperature difference to determine heat exchanger area. The overall heat transfer coefficient was given in each case. A more complete and much more time-consuming problem is encountered when one must determine U according to the procedures presented in Chapter 5 and use equation (7-11) to determine the required area. Certain of the problems at the end of this chapter will require this more extensive solution.

For a more complete discussion of the subject of compact heat exchangers, the excellent book by Kays and London[4] is recommended.

7.4 THE NUMBER-OF-TRANSFER-UNITS (NTU) METHOD OF HEAT EXCHANGER DESIGN AND ANALYSIS

The concept of a *heat exchanger effectiveness* was first proposed by Nusselt[5] in 1930. The effectiveness $\mathscr{E}$ is defined as the ratio of actual heat transfer accomplished in a heat exchanger to the maximum possible—if infinite heat transfer area were available.

Referring to Figure 7.10, which depicts typical temperature profiles for the counterflow and parallel flow single-pass configurations, it is apparent that, in general, one fluid experiences a greater temperature change than the other. The relative temperature change of the two fluids is inversely related to their capacity rates, the one with a smaller value of C experiencing the greater change in temperature.

The larger capacity rate is designated C_{max} and the smaller C_{min}. In the case of counterflow, it is apparent that, as the heat exchanger area is increased, the outlet temperature of the minimum fluid (the one having C_{min}) will approach the inlet temperature of the maximum fluid in the limit as the area approaches infinity: $T_{min\ out} \rightarrow T_{max\ in}$. In the parallel flow case, an infinite area will mean simply that the temperature of both fluids, in such a case, would be that which would be achieved if both were allowed to mix freely in an open-type exchanger.

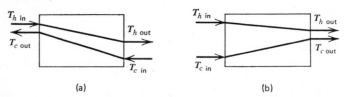

(a)   (b)

Figure 7.10   Typical temperature profiles for counterflow and parallel flow heat exchangers. (a) Counterflow, $C_h > C_c$. (b) Parallel flow, $C_h > C_c$.

[4] Kays and London, op. cit.
[5] W. Nusselt, *Tech. Mechanik and Thermodynamik* **12** (1930).

Referring once more to the counterflow case, we see that for $C_c = C_{min}$, as shown in Figure 7.10(a), and as the heat transfer area approaches infinity, $T_{c\ out} \rightarrow T_{h\ in}$. The effectiveness in this case becomes

$$\mathscr{E} = \frac{C_h(T_{h\,in} - T_{h\,out})}{C_c(T_{c\,out} - T_{c\,in})|_{max}} = \frac{C_{max}(T_{h\,in} - T_{h\,out})}{C_{min}(T_{h\,in} - T_{c\,in})} \qquad (7\text{-}19)$$

If the hot fluid is the minimum fluid, then as the area approaches infinity, $T_{h\ out} \rightarrow T_{c\ in}$, and the effectiveness may be written

$$\mathscr{E} = \frac{C_c(T_{c\,out} - T_{c\,in})}{C_h(T_{h\,in} - T_{c\,in})|_{max}} = \frac{C_{max}(T_{c\,out} - T_{c\,in})}{C_{min}(T_{h\,in} - T_{c\,in})} \qquad (7\text{-}20)$$

In each of these two expressions the numerator represents actual heat transfer, the denominator represents maximum heat transfer, and the denominator in each case is the same!

Rewriting these equations, we have an additional expression for heat transfer

$$q = \mathscr{E} C_{min}(T_{h\ in} - T_{c\ in}) \qquad (7\text{-}21)$$

which may be used in analyzing and designing heat exchangers. Equation (7-21) differs from the other design expressions considered in that the driving force is the difference between the *inlet* temperatures of the fluid streams, and it alone contains the effectiveness. We will now address ourselves to the task of evaluating the heat exchanger effectiveness $\mathscr{E}$.

7.4-1 Heat Exchanger Effectiveness

The analysis to follow is based upon equation (7-21). Counterflow operation of a single-pass exchanger is considered; subscripts used refer to the conditions shown in Figure 7.6. The cold fluid is assumed to be the minimum fluid. With these conditions, equations (7-3) and (7-15) are written as

$$q = C_c(T_{c1} - T_{c2}) = UA \frac{(T_{h1} - T_{c1}) - (T_{h2} - T_{c2})}{\ln\left[(T_{h1} - T_{c1})/(T_{h2} - T_{c2})\right]} \qquad (7\text{-}22)$$

The entering temperature of the hot fluid T_{h1} may be expressed in terms of $\mathscr{E}$, using equation (7-21), as

$$T_{h1} = T_{c2} + q/\mathscr{E} C_{min}$$

$$= T_{c2} + \frac{T_{c1} - T_{c2}}{\mathscr{E}} \qquad (7\text{-}23)$$

The temperature difference at (1) between hot and cold fluid streams thus becomes

$$T_{h1} - T_{c1} = T_{c2} - T_{c1} + \frac{T_{c1} - T_{c2}}{\mathcal{E}}$$

$$= \left(\frac{1}{\mathcal{E}} - 1\right)(T_{c1} - T_{c2}) \qquad (7\text{-}24)$$

The other temperature difference $T_{h2} - T_{c2}$ needed for substitution into equation (7-22) is obtained by solving equation (7-5) for T_{h2}, yielding

$$T_{h2} = T_{h1} - \frac{C_{\min}}{C_{\max}}(T_{c1} - T_{c2})$$

thus

$$T_{h2} - T_{c2} = T_{h1} - T_{c2} - \frac{C_{\min}}{C_{\max}}(T_{c1} - T_{c2}) \qquad (7\text{-}25)$$

Using equation (7-23), we rearrange this expression as

$$T_{h2} - T_{c2} = \frac{T_{c1} - T_{c2}}{\mathcal{E}} - \frac{C_{\min}}{C_{\max}}(T_{c1} - T_{c2})$$

or

$$T_{h2} - T_{c2} = \left(\frac{1}{\mathcal{E}} - \frac{C_{\min}}{C_{\max}}\right)(T_{c1} - T_{c2}) \qquad (7\text{-}26)$$

Equations (7-24) and (7-26) are now substituted into equation (7-22). With some rearrangement we obtain

$$\ln \frac{\dfrac{1}{\mathcal{E}} - \dfrac{C_{\min}}{C_{\max}}}{\dfrac{1}{\mathcal{E}} - 1} = \frac{UA}{C_{\min}}\left(1 - \frac{C_{\min}}{C_{\max}}\right)$$

By taking the antilog of both sides of this expression, we evaluate $\mathcal{E}$ as

$$\mathcal{E} = \frac{1 - \exp\left[-\dfrac{UA}{C_{\min}}\left(1 - \dfrac{C_{\min}}{C_{\max}}\right)\right]}{1 - \dfrac{C_{\min}}{C_{\max}}\exp\left[-\dfrac{UA}{C_{\min}}\left(1 - \dfrac{C_{\min}}{C_{\max}}\right)\right]} \qquad (7\text{-}27)$$

Recall that equation (7-27) was derived assuming the cold fluid as minimum. It is a simple exercise to show that this identical expression is obtained for the hot fluid as the minimum fluid. Thus with $UA/C_{\min}$ designated NTU, the *number of transfer units*, we have, as a general expression for counterflow operation, the equation

$$\mathscr{E} = \frac{1 - \exp\left[-\text{NTU}\left(1 - C_{\min}/C_{\max}\right)\right]}{1 - \dfrac{C_{\min}}{C_{\max}}\exp\left[-\text{NTU}\left(1 - C_{\min}/C_{\max}\right)\right]} \tag{7-28}$$

In the case of parallel flow, an analysis similar to that just gone through will yield, for $\mathscr{E}$, the expression

$$\mathscr{E} = \frac{1 - \exp\left[-\text{NTU}\left(1 + C_{\min}/C_{\max}\right)\right]}{1 + C_{\min}/C_{\max}} \tag{7-29}$$

The last two equations for single-pass heat exchangers relate the effectiveness $\mathscr{E}$ in terms of the two parameters NTU and $C_{\min}/C_{\max}$. It seems reasonable that these expressions be represented graphically for easy use. Kays and London[6] have generated plots of heat exchanger effectiveness for these single-pass cases and for various shell-and-tube and crossflow arrangements. Figures 7.11 and 7.12 show plots of $\mathscr{E}$ for counterflow, parallel flow, and selected compact configurations.

Equations (7-21) may be used, along with effectiveness plots, both to design heat exchangers and to evaluate existing equipment. The examples which follow illustrate both applications.

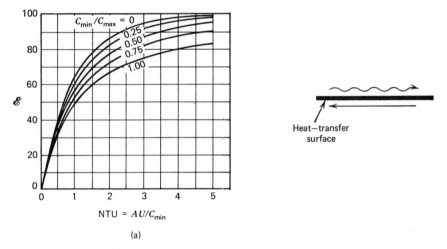

(a)

Figure 7.11 Heat exchanger effectiveness for single-pass and shell-and-tube configurations. (a) Single-pass counterflow. [From W. A. Kays and A. L. London, *Compact Heat Exchangers*, 2nd ed. (New York: McGraw-Hill Book Company, 1964). By permission of the publishers.]

[6] Kays and London, op. cit.

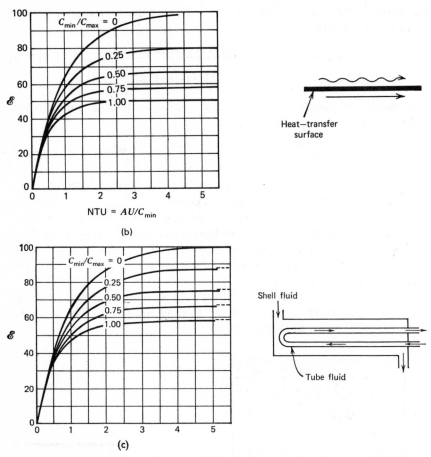

Figure 7.11 (b) Single-pass parallel flow. (c) Shell-and-tube with one shell pass and two, or a multiple of two, tube passes. [From W. A. Kays and A. L. London, *Compact Heat Exchangers*, 2nd ed. (New York: McGraw-Hill Book Company, 1964). By permission of the publishers.]

Example 7.3

Rework Example 7.2, using the heat exchanger effectiveness approach to determine the area required. From previous calculations the following values have been established.

$$q = 185,000 \text{ Btu/hr} \qquad C_{H_2O} = 5000 \text{ Btu/hr-}°F$$

$$T_{H_2O \text{ exit}} = 92°F \qquad C_{Benz} = 3360 \text{ Btu/hr-}°F$$

Benzene is seen to be the minimum fluid.

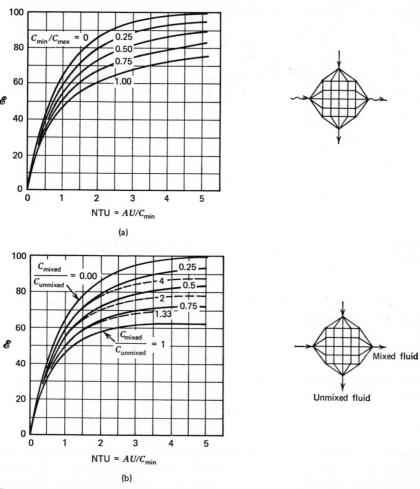

Figure 7.12   Heat exchanger effectiveness for three crossflow configurations. (a) Crossflow, both fluids unmixed. (b) Crossflow, one fluid mixed. [From W. A. Kays and A. L. London, *Compact Heat Exchangers*, 2nd ed. (New York: McGraw-Hill Book Company, 1964). By permission of the publishers.]

The required value of $\mathscr{E}$ may be determined, using equation (7-21).

$$\mathscr{E} = \frac{q}{C_{min}(T_{h\ in} - T_{c\ in})} = \frac{185{,}000\ \text{Btu/hr}}{(3360\ \text{Btu/hr-}^\circ\text{F})(170 - 55)^\circ\text{F}} = 0.479$$

This value and the numerical value $C_{min}/C_{max} = (3360/5000) = 0.672$ allow NTU to be determined for each of the configurations in question.

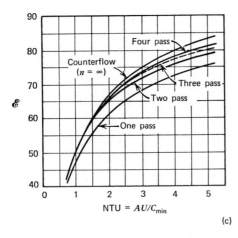
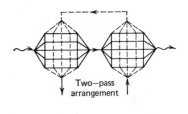

NTU = AU/C_{min}

(c)

Figure 7.12 (c)   Crossflow, multiple passes, both fluids unmixed. [From W. A. Kays and and A. L. London, *Compact Heat Exchangers*, 2nd ed. (New York: McGraw-Hill Book Company, 1964). By permission of the publishers.]

For the counterflow, single-pass case, NTU = $UA/C_{min} \simeq 0.8$, and the area required is determined as

$$A = \frac{0.8C_{min}}{U} = \frac{0.8(3360 \text{ Btu/hr-}^\circ\text{F})}{55 \text{ Btu/hr ft}^2\text{-}^\circ\text{F}} = 48.9 \text{ ft}^2$$

For (b), with one shell pass and four tube passes, Figure 7.11(c) yields the value NTU = $UA/C_{min} \simeq 0.85$, and

$$A = \frac{0.85 \, C_{min}}{U} = \frac{0.85(3360 \text{ Btu/hr-}^\circ\text{F})}{55 \text{ Btu/hr-ft}^2\text{-}^\circ\text{F}} = 51.9 \text{ ft}^2$$

and for crossflow single-pass with the water mixed and benzene unmixed, Figure 7.12(b) gives NTU = $UA/C_{min} \simeq 0.85$, for which the required area will be the same as in (b), 51.9 ft².

In this example the work required to achieve the answer is approximately the same as in the $\Delta T_{\text{log mean}}$ method. One situation where the approach using heat exchanger effectiveness is clearly superior is illustrated in the next example.

Example 7.4

A shell-and-tube heat exchanger with one shell pass and four tube passes has 52 ft² of heat transfer area. The effective overall heat transfer coefficient of this unit is 55 Btu/hr-ft²-°F. This exchanger was designed for use with water and benzene. It is now proposed to cool a stream of oil ($c_p = 0.53$ Btu/lb$_m$-°F) at 250°F, flowing at 12,000 lb$_m$/hr, with cooling water, available at 55°F and a flow of 5000 lb$_m$/hr.

In this application what will be the exciting temperatures of the two fluid streams? The heat transfer may be written, using four expressions. The first law of thermodynamics, equations (7-3) and (7-4), will yield

$$q = C_c(T_{c\ out} - T_{c\ in})$$
$$= (5000\ lb_m/hr)(1\ Btu/lb_m\text{-}°F)(T_{H_2O\ out} - 55°F)$$
$$= 5000(T_{H_2O\ out} - 55)\ Btu/hr \tag{1}$$
$$q = C_H(T_{h\ in} - T_{h\ out})$$
$$= (12,000\ lb_m/hr)(0.53\ Btu/lb_m\text{-}°F)(250 - T_{oil\ out})$$
$$= 6360(250 - T_{oil\ out}) \tag{2}$$

Equation (7-18) is the proper ΔT_{lm} expression. Written in appropriate form for this problem, equation (7-18) becomes

$$q = UAF\ \Delta T_{lm}$$
$$= (55\ Btu/hr\text{-}ft^2\text{-}°F)(52\ ft^2)F\left[\frac{(T_{oil} - 55) - (250 - T_{H_2O})}{\ln \dfrac{T_{oil} - 55}{250 - T_{H_2O}}}\right] \tag{3}$$

The last of the equations to be written is (7-21).

$$q = \mathscr{E}C_{min}(T_{oil\ in} - T_{H_2O\ in})$$
$$= \mathscr{E}(5000\ Btu/hr\text{-}°F)(250 - 50)°F$$
$$= \mathscr{E}(10^6)\ Btu/hr \tag{4}$$

Equations (1), (2), (3), and (4) must all be satisfied; among them they contain three unknowns: q, $T_{H_2O\ out}$, and $T_{oil\ out}$.

Were equations (1), (2), and (3) used to solve this problem, a trial-and-error procedure would be required. The unknown temperatures are tied up in the argument of a logarithm in equation (3); moreover, the factor F is a function of the unknown temperatures as well, and is determined from graphs. While the procedure to solve the problem this way is not extremely difficult, such a prospect is not overly attractive.

Equations (1) and (2) may be used along with (4) in a straightforward fashion involving no trial-and-error methods. This approach is definitely more attractive than the one discussed above. Carrying out the necessary calculations,

$$\frac{C_{min}}{C_{max}} = \frac{5000}{6360} = 0.786$$

$$\frac{UA}{C_{min}} = \frac{(55\ Btu/hr\text{-}ft^2\text{-}°F)(52\ ft^2)}{5000\ Btu/hr\text{-}°F} = 0.572$$

$$\mathscr{E} \simeq 0.36 \qquad \text{[from Figure 7.11(c)]}$$

The heat transferred is now evaluated from (4) as

$$q = 0.36(10^6)\ Btu/hr = 360,000\ Btu/hr$$

and the sought-for exit temperatures may be evaluated from (1) and (2), yielding

$$T_{H_2O \text{ out}} = 55°F + \frac{360,000 \text{ Btu/hr}}{5000 \text{ Btu/hr-°F}}$$

$$= 55°F + 72°F = 127°F$$

$$T_{\text{oil out}} = 250°F - \frac{360,000 \text{ Btu/hr}}{6360 \text{ Btu/hr-°F}} = 193°F$$

Some comments are in order concerning the plots of heat exchanger effectiveness. In every case, the value of $\mathscr{E}$ reaches a maximum at some modest value of NTU, on the order of 2 to 5. For greater heat transfer areas, the value of NTU will increase; however, the heat exchanger effectiveness, once the maximum value is approached, will not show any significant increase. Thus the idea that any increase in heat transfer area in an exchanger is accompanied by a corresponding increase in exchanger capacity is false.

The reader may also note that certain values of $\mathscr{E}$ may never be achieved for certain equipment configurations and operating conditions. For example, in a single-pass counterflow situation with $C_{\min}/C_{\max} = 1$, a value of $\mathscr{E} = 0.90$ is impractical.

Example 7.5 represents a typical approach to heat exchanger design.

Example 7.5

Water is available as a coolant at 60°F with a flow of 150 lb_m/min. It is to reach 140°F by exchanging heat with an oil ($c_p = 0.45$ Btu/lb_m-°F). The oil will enter the exchanger at 240°F and leave at 80°F. The overall heat transfer coefficient is 50 Btu/hr-ft^2-°F. Determine
(a)   the required area in a single-pass counterflow exchanger.
(b)   the required area if a shell-and-tube exchanger is to be used, the water making one shell pass and the oil two tube passes.
(c)   the exit temperatures if, for the exchanger of (a), the flow rate of water is decreased to 120 lb_m/min.
Solving this problem, we first determine the required heat transfer from the water flow rate.

$$q = C_{H_2O} \, \Delta T_{H_2O}$$

$$= (150 \text{ lb}_m/\text{min})(1 \text{ Btu/lb}_m\text{-°F})(80°F)$$

$$= 12,000 \text{ Btu/min}$$

Either the ΔT_{lm} or effectiveness approach may be used to solve for the area in (a). We will use the NTU/effectiveness approach; the necessary quantities are determined as

$$q = C_{H_2O} \, \Delta T_{H_2O} = C_{\text{oil}} \, \Delta T_{\text{oil}}$$

$$C_{H_2O} = 150 \text{ Btu/min-°F}$$

$$C_{\text{oil}} = \frac{12,000 \text{ Btu/min}}{160°F} = 75 \text{ Btu/min-°F}$$

thus

$$C_{oil} = C_{min}$$

$$NTU = \frac{UA}{C_{min}} = \frac{(50 \text{ Btu/hr-ft}^2\text{-}^\circ\text{F})(A \text{ ft}^2)}{(75 \text{ Btu/min-}^\circ\text{F})(60 \text{ min/hr})}$$

$$= 0.0111 A \text{ ft}^2$$

$$\mathscr{E} = \frac{q_{actual}}{q_{max}} = \frac{12,000 \text{ Btu/min}}{C_{min}(T_{h \text{ in}} - T_{c \text{ in}})}$$

$$= \frac{12,000 \text{ Btu/min}}{(75 \text{ Btu/min-}^\circ\text{F})(240^\circ\text{F} - 60^\circ\text{F})} = 0.889$$

$$C_{min}/C_{max} = 0.5$$

From Figure 7.11(a), NTU $\simeq 3.3$, and the required area is

$$A = 3.3/0.0111 = 297 \text{ ft}^2$$

For (b), the NTU/effectiveness approach will involve the same values for the parameters NTU, $\mathscr{E}$ and C_{min}/C_{max}.

From Figure 7.11(c), the conditions of this problem are seen to be impossible as far as a shell-and-tube configuration is concerned. In order to use a shell-and-tube exchanger for the streams with flow rates and entering temperatures specified, *less* heat transfer would be accomplished and the temperature changes of the two fluid streams would be less than in the problem statement.

Working (c) of this example, we now have

$$A = 297 \text{ ft}^2$$

$$C_{H_2O} = \left(120 \frac{\text{lb}_m}{\text{min}}\right)(1 \text{ Btu/lb}_m\text{-}^\circ\text{F}) = 120 \frac{\text{Btu}}{\text{min-}^\circ\text{F}}$$

$$C_{oil} = 75 \text{ Btu/min-}^\circ\text{F}, \qquad \text{thus} \qquad C_{oil} \text{ is } C_{min}$$

$$NTU = \frac{UA}{C_{min}} = 3.3$$

$$\frac{C_{min}}{C_{max}} = \frac{75}{120} = 0.625$$

and, reading Figure 7.11(a), we obtain $\mathscr{E} = 0.86$. We may now determine q as

$$q = \mathscr{E} C_{min}(T_{h \text{ in}} - T_{c \text{ in}})$$

$$= 0.86(75 \text{ Btu/min-}^\circ\text{F})(240^\circ\text{F} - 60^\circ\text{F})$$

$$= 11,610 \text{ Btu/min}$$

Each of the exiting stream temperatures is determined easily from this result.

$$q = C_{H_2O}(T_{H_2O \text{ out}} - T_{H_2O \text{ in}}) = C_{oil}(T_{oil \text{ in}} - T_{oil \text{ out}})$$

$$T_{H_2O \text{ out}} = 60 + \frac{11,610 \text{ Btu/min}}{120 \text{ Btu/min-}^\circ\text{F}} = 157^\circ\text{F}$$

$$T_{oil \text{ out}} = 240^\circ\text{F} - \frac{11,610 \text{ Btu/min}}{75 \text{ Btu/min-}^\circ\text{F}} = 85.2^\circ\text{F}$$

In this example, (c) was a relatively simple evaluation, using the NTU/ effectiveness approach, while a laborious trial-and-error procedure would have been necessary had the ΔT_{lm} approach been used.

7.5 ADDITIONAL CONSIDERATIONS IN HEAT EXCHANGER ANALYSIS AND DESIGN

7.5-1 Heat Exchanger Fouling

When a heat exchanger has been in operation over some extended period of time, there will likely be a build-up of scale on tube surfaces or a deterioration of the surface itself due to corrosion. These effects, in time, will alter the performance of the heat exchanger. A heat transfer surface which has been affected in such a manner is said to be "fouled."

A fouled surface is normally considered to be one which presents some additional heat transfer resistance due to the buildup of foreign material, or "scale." This additional thermal resistance will naturally cause the heat transfer to be less than in the case of no fouling resistance.

The prediction of scale build-up or the corresponding effect on heat transfer is a most difficult task. The actual performance of a heat exchanger can be evaluated after some period of service and the fouling resistance determined. With a clean surface,

$$q_o = U_o A \, \Delta T_{lm} = \frac{\Delta T_{lm}}{\sum R_{to}} \tag{7-30}$$

Similarly, for a fouled surface,

$$q_f = U_f A \, \Delta T_{lm} = \frac{\Delta T_{lm}}{\sum R_{tf}} \tag{7-31}$$

where $\sum R_{to}$ and $\sum R_{tf}$ represent the total thermal resistance for clean and fouled surfaces, respectively.

In the case of a clean surface,

$$U_o = \frac{1}{\dfrac{1}{h_o} + \dfrac{A_o \ln (r_o/r_i)}{2\pi k} + \dfrac{A_o}{A_i h_i}} \tag{7-32}$$

where the terms in the denominator are thermal resistances due to convection on the outside tube surface, conduction through the tube wall, and convection at the inside tube surface, respectively.

For a fouled surface, the expression for U is

$$U_f = \frac{1}{\dfrac{1}{h_o} + R_o + \dfrac{A_o \ln (r_o/r_i)}{2\pi k} + R_i + \dfrac{A_o}{A_i h_i}} \quad (7\text{-}33)$$

with the additional terms R_o and R_i representing the fouling resistances on the outside and inside tube surfaces, respectively. Some typical values of fouling resistance to be used in equation (7-33) are given in Table 7.1. These values have been suggested by the Tubular Exchanger Manufacturers Association.[7]

Table 7.1 Fouling Resistance for Certain Fluids in Heat Exchangers

Fluid	Fouling Resistance (hr-ft^2-°F/Btu)
Distilled water	0.0005
Sea water, below 125°F	0.0005
above 125°F	0.001
Boiler feedwater, treated	0.001
City or well water, below 125°F	0.001
above 125°F	0.002
Refrigerating liquids	0.001
Refrigerating vapors	0.002
Liquid gasoline and organic vapors	0.0005
Fuel oil	0.005
Quenching oil	0.004
Steam, non-oil-bearing	0.0005
Industrial air	0.002

Table 7.2 gives some representative values of the overall heat transfer coefficient for different fluid combinations. The values listed were suggested by Mueller.[8] These values are not to be taken as exact, merely representative of the magnitude to be expected in a heat exchanger with the fluid combinations listed.

[7] TEMA, op. cit.
[8] A. C. Mueller, *Purdue Univ. Eng. Expt. Sta., Engr. Bulletin Res. Ser.* 121 (1954).

Table 7.2   Overall   Heat   Transfer   Coefficients— Approximate Values

Fluid Combination	$U(\text{Btu/hr-ft}^2 {}^\circ\text{F})$
Water to compressed air	10–30
Water to water, jacket water coolers	150–275
Water to brine	100–200
Water to gasoline	60–90
Water to gas oil or distillate	35–60
Water to organic solvents, alcohol	50–150
Water to condensing alcohol	45–120
Water to lubricating oil	20–60
Water to condensing oil vapors	40–100
Water to condensing or boiling Freon-12	50–150
Water to condensing ammonia	150–250
Steam to water, instantaneous heater	400–600
storage-tank heater	175–300
Steam to oil, heavy fuel	10–30
light fuel	30–60
light petroleum distillate	50–200
Steam to aqueous solutions	100–600
Steam to gases	5–50
Light organics to light organics	40–75
Medium organics to medium organics	20–60
Heavy organics to heavy organics	10–40
Heavy organics to light organics	10–60
Crude oil to gas oil	30–55

7.6   CLOSURE

This chapter has introduced the classification, standard nomenclature, and quantitative techniques for heat exchanger design and analysis. The following equations provide the basis for all such analysis:

$$q = C_c(T_{c\ \text{out}} - T_{c\ \text{in}}) \qquad (7\text{-}3)$$

$$q = C_H(T_{h\ \text{in}} - T_{h\ \text{out}}) \qquad (7\text{-}4)$$

$$q = UAF\,\Delta T_{\text{lm}} \qquad (7\text{-}18)$$

$$q = \mathscr{E}C_{\text{min}}(T_{h\ \text{in}} - T_{c\ \text{in}}) \qquad (7\text{-}21)$$

Charts were included to provide a means of obtaining F and $\mathscr{E}$ for single-pass and compact heat exchanger configurations.

The ΔT_{lm} approach to exchanger design and analysis is represented by equation (7-18). The NTU/effectiveness approach utilizes equation (7-21). Either technique is valid; however, the second approach is simpler to use when an exchanger is operated at other than design conditions.

Fouling resistances were listed to provide an approximate means of predicting heat exchanger performance when scale build-up has occurred over an extended period of equipment operation.

APPENDIX A

MATERIAL PROPERTIES

A-1 Physical Properties of Solids

Material	ρ (lb$_m$/ft^3) (68°F)	c_p (Btu/lb$_m$°F) (68°F)	α (ft^2/hr) (68°F)	k (Btu/hr-ft°F) (68°F)	(212°F)	(572°F)
Metals						
Aluminum	168.6	0.224	3.55	132	132	133
Copper	555	0.092	3.98	223	219	213
Gold	1206	0.031	4.52	169	170	172
Iron	492	0.122	0.83	42.3	39.0	31.6
Lead	708	0.030	0.80	20.3	19.3	17.2
Magnesium	109	0.248	3.68	99.5	96.8	91.4
Nickel	556	0.111	0.87	53.7	47.7	36.9
Platinum	1340	0.032	0.09	40.5	41.9	43.5
Silver	656	0.057	6.42	240	237	209
Tin	450	0.051	1.57	36	34	...
Tungsten	1206	0.032	2.44	94	87	77
Uranium α	1167	0.027	0.53	16.9	17.2	19.6
Zinc	446	0.094	1.55	65	63	58
Alloys						
Aluminum 2024	173	0.23	1.76	70.2		
Brass (70% Cu, 30% Zn)	532	0.091	1.27	61.8	73.9	85.3

(continued)

Material	ρ $(\mathrm{lb}_m/\mathrm{ft}^3)$ (68°F)	c_p $(\mathrm{Btu}/\mathrm{lb}_m\text{°F})$ (68°F)	α $(\mathrm{ft}^2/\mathrm{hr})$ (68°F)	k $(\mathrm{Btu}/\mathrm{hr\text{-}ft\text{°}F})$ (68°F)	(212°F)	(572°F)
Alloys						
Constantan						
(60% Cu, 40% Ni)	557	0.098	0.24	13.1	15.4	
Iron, cast	455	0.100	0.65	29.6	26.8	
Nichrome V	530	0.106	0.12	7.06	7.99	9.94
Stainless steel	488	0.110	0.17	9.4	10.0	13
Steel, mild (1%C)	488	0.113	0.45	24.8	24.8	22.9
Nonmetals						
Asbestos	36	0.25		0.092	0.11	0.125
Brick (fire clay)	144	0.22			0.65	
Brick (masonry)	106	0.20		0.38		
Brick (chrome)	188	0.20			0.67	
Concrete	144	0.21		0.70		
Corkboard	10	0.4		0.025		
Diatomaceous earth, powdered	14	0.2		0.03		
Glass, window	170	0.2		0.45		
Glass, Pyrex	140	0.2		0.63	0.67	0.84
Kaolin firebrick	19					0.052
85% magnesia	17			0.038	0.041	
Sandy loam, 4% H_2O	104	~0.4		0.54		
Sandy loam, 10% H_2O	121			1.08		
Rock wool	~10	0.2		0.023	0.033	
Wood, oak, ⊥ to grain	51	0.57		0.12		
Wood, oak, ‖ to grain	51	0.57		0.23		

A-2 Physical Properties of Liquids

Liquids

Water

T (°F)	ρ (lb_m/ft^3)	c_p (Btu/lb_m-°F)	$\mu \times 10^3$ (lb_m/ft-sec)	$\nu \times 10^5$ (ft^2/sec)	k (Btu/hr-ft-°F)	$\alpha \times 10^3$ (ft^2/hr)	Pr	$\beta \times 10^4$ (1/°F)	$g\beta\rho^2/\mu^2 \times 10^{-6}$ (1/°F-ft^3)
32	62.4	1.01	1.20	1.93	0.319	5.06	13.7	−0.350	
60	62.3	1.00	0.760	1.22	0.340	5.45	8.07	0.800	17.2
80	62.2	0.999	0.578	0.929	0.353	5.67	5.89	1.30	48.3
100	62.1	0.999	0.458	0.736	0.364	5.87	4.51	1.80	107
150	61.3	1.00	0.290	0.474	0.383	6.26	2.72	2.80	403
200	60.1	1.01	0.206	0.342	0.392	6.46	1.91	3.70	1,010
250	58.9	1.02	0.160	0.272	0.395	6.60	1.49	4.70	2,045
300	57.3	1.03	0.130	0.227	0.395	6.70	1.22	5.60	3,510
400	53.6	1.08	0.0930	0.174	0.382	6.58	0.950	7.80	8,350
500	49.0	1.19	0.0700	0.143	0.349	5.98	0.859	11.0	17,350
600	42.4	1.51	0.0579	0.137	0.293	4.58	1.07	17.5	30,300

(continued through page 419)

Aniline

T (°F)	ρ (lb_m/ft³)	c_p (Btu/lb_m-°F)	$\mu \times 10^5$ (lb_m/ft-sec)	$\nu \times 10^5$ (ft²/sec)	k (Btu/hr-ft-°F)	$\alpha \times 10^3$ (ft²/hr)	Pr	$\beta \times 10^3$ (1/°F)	$g\beta\rho^2/\mu^2 \times 10^{-6}$ (1/°F-ft³)
60	64.0	0.480	305	4.77	0.101	3.29	52.3		
80	63.5	0.485	240	3.78	0.100	3.25	41.8		
100	63.0	0.490	180	2.86	0.100	3.24	31.8	0.45	17.7
150	61.6	0.503	100	1.62	0.0980	3.16	18.4		
200	60.2	0.515	62	1.03	0.0962	3.10	12.0		
250	58.9	0.527	42	0.714	0.0947	3.05	8.44		
300	57.5	0.540	30	0.522	0.0931	2.99	6.28		

Ammonia

T (°F)	ρ (lb_m/ft³)	c_p (Btu/lb_m-°F)	$\mu \times 10^5$ (lb_m/ft-sec)	$\nu \times 10^5$ (ft²/sec)	k (Btu/hr-ft-°F)	$\alpha \times 10^3$ (ft²/hr)	Pr	$\beta \times 10^3$ (1/°F)	$\beta g\rho^2/\mu^2 \times 10^{-7}$ (1/°F-ft³)
−60	43.9	1.07	20.6	0.471	0.316	6.74	2.52	0.94	132
−30	42.7	1.07	18.2	0.426	0.317	6.93	2.22	1.02	265
0	41.3	1.08	16.9	0.409	0.315	7.06	2.08	1.1	467
30	40.0	1.11	16.2	0.402	0.312	7.05	2.05	1.19	757
60	38.5	1.14	15.0	0.391	0.304	6.92	2.03	1.3	1130
80	37.5	1.16	14.2	0.379	0.296	6.79	2.01	1.4	1650
100	36.4	1.19	13.5	0.368	0.287	6.62	2.00	1.5	2200
120	35.3	1.22	12.6	0.356	0.275	6.43	2.00	1.68	3180

Freon-12

T (°F)	ρ (lb$_m$/ft³)	c_p (Btu/lb$_m$-°F)	$\mu \times 10^5$ (lb$_m$/ft-sec)	$\nu \times 10^5$ (ft²/sec)	k (Btu/hr-ft-°F)	$\alpha \times 10^3$ (ft²/hr)	Pr	$\beta \times 10^4$ (1/°F)	$g\beta\rho^2/\mu^2 \times 10^{-6}$ (1/°F-ft³)
−40	94.5	0.202	125	1.32	0.0650	3.40	14.0	9.10	168
−30	93.5	0.204	123	1.32	0.0640	3.35	14.1	9.60	179
0	90.9	0.212	116	1.28	0.0578	3.00	15.4	11.4	225
30	87.4	0.221	108	1.24	0.0564	2.92	15.3	13.1	277
60	84.0	0.230	99.6	1.19	0.0528	2.74	15.6	14.9	341
80	81.3	0.238	94.0	1.16	0.0504	2.60	16.0	16.0	384
100	78.7	0.246	88.4	1.12	0.0480	2.48	16.3	17.2	439
150	71.0	0.271	74.8	1.05	0.0420	2.18	17.4	19.5	625

n-Butyl Alcohol

T (°F)	ρ (lb$_m$/ft³)	c_p (Btu/lb$_m$-°F)	$\mu \times 10^5$ (lb$_m$/ft-sec)	$\nu \times 10^5$ (ft²/sec)	k (Btu/hr-ft-°F)	$\alpha \times 10^3$ (ft²/hr)	Pr	$\beta \times 10^3$ (1/°F)	$g\beta\rho^2/\mu^2 \times 10^{-6}$ (1/°F-ft³)
60	50.5	0.55	225.	4.46	0.100	3.59	44.6		
80	50.0	0.58	180	3.60	0.099	3.41	38.0		
100	49.6	0.61	130	2.62	0.098	3.25	29.1	0.25	6.23
150	48.5	0.68	68	1.41	0.098	2.97	17.1	0.43	2.02

Benzene

T (°F)	ρ (lb$_m$/ft³)	c_p (Btu/lb$_m$-°F)	$\mu \times 10^5$ (lb$_m$/ft-sec)	$\nu \times 10^5$ (ft²/sec)	k (Btu/hr-ft-°F)	$\alpha \times 10^3$ (ft²/hr)	Pr	$\beta \times 10^4$ (1/°F)	$g\beta\rho^2/\mu^2 \times 10^{-6}$ (1/°F-ft³)
60	55.2	0.395	44.5	0.806	0.0856	3.93	7.39		
80	54.6	0.410	38	0.695	0.0836	3.73	6.70	7.5	498
100	53.6	0.420	33	0.615	0.0814	3.61	6.13	7.2	609
150	51.8	0.450	24.5	0.473	0.0762	3.27	5.21	6.8	980
200	49.9	0.480	19.4	0.390	0.0711	2.97	4.73		

Hydraulic Fluid (MIL-M-5606)

T (°F)	ρ (lb$_m$/ft³)	c_p (Btu/lb$_m$-°F)	$\mu \times 10^5$ (lb$_m$/ft-sec)	$\nu \times 10^5$ (ft²/sec)	k (Btu/hr-ft-°F)	$\alpha \times 10^3$ (ft²/hr)	Pr	$\beta \times 10^3$ (1/°F)	$g\beta\rho^2/\mu^2 \times 10^{-4}$ (1/°F-ft³)
0	55.0	0.400	5550	101	0.0780	3.54	1030	0.76	2.39
30	54.0	0.420	2220	41.1	0.0755	3.32	446	0.68	13.0
60	53.0	0.439	1110	20.9	0.0732	3.14	239	0.60	44.1
80	52.5	0.453	695	13.3	0.0710	3.07	155	0.52	95.7
100	52.0	0.467	556	10.7	0.0690	2.84	136	0.47	132
150	51.0	0.499	278	5.45	0.0645	2.44	80.5	0.32	346
200	50.0	0.530	250	5.00	0.0600	2.27	79.4	0.20	258

Glycerin

T (°F)	ρ (lb$_m$/ft^3)	c_p (Btu/lb$_m$-°F)	μ (lb$_m$/ft-sec)	$v \times 10^2$ (ft^2/sec)	k (Btu/hr-ft-°F)	$\alpha \times 10^3$ (ft^2/hr)	$Pr \times 10^{-2}$	$\beta \times 10^3$ (1/°F)	$g\beta\rho^2/\mu^2$ (1/°F-ft^3)
30	79.7	0.540	7.2	9.03	0.168	3.91	832		
60	79.1	0.563	1.4	1.77	0.167	3.75	170	0.30	166
80	78.7	0.580	0.6	0.762	0.166	3.64	75.3		
100	78.2	0.598	0.1	0.128	0.165	3.53	13.1		

Kerosene

T (°F)	ρ (lb$_m$/ft^3)	c_p (Btu/lb$_m$-°F)	$\mu \times 10^5$ (lb$_m$/ft-sec)	$v \times 10^5$ (ft^2/sec)	k (Btu/hr-ft-°F)	$\alpha \times 10^3$ (ft^2/hr)	Pr	$\beta \times 10^3$ (1/°F)	$g\beta\rho^2/\mu^2 \times 10^{-4}$ (1/°F-ft^3)
30	48.8	0.456	800	16.4	0.0809	3.63	163		
60	48.1	0.474	600	12.5	0.0805	3.53	127	0.58	120
80	47.6	0.491	490	10.3	0.0800	3.42	108	0.48	146
100	47.2	0.505	420	8.90	0.0797	3.35	95.7	0.47	192
150	46.1	0.540	320	6.83	0.0788	3.16	77.9		

Liquid Hydrogen

T (°F)	ρ (lb$_m$/ft^3)	c_p (Btu/lb$_m$-°F)	$\mu \times 10^5$ (lb$_m$/ft-sec)	$\nu \times 10^5$ (ft^2/sec)	k (Btu/hr-ft-°F)	$\alpha \times 10^3$ (ft^2/hr)	Pr	$\beta \times 10^3$ (1/°F)	$g\beta\rho^2/\mu^2 \times 10^{-4}$ (1/°F-ft^3)
−435	4.84	1.69	1.63	0.337	0.0595	7.28	1.67		
−433	4.77	1.78	1.52	0.319	0.0610	7.20	1.59		
−431	4.71	1.87	1.40	0.297	0.0625	7.09	1.51	7.1	2.59
−429	4.64	1.96	1.28	0.276	0.0640	7.03	1.41		
−427	4.58	2.05	1.17	0.256	0.0655	6.97	1.32		
−425	4.51	2.15	1.05	0.233	0.0670	6.90	1.21		

Liquid Oxygen

T (°F)	ρ (lb$_m$/ft^3)	c_p (Btu/lb$_m$-°F)	$\mu \times 10^5$ (lb$_m$/ft-sec)	$\nu \times 10^5$ (ft^2/sec)	$k \times 10^3$ (Btu/hr-ft-°F)	$\alpha \times 10^5$ (ft^2/hr)	Pr	$\beta \times 10^3$ (1/°F)	$g\beta\rho^2/\mu^2 \times 10^{-8}$ (1/°F-ft^3)
−350	80.1	0.400	38.0	0.474	3.1	9.67	172		
−340	78.5	0.401	28.0	0.356	3.4	10.8	109		
−330	76.8	0.402	21.8	0.284	3.7	12.0	85.0	3.19	186
−320	75.1	0.404	17.4	0.232	4.0	12.2	63.5		
−310	73.4	0.405	14.8	0.202	4.3	14.5	50.1		
−300	71.7	0.406	13.0	0.181	4.6	15.8	41.2		

T ($^\circ$F)	ρ (lb$_m$/ft^3)	c_p (Btu/lb$_m$-$^\circ$F)	$\mu \times 10^3$ (lb$_m$/ft-sec)	$\nu \times 10^6$ (ft^2/sec)	k (Btu/hr-ft-$^\circ$F)	α (ft^2/hr)	Pr	$\beta \times 10^3$ (1/$^\circ$F)	$g\beta\rho^2/\mu^2 \times 10^{-9}$ (1/$^\circ$F-ft^3)
					Bismuth				
600	625	0.0345	1.09	1.75	8.58	0.397	0.0159		
700	622	0.0353	0.990	1.59	8.87	0.405	0.0141	0.062	0.786
800	618	0.0361	0.900	1.46	9.16	0.408	0.0129	0.065	0.985
900	613	0.0368	0.830	1.35	9.44	0.418	0.0116	0.068	1.19
1000	608	0.0375	0.765	1.26	9.74	0.427	0.0106	0.071	1.45
1100	604	0.0381	0.710	1.17	10.0	0.435	0.00970	0.074	1.72
1200	599	0.0386	0.660	1.10	10.3	0.446	0.00895	0.077	2.04
1300	595	0.0391	0.620	1.04	10.6	0.456	0.00820		

Mercury

T (°F)	ρ (lb$_m$/ft^3)	c_p (Btu/lb$_m$-°F)	$\mu \times 10^3$ (lb$_m$/ft-sec)	$\nu \times 10^6$ (ft^2/sec)	k (Btu/hr-ft-°F)	α (ft^2/hr)	Pr	$\beta \times 10^3$ (1/°F)	$g\beta\rho^2/\mu^2 \times 10^{-9}$ (1/°F-ft^3)
40	848	0.0334	1.11	1.31	4.55	0.161	0.0292		1.57
60	847	0.0333	1.05	1.24	4.64	0.165	0.0270		1.76
80	845	0.0332	1.00	1.18	4.72	0.169	0.0252		1.94
100	843	0.0331	0.960	1.14	4.80	0.172	0.0239		2.09
150	839	0.0330	0.893	1.06	5.03	0.182	0.0210		2.38
200	835	0.0328	0.850	1.02	5.25	0.192	0.0191		2.62
250	831	0.0328	0.806	0.970	5.45	0.200	0.0175		2.87
300	827	0.0328	0.766	0.928	5.65	0.209	0.0160	0.084	3.16
400	819	0.0328	0.700	0.856	6.05	0.225	0.0137		3.70
500	811	0.0328	0.650	0.803	6.43	0.243	0.0119		4.12
600	804	0.0328	0.606	0.754	6.80	0.259	0.0105		4.80
800	789	0.0329	0.550	0.698	7.45	0.289	0.0087		5.54

Sodium

T (°F)	ρ (lb$_m$/ft³)	c_p (Btu/lb$_m$-°F)	$\mu \times 10^3$ (lb$_m$/ft-sec)	$\nu \times 10^6$ (ft²/sec)	k (Btu/hr-ft-°F)	α (ft²/hr)	Pr	$\beta \times 10^3$ (1/°F)	$g\beta\rho^2/\mu^2 \times 10^{-6}$ (1/°F-ft³)
200	58.1	0.332	0.489	8.43	49.8	2.58	0.0118		68.0
250	57.6	0.328	0.428	7.43	49.3	2.60	0.0103		87.4
300	57.2	0.324	0.378	6.61	48.8	2.64	0.00903		110
400	56.3	0.317	0.302	5.36	47.3	2.66	0.00725		168
500	55.5	0.309	0.258	4.64	45.5	2.64	0.00633	0.15	224
600	54.6	0.305	0.224	4.11	43.1	2.58	0.00574		287
800	52.9	0.304	0.180	3.40	38.8	2.41	0.00510		418
1000	51.2	0.304	0.152	2.97	36.0	2.31	0.00463		548
1300	48.7	0.305	0.120	2.47	34.2	2.31	0.00385		795

A-3 Physical Properties of Gases

T (°F)	c_p (Btu/lb$_m$-°F)	ρ (lb$_m$/ft³)	$\mu \times 10^5$ (lb$_m$/ft-sec)	$\nu \times 10^3$ (ft²/sec)	k (Btu/hr-ft-°F)	α (ft²/hr)	Pr	$\beta \times 10^3$ (1/°F)	$g\beta\rho^2/\mu^2$ (1/°F-ft³)
					Air				
0	0.240	0.0862	1.09	0.126	0.0132	0.639	0.721	2.18	4.39×10^6
30	0.240	0.0810	1.15	0.142	0.0139	0.714	0.716	2.04	3.28
60	0.240	0.0764	1.21	0.159	0.0146	0.798	0.711	1.92	2.48
80	0.240	0.0735	1.24	0.169	0.0152	0.855	0.708	1.85	2.09
100	0.240	0.0710	1.28	0.181	0.0156	0.919	0.703	1.79	1.76
150	0.241	0.0651	1.36	0.209	0.0167	1.06	0.698	1.64	1.22
200	0.241	0.0602	1.45	0.241	0.0179	1.24	0.694	1.52	0.840
250	0.242	0.0559	1.53	0.274	0.0191	1.42	0.690	1.41	0.607
300	0.243	0.0523	1.60	0.306	0.0203	1.60	0.686	1.32	0.454
400	0.245	0.0462	1.74	0.377	0.0225	2.00	0.681	1.16	0.264
500	0.247	0.0413	1.87	0.453	0.0246	2.41	0.680	1.04	0.163
600	0.251	0.0374	2.00	0.535	0.0270	2.88	0.680	0.944	79.4×10^3
800	0.257	0.0315	2.24	0.711	0.0303	3.75	0.684	0.794	50.6
1000	0.263	0.0272	2.46	0.906	0.0337	4.72	0.689	0.685	27.0
1500	0.277	0.0203	2.92	1.44	0.0408	7.27	0.705	0.510	7.96

Note: All gas properties are for atmospheric pressure.

(continued through page 429)

Steam

T (°F)	ρ (lb$_m$/ft^3)	c_p (Btu/lb$_m$-°F)	$\mu \times 10^5$ (lb$_m$/ft-sec)	$\nu \times 10^3$ (ft^2/sec)	k (Btu/hr-ft-°F)	α (ft^2/hr)	Pr	$\beta \times 10^3$ (1/°F)	$g\beta\rho^2/\mu^2$ (1/°F-ft^3)
212	0.0372	0.493	0.870	0.234	0.0145	0.794	1.06	1.49	0.873×10^6
250	0.0350	0.483	0.890	0.254	0.0155	0.920	0.994	1.41	0.698
300	0.0327	0.476	0.960	0.294	0.0171	1.10	0.963	1.32	0.493
400	0.0289	0.472	1.09	0.377	0.0200	1.47	0.924	1.16	0.262
500	0.0259	0.477	1.23	0.474	0.0228	1.85	0.922	1.04	0.148
600	0.0234	0.483	1.37	0.585	0.0258	2.29	0.920	0.944	88.9×10^3
800	0.0197	0.498	1.63	0.828	0.0321	3.27	0.912	0.794	37.8
1000	0.0170	0.517	1.90	1.12	0.0390	4.44	0.911	0.685	17.2
1500	0.0126	0.564	2.57	2.05	0.0580	8.17	0.906	0.510	3.97

T (°F)	ρ (lb$_m$/ft³)	c_p (Btu/lb$_m$-°F)	$\mu \times 10^5$ (lb$_m$/ft-sec)	$\nu \times 10^3$ (ft²/sec)	k (Btu/hr-ft-°F)	α (ft²/hr)	Pr	$\beta \times 10^3$ (1/°F)	$g\beta\rho^2/\mu^2$ (1/°F-ft³)
					Nitrogen				
0	0.0837	0.249	1.06	0.127	0.0132	0.633	0.719	2.18	4.38 × 10⁶
30	0.0786	0.249	1.12	0.142	0.0139	0.710	0.719	2.04	3.29
60	0.0740	0.249	1.17	0.158	0.0146	0.800	0.716	1.92	2.51
80	0.0711	0.249	1.20	0.169	0.0151	0.853	0.712	1.85	2.10
100	0.0685	0.249	1.23	0.180	0.0154	0.915	0.708	1.79	1.79
150	0.0630	0.249	1.32	0.209	0.0168	1.07	0.702	1.64	1.22
200	0.0580	0.249	1.39	0.240	0.0174	1.25	0.690	1.52	0.854
250	0.0540	0.249	1.47	0.271	0.0192	1.42	0.687	1.41	0.616
300	0.0502	0.250	1.53	0.305	0.0202	1.62	0.685	1.32	0.457
400	0.0443	0.250	1.67	0.377	0.0212	2.02	0.684	1.16	0.263
500	0.0397	0.253	1.80	0.453	0.0244	2.43	0.683	1.04	0.163
600	0.0363	0.256	1.93	0.532	0.0252	2.81	0.686	0.944	0.108
800	0.0304	0.262	2.16	0.710	0.0291	3.71	0.691	0.794	0.0507
1000	0.0263	0.269	2.37	0.901	0.0336	4.64	0.700	0.685	0.0272
1500	0.0195	0.283	2.82	1.45	0.0423	7.14	0.732	0.510	0.00785

T (°F)	ρ (lb$_m$/ft^3)	c_p (Btu/lb$_m$-°F)	$\mu \times 10^5$ (lb$_m$/ft-sec)	$\nu \times 10^3$ (ft^2/sec)	k (Btu/hr-ft-°F)	α (ft^2/hr)	Pr	$\beta \times 10^3$ (1/°F)	$g\beta\rho^2/\mu^2$ (1/°F-ft^3)
					Oxygen				
0	0.0955	0.219	1.22	0.128	0.0134	0.641	0.718	2.18	4.29×10^6
30	0.0897	0.219	1.28	0.143	0.0141	0.718	0.716	2.04	3.22
60	0.0845	0.219	1.35	0.160	0.0149	0.806	0.713	1.92	2.43
80	0.0814	0.220	1.40	0.172	0.0155	0.866	0.713	1.85	2.02
100	0.0785	0.220	1.43	0.182	0.0160	0.925	0.708	1.79	1.74
150	0.0720	0.221	1.52	0.211	0.0172	1.08	0.703	1.64	1.19
200	0.0665	0.223	1.62	0.244	0.0185	1.25	0.703	1.52	0.825
250	0.0618	0.225	1.70	0.276	0.0197	1.42	0.700	1.41	0.600
300	0.0578	0.227	1.79	0.310	0.0209	1.60	0.700	1.32	0.442
400	0.0511	0.230	1.95	0.381	0.0233	1.97	0.698	1.16	0.257
500	0.0458	0.234	2.10	0.458	0.0254	2.37	0.696	1.04	0.160
600	0.0414	0.239	2.25	0.543	0.0281	2.84	0.688	0.944	0.103
800	0.0349	0.246	2.52	0.723	0.0324	3.77	0.680	0.794	49.4×10^3
1000	0.0300	0.252	2.79	0.930	0.0366	4.85	0.691	0.685	25.6
1500	0.0224	0.264	3.39	1.52	0.0465	7.86	0.696	0.510	7.22

T ($°F$)	ρ (lb_m/ft^3)	c_p (Btu/lb_m-$°F$)	$\mu \times 10^5$ (lb_m/ft-sec)	$\nu \times 10^3$ (ft^2/sec)	k (Btu/hr-ft-$°F$)	α (ft^2/hr)	Pr	$\beta \times 10^3$ ($1/°F$)	$g\beta\rho^2/\mu^2$ ($1/°F$-ft^3)
					Carbon Dioxide				
0	0.132	0.193	0.865	0.0655	0.00760	0.298	0.792	2.18	16.3×10^6
30	0.124	0.198	0.915	0.0739	0.00830	0.339	0.787	2.04	12.0
60	0.117	0.202	0.965	0.0829	0.00910	0.387	0.773	1.92	9.00
80	0.112	0.204	1.00	0.0891	0.00960	0.421	0.760	1.85	7.45
100	0.108	0.207	1.03	0.0953	0.0102	0.455	0.758	1.79	6.33
150	0.100	0.213	1.12	0.113	0.0115	0.539	0.755	1.64	4.16
200	0.092	0.219	1.20	0.131	0.0130	0.646	0.730	1.52	2.86
250	0.0850	0.225	1.32	0.155	0.0148	0.777	0.717	1.41	2.04
300	0.0800	0.230	1.36	0.171	0.0160	0.878	0.704	1.32	1.45
400	0.0740	0.239	1.45	0.196	0.0180	1.02	0.695	1.16	1.11
500	0.0630	0.248	1.65	0.263	0.0210	1.36	0.700	1.04	0.485
600	0.0570	0.256	1.78	0.312	0.0235	1.61	0.700	0.944	0.310
800	0.0480	0.269	2.02	0.420	0.0278	2.15	0.702	0.794	0.143
1000	0.0416	0.280	2.25	0.540	0.0324	2.78	0.703	0.685	75.3×10^3
1500	0.0306	0.301	2.80	0.913	0.0340	4.67	0.704	0.510	19.6

T (°F)	ρ (lb$_m$/ft^3)	c_p (Btu/lb$_m$-°F)	$\mu \times 10^5$ (lb$_m$/ft-sec)	$\nu \times 10^3$ (ft^2/sec)	k (Btu/hr-ft-°F)	α (ft^2/hr)	Pr	$\beta \times 10^3$ (1/°F)	$g\beta\rho^2/\mu^2$ (1/°F-ft^3)
					Hydrogen				
0	0.00597	3.37	0.537	0.900	0.092	4.59	0.713	2.18	87,000
30	0.00562	3.39	0.562	1.00	0.097	5.09	0.709	2.04	65,700
60	0.00530	3.41	0.587	1.11	0.102	5.65	0.707	1.92	50,500
80	0.00510	3.42	0.602	1.18	0.105	6.04	0.705	1.85	42,700
100	0.00492	3.42	0.617	1.25	0.108	6.42	0.700	1.79	36,700
150	0.00450	3.44	0.653	1.45	0.116	7.50	0.696	1.64	25,000
200	0.00412	3.45	0.688	1.67	0.123	8.64	0.696	1.52	17,500
250	0.00382	3.46	0.723	1.89	0.130	9.85	0.690	1.41	12,700
300	0.00357	3.46	0.756	2.12	0.137	11.1	0.687	1.32	9,440
400	0.00315	3.47	0.822	2.61	0.151	13.8	0.681	1.16	5,470
500	0.00285	3.47	0.890	3.12	0.165	16.7	0.675	1.04	3,430
600	0.00260	3.47	0.952	3.66	0.179	19.8	0.667	0.944	2,270
800	0.00219	3.49	1.07	4.87	0.205	26.8	0.654	0.794	1,080
1000	0.00189	3.52	1.18	6.21	0.224	33.7	0.664	0.685	571
1500	0.00141	3.62	1.44	10.2	0.265	51.9	0.708	0.510	158

T (°F)	ρ (lb$_m$/ft³)	c_p (Btu/lb$_m$-°F)	$\mu \times 10^5$ (lb$_m$/ft-sec)	$\nu \times 10^3$ (ft²/sec)	k (Btu/hr-ft-°F)	α (ft²/hr)	Pr	$\beta \times 10^3$ (1/°F)	$g\beta\rho^2/\mu^2$ (1/°F-ft³)
				Carbon Monoxide					
0	0.0832	0.249	1.05	0.126	0.0128	0.620	0.749	2.18	4.40×10^6
30	0.0780	0.249	1.11	0.142	0.0134	0.691	0.744	2.04	3.32
60	0.0736	0.249	1.16	0.157	0.0142	0.775	0.740	1.92	2.48
80	0.0709	0.249	1.20	0.169	0.0146	0.828	0.737	1.85	2.09
100	0.0684	0.249	1.23	0.180	0.0150	0.884	0.735	1.79	1.79
150	0.0628	0.249	1.32	0.210	0.0163	1.04	0.730	1.64	1.19
200	0.0580	0.250	1.40	0.241	0.0174	1.20	0.726	1.52	0.842
250	0.0539	0.250	1.48	0.275	0.0183	1.36	0.722	1.41	0.604
300	0.0503	0.251	1.56	0.310	0.0196	1.56	0.720	1.32	0.442
400	0.0445	0.253	1.73	0.389	0.0217	1.92	0.726	1.16	0.248
500	0.0399	0.256	1.85	0.463	0.0234	2.30	0.729	1.04	0.156
600	0.0361	0.259	1.97	0.545	0.0253	2.71	0.726	0.944	0.101
800	0.0304	0.266	2.21	0.728	0.0288	3.57	0.735	0.794	48.2×10^3
1000	0.0262	0.273	2.43	0.929	0.0324	4.54	0.740	0.685	25.6
1500	0.0195	0.286	3.00	1.54	0.0410	7.35	0.756	0.510	6.93

T (°F)	ρ ($\text{lb}_\text{m}/\text{ft}^3$)	c_p ($\text{Btu}/\text{lb}_\text{m}\text{-°F}$)	$\mu \times 10^6$ ($\text{lb}_\text{m}/\text{ft-sec}$)	$\nu \times 10^3$ (ft^2/sec)	k ($\text{Btu}/\text{hr-ft-°F}$)	α (ft^2/hr)	Pr	$\beta \times 10^3$ ($1/°F$)	$g\beta\rho^2/\mu^2 \times 10^{-6}$ ($1/°F\text{-ft}^3$)
				Chlorine					
0	0.211	0.113	8.06	0.0381	0.00418	0.175	0.785	2.18	48.3
30	0.197	0.114	8.40	0.0426	0.00450	0.201	0.769	2.04	36.6
60	0.187	0.114	8.80	0.0470	0.00480	0.225	0.753	1.92	28.1
80	0.180	0.115	9.07	0.0504	0.00500	0.242	0.753	1.85	24.3
100	0.173	0.115	9.34	0.0540	0.00520	0.261	0.748	1.79	19.9
150	0.159	0.117	10.0	0.0629	0.00570	0.306	0.739	1.64	13.4

427

T (°F)	ρ (lb$_m$/ft³)	c_p (Btu/lb$_m$-°F)	$\mu \times 10^7$ (lb$_m$/ft-sec)	$\nu \times 10^3$ (ft²/sec)	k (Btu/hr-ft-°F)	α (ft²/hr)	Pr	$\beta \times 10^3$ (1/°F)	$g\beta\rho^2/\mu^2$ (1/°F-ft³)
					Helium				
0	0.0119	1.24	122	1.03	0.0784	5.30	0.698	2.18	66,800
30	0.0112	1.24	127	1.14	0.0818	5.89	0.699	2.04	51,100
60	0.0106	1.24	132	1.25	0.0852	6.46	0.700	1.92	40,000
80	0.0102	1.24	135	1.32	0.0872	6.88	0.701	1.85	33,900
100	0.00980	1.24	138	1.41	0.0892	7.37	0.701	1.79	29,000
150	0.00900	1.24	146	1.63	0.0937	8.36	0.703	1.64	20,100
200	0.00829	1.24	155	1.87	0.0977	9.48	0.705	1.52	14,000
250	0.00772	1.24	162	2.09	0.102	10.7	0.707	1.41	10,400
300	0.00722	1.24	170	2.36	0.106	11.8	0.709	1.32	7,650
400	0.00637	1.24	185	2.91	0.114	14.4	0.714	1.16	4,410
500	0.00572	1.24	198	3.46	0.122	17.1	0.719	1.04	2,800
600	0.00517	1.24	209	4.04	0.130	20.6	0.720	0.944	1,850
800	0.00439	1.24	232	5.28	0.145	27.6	0.722	0.794	915
1000	0.00376	1.24	255	6.78	0.159	35.5	0.725	0.685	480
1500	0.00280	1.24	309	11.1	0.189	59.7	0.730	0.510	135

T (°F)	ρ (lb_m/ft^3)	c_p (Btu/lb_m-°F)	$\mu \times 10^5$ (lb_m/ft-sec)	$\nu \times 10^3$ (ft^2/sec)	k (Btu/hr-ft-°F)	α (ft^2/hr)	Pr	$\beta \times 10^3$ (1/°F)	$g\beta\rho^2/\mu^2$ (1/°F-ft^3)
				Sulfur Dioxide					
0	0.195	0.142	0.700	3.59	0.00460	0.166	0.778	2.03	50.6×10^6
100	0.161	0.149	0.890	5.52	0.00560	0.233	0.854	1.79	19.0
200	0.136	0.157	1.05	7.74	0.00670	0.313	0.883	1.52	8.25
300	0.118	0.164	1.20	10.2	0.00790	0.407	0.898	1.32	4.12
400	0.104	0.170	1.35	13.0	0.00920	0.520	0.898	1.16	2.24
500	0.0935	0.176	1.50	16.0	0.00990	0.601	0.958	1.04	1.30
600	0.0846	0.180	1.65	19.5	0.0108	0.711	0.987	0.994	0.795

429

B_____

CHARTS FOR SOLUTION OF TRANSIENT CONDUCTION PROBLEMS

Table B.1   Symbols for Unsteady-State Charts

	Parameter Symbol	Heat Conduction
Dimensionless temperature	Y	$\dfrac{T - T_\infty}{T_0 - T_\infty}$
Relative time	X	$\dfrac{\alpha t}{x_1^2}$
Relative position	n	$\dfrac{x}{x_1}$
Relative resistance	m	$\dfrac{k}{hx_1}$

T = temperature
x = distance from center to
. any point
t = time
k = thermal conductivity
h = convective transfer
   coefficient
α = thermal diffusivity

Subscripts:
0 = initial condition at time
   $t = 0$
1 = boundary
∞ = reference fluid condition

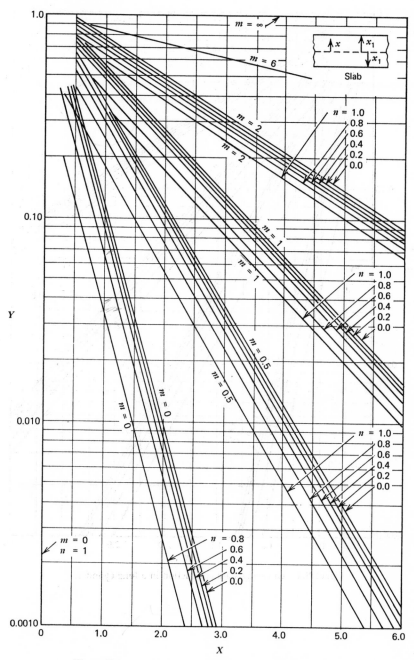

Figure B.1 Unsteady-state conduction in a large flat slab.

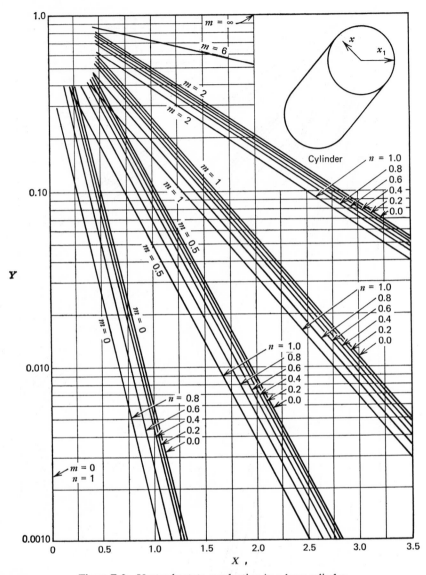

Figure B.2 Unsteady-state conduction in a long cylinder.

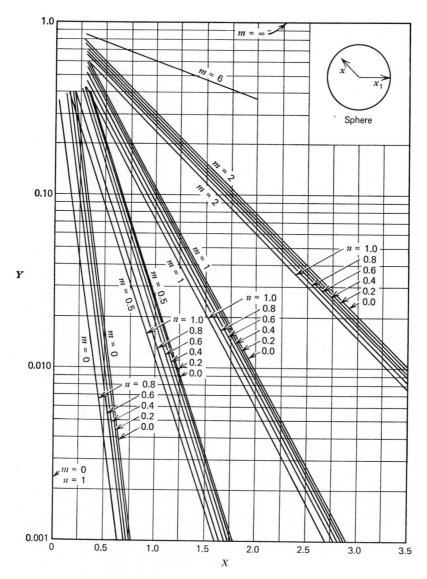

Figure B.3 Unsteady-state conduction in a sphere.

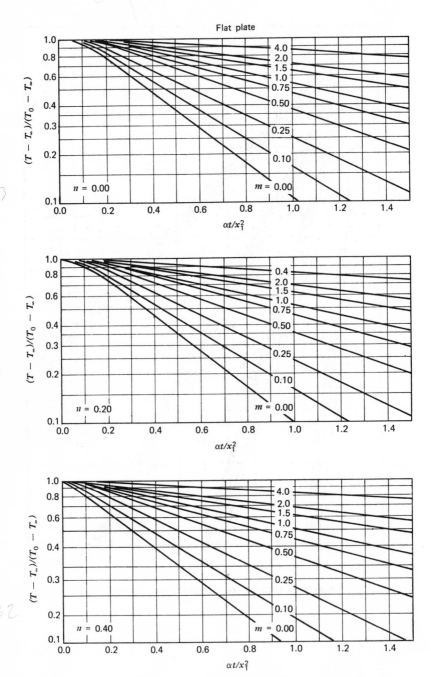

Figure B.4 Charts for solution of unsteady conduction problems: flat plate.

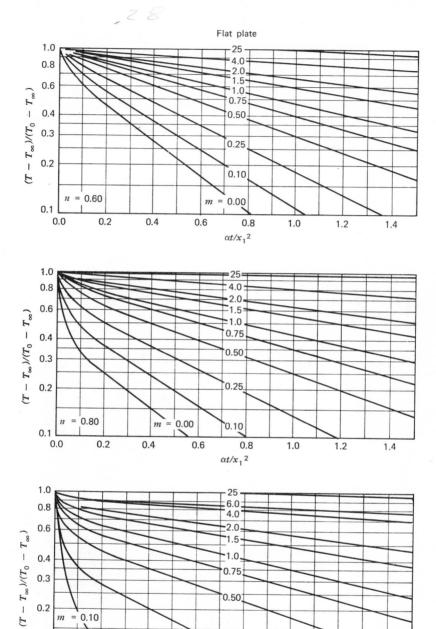

Figure B.4 (continued)

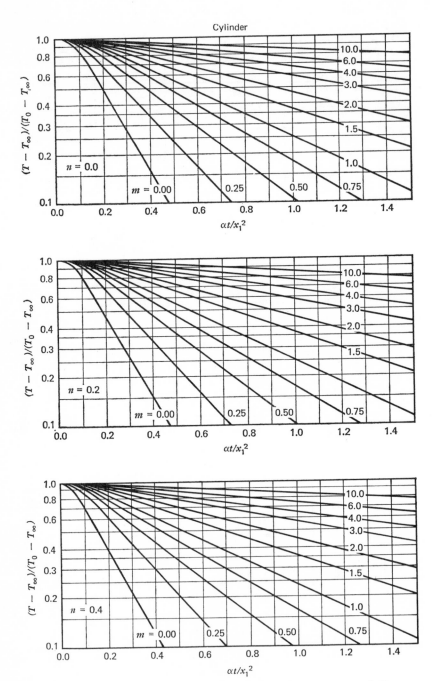

Figure B.5 Charts for solution of unsteady conduction problems: cylinder.

436

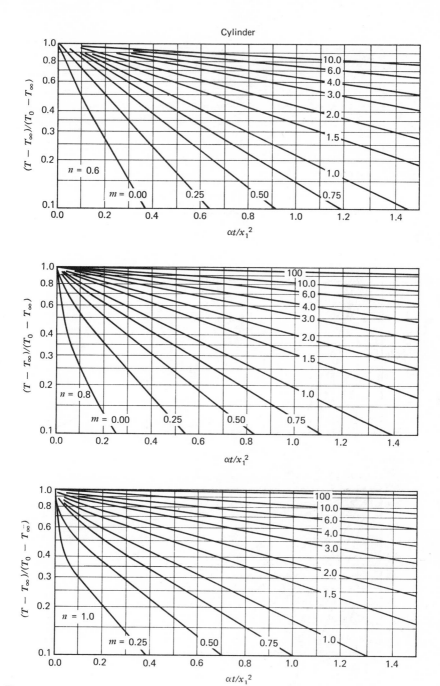

Figure B.5 (continued)

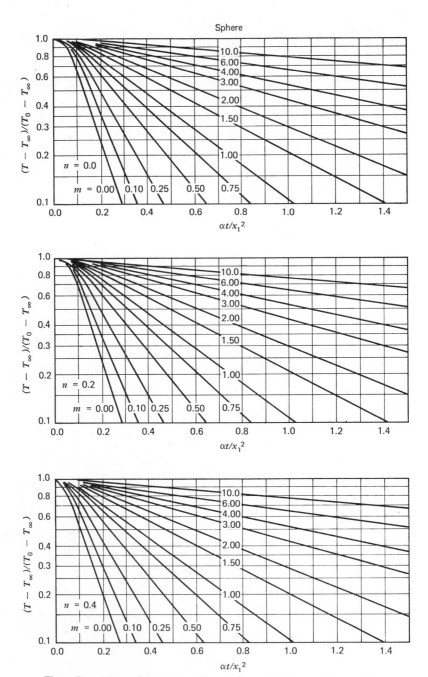

Figure B.6 Charts for solution of unsteady conduction problems: sphere.

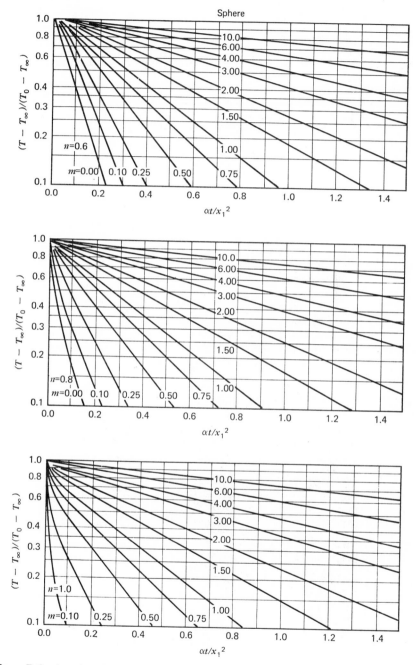

Figure B.6 (continued)

439

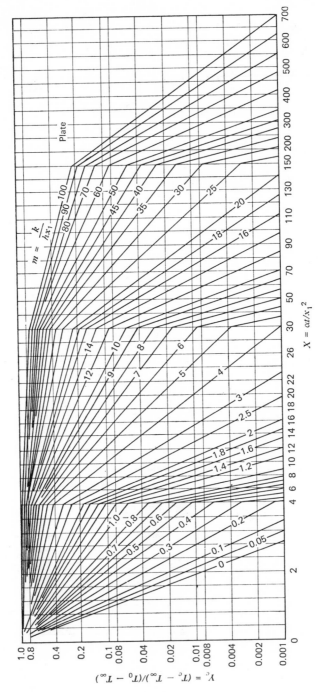

Figure B.7 Center temperature history for an infinite plate.

440

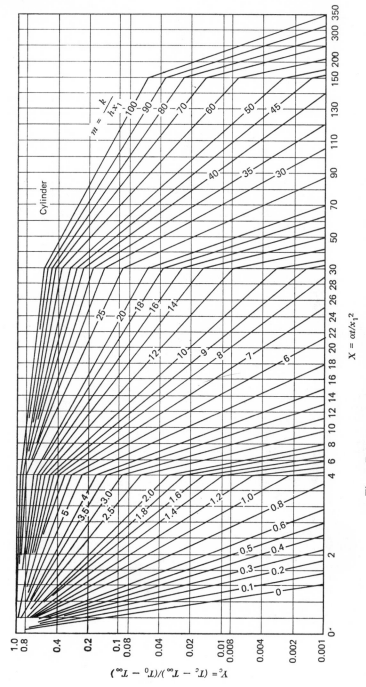

Figure B.8 Center temperature history for an infinite cylinder.

$X = \alpha t / x_1^2$

$Y_c = (T_c - T_\infty)/(T_0 - T_\infty)$

441

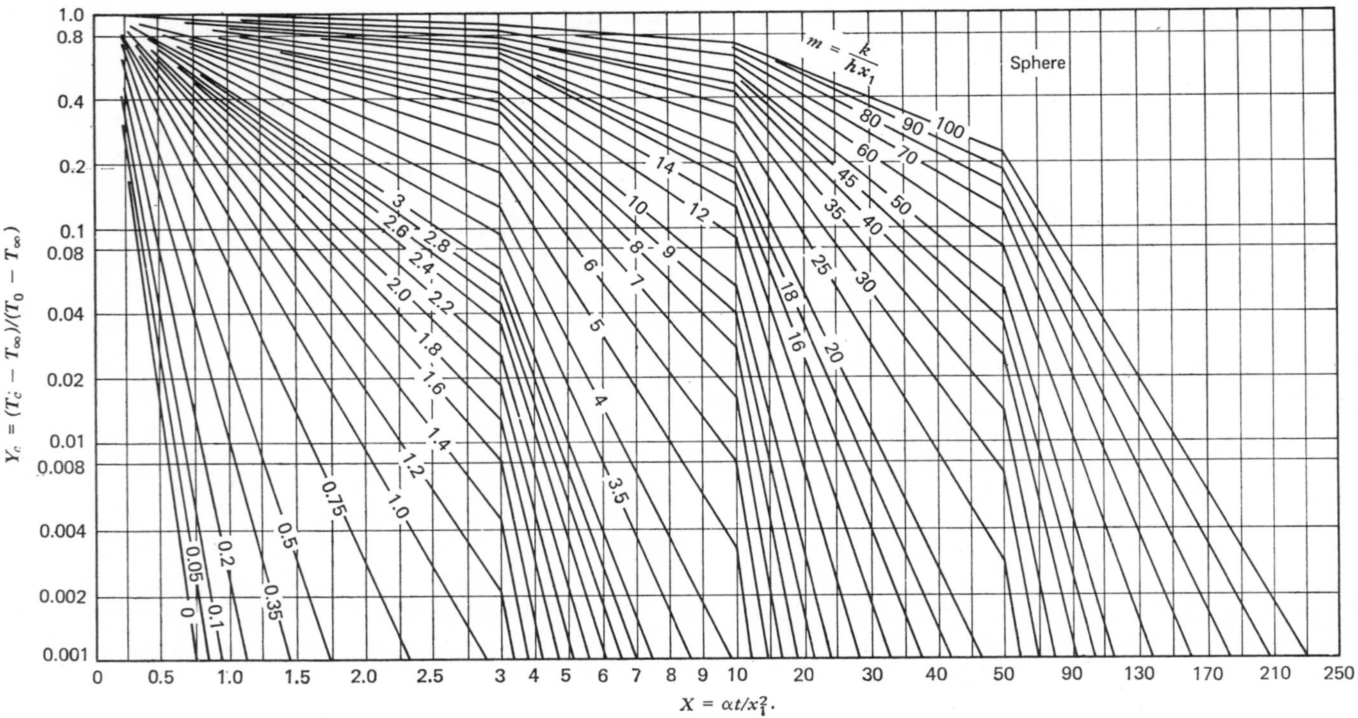
Sphere
m = k/hx1
100
90
80
70
60
50
45
40
35
30
25
20
18
16
14
12
10
9
8
7
6
5
4
3.5
3
2.8
2.6
2.4
2.2
2.0
1.8
1.6
1.4
1.2
1.0
0.75
0.5
0.35
0.2
0.1
0.05
0
1.0
0.8
0.4
0.2
0.1
0.08
0.04
0.02
0.01
0.008
0.004
0.002
0.001
0
0.5
1.0
1.5
2.0
2.5
3
4
5
6
7
8
9
10
20
30
40
50
90
130
170
210
250
Yc = (Tc − T∞)/(T0 − T∞)
X = αt/x1².

Figure B.9 Center temperature history for a sphere.

C

THE ERROR FUNCTION

$$\text{erf } \phi \equiv \frac{2}{\sqrt{\pi}} \int_0^{\phi} e^{-\eta^2} \, d\eta$$

ϕ	erf ϕ	ϕ	erf ϕ
0	0.0	0.85	0.7707
0.025	0.0282	0.90	0.7970
0.05	0.0564	0.95	0.8209
0.10	0.1125	1.0	0.8427
0.15	0.1680	1.1	0.8802
0.20	0.2227	1.2	0.9103
0.25	0.2763	1.3	0.9340
0.30	0.3286	1.4	0.9523
0.35	0.3794	1.5	0.9661
0.40	0.4284	1.6	0.9763
0.45	0.4755	1.7	0.9838
0.50	0.5205	1.8	0.9891
0.55	0.5633	1.9	0.9928
0.60	0.6039	2.0	0.9953
0.65	0.6420	2.2	0.9981
0.70	0.6778	2.4	0.9993
0.75	0.7112	2.6	0.9998
0.80	0.7421	2.8	0.9999

NORMAL TOTAL EMISSIVITIES
OF VARIOUS SURFACES

Surface	$T(^\circ F)$	Emissivity
Metals and Their Oxides		
Aluminum		
Highly polished plate, 98.3% pure	440–1070	0.039–0.057
Commercial sheet	212	0.09
Oxidized at 1110°F	390–1110	0.11–0.19
Heavily oxidized	200–940	0.20–0.31
Brass		
Polished	100–600	0.10
Oxidized by heating at 1110°F	390–1110	0.61–0.59
Chromium (see nickel alloys for Ni–Cr steels)		
Polished	100–2000	0.08–0.36
Copper		
Polished	212	0.052
Plate heated at 1110°F	390–1110	0.57
Cuprous oxide	1470–2010	0.66–0.54
Molten copper	1970–2330	0.16–0.13
Gold		
Pure, highly polished	440–1160	0.018–0.035
Graphite	100–5000	0.41–0.73

(continued through page 447)

444

Surface	$T(^\circ F)$	Emissivity
Metals and Their Oxides		

Iron and steel (not including stainless)
  Metallic surfaces (or very thin oxide layer)

Iron, polished	800–1880	0.14–0.38
Cast iron, polished	392	0.21
Wrought iron, highly polished	100-480	0.28
Oxidized surfaces		
Iron plate, completely rusted	67	0.69
Steel plate, rough	100–700	0.94–0.97
Molten surfaces		
Cast iron	2370–2550	0.29
Mild steel	2910–3270	0.28
Lead		
Pure (99.96%), unoxidized	260–440	0.057–0.075
Gray oxidized	75	0.28
Magnesium	100–1000	0.07–0.18
Molybdenum	1000–5000	0.08–0.29
Nickel		
Pure, polished	500–1000	0.07–0.10
Pure, oxidized	100–1000	0.39–0.67
Nickel alloys		
Chromnickel	125–1894	0.64–0.76
Copper-nickel, polished	212	0.059
Nichrome wire, bright	120–1830	0.65–0.79
Nichrome wire, oxidized	120–930	0.95–0.98
Platinum		
Pure, polished plate	440–1160	0.054–0.104
Strip	1700–2960	0.12–0.17
Filament	80–2240	0.036–0.192
Wire	440–2510	0.073–0.182
Silver		
Polished, pure	440–1160	0.020–0.032
Polished	100–700	0.022–0.031
Stainless steels		
Inconel X, polished	−300–900	0.19–0.20
Inconel B, polished	−300–900	0.19–0.22
Type 301, polished	75	0.16
Type 310, smooth	1500	0.39
Type 316, polished	400–1900	0.24–0.31
Tantalum	2500–5000	0.20–0.30
Tungsten		
Filament, aged	80–6000	0.032–0.35
Filament	6000	0.39
Polished coat	212	0.066

Surface	$T(°F)$	Emissivity
Metals and Their Oxides		
Zinc		
Commercial 99.1% pure, polished	440–620	0.045–0.053
Galvanized sheet, fairly bright	100	0.23
Oxidized by heating at 750°F	750	0.11
Refractories, Building Materials, and Miscellaneous		
Asbestos		
Board	74	0.96
Paper	100–700	0.93–0.94
Asphalt	solar	0.85
Brick		
Red, rough, but no gross irregularities	70	0.93
Glazed	2012	0.75
Building	1832	0.45
Fireclay	1832	0.75
Carbon		
Filament	1900–2560	0.526
Lampblack-waterglass coating	209–440	0.96–0.95
Thin layer of same on iron plate	69	0.927
Clay tiles		
Red	solar	0.71
Dark purple	solar	0.82
Concrete, rough	100	0.94
Concrete tiles		
Uncolored	solar	0.65
Brown	solar	0.85

Surface	$T(°F)$	Emissivity
Refractories, Building Materials, and Miscellaneous		
Glass		
Smooth	72	0.94
Pyrex, lead, and soda	500–1000	0.95–0.85
Gypsum, 0.02 in. thick on smooth or blackened		
plate	70	0.903
Magnesite refractory brick	1832	0.38
Marble, light gray, polished	72	0.93
Oak, planed	70	0.90
Paints, lacquers, varnishes		
Black or white lacquer	100–200	0.80–0.95
Flat black lacquer	100–200	0.96–0.98
Oil paints, 16 different, all colors	212	0.92–0.96
A1 paint, after heating to 620°F	300–600	0.35
Plaster, rough lime	50–190	0.91
Porcelain, glazed	70	0.92
Roofing paper	69	0.91
Rubber		
Hard, glossy plate	74	0.94
Soft, gray, rough (reclaimed)	76	0.86
Slate	100	0.67–0.80
Water	32–212	0.95–0.963
Ice	32	0.97

PLANCK RADIATION FUNCTIONS

$\lambda T [\mu^\circ R]$	$\dfrac{E_{b\lambda} \times 10^5}{\sigma T^5}$	$\dfrac{F_{0-\lambda T}}{\sigma T^4}$	$\lambda T [\mu^\circ R]$	$\dfrac{E_{b\lambda} \times 10^5}{\sigma T^5}$	$\dfrac{F_{0-\lambda T}}{\sigma T^4}$
1000.0	0.000039	0.0000	10400.0	5.142725	0.7183
1200.0	0.001191	0.0000	10600.0	4.921745	0.7284
1400.0	0.012008	0.0000	10800.0	4.710716	0.7380
1600.0	0.062118	0.0000	11000.0	4.509291	0.7472
1800.0	0.208018	0.0003	11200.0	4.317109	0.7561
2000.0	0.517405	0.0010	11400.0	4.133804	0.7645
2200.0	1.041926	0.0025	11600.0	3.959010	0.7726
2400.0	1.797651	0.0053	11800.0	3.792363	0.7803
2600.0	2.761875	0.0098	12000.0	3.633505	0.7878
2800.0	3.882650	0.0164	12200.0	3.482084	0.7949
3000.0	5.093279	0.0254	12400.0	3.337758	0.8017
3200.0	6.325614	0.0368	12600.0	3.200195	0.8082
3400.0	7.519353	0.0507	12800.0	3.069073	0.8145
3600.0	8.626936	0.0668	13000.0	2.944084	0.8205
3800.0	9.614973	0.0851	13200.0	2.824930	0.8263
4000.0	10.463377	0.1052	13400.0	2.711325	0.8318
4200.0	11.163315	0.1269	13600.0	2.602997	0.8371
4400.0	11.714711	0.1498	13800.0	2.499685	0.8422
4600.0	12.123821	0.1736	14000.0	2.401139	0.8471
4800.0	12.401105	0.1982	14200.0	2.307123	0.8518

(continued)

$\lambda T[\mu\,^\circ\text{R}]$	$\dfrac{E_{b\lambda} \times 10^5}{\sigma T^5}$	$\dfrac{F_{0-\lambda T}}{\sigma T^4}$	$\lambda T[\mu\,^\circ\text{R}]$	$\dfrac{E_{b\lambda} \times 10^5}{\sigma T^5}$	$\dfrac{F_{0-\lambda T}}{\sigma T^4}$
5000.0	12.559492	0.2232	14400.0	2.217411	0.8564
5200.0	12.613057	0.2483	14600.0	2.131788	0.8607
5400.0	12.576066	0.2735	14800.0	2.050049	0.8649
5600.0	12.462308	0.2986	15000.0	1.972000	0.8689
5800.0	12.284687	0.3234	16000.0	1.630989	0.8869
6000.0	12.054971	0.3477	17000.0	1.358304	0.9018
6200.0	11.783688	0.3715	18000.0	1.138794	0.9142
6400.0	11.480102	0.3948	19000.0	0.960883	0.9247
6600.0	11.152254	0.4174	20000.0	0.815714	0.9335
6800.0	10.807041	0.4394	21000.0	0.696480	0.9411
7000.0	10.450309	0.4607	22000.0	0.597925	0.9475
7200.0	10.086964	0.4812	23000.0	0.515964	0.9531
7400.0	9.721078	0.5010	24000.0	0.447405	0.9579
7600.0	9.355994	0.5201	25000.0	0.389739	0.9621
7800.0	8.994419	0.5384	26000.0	0.340978	0.9657
8000.0	8.638524	0.5561	27000.0	0.299540	0.9689
8200.0	8.290014	0.5730	28000.0	0.264157	0.9717
8400.0	7.950202	0.5892	29000.0	0.233807	0.9742
8600.0	7.620072	0.6048	30000.0	0.207663	0.9764
8800.0	7.300336	0.6197	40000.0	0.074178	0.9891
9000.0	6.991475	0.6340	50000.0	0.032617	0.9941
9200.0	6.693786	0.6477	60000.0	0.016479	0.9965
9400.0	6.407408	0.6608	70000.0	0.009192	0.9977
9600.0	6.132361	0.6733	80000.0	0.005521	0.9984
9800.0	5.868560	0.6853	90000.0	0.003512	0.9989
10000.0	5.615844	0.6968	100000.0	0.002339	0.9991
10200.0	5.373989	0.7078			

F

STANDARD PIPE AND TUBING

GAGES

F-1 Standard Pipe Sizes

Nominal Pipe Size (in.)	Outside Diam. (in.)	Schedule No.	Wall Thickness (in.)	Inside Diam. (in.)	Cross-Sectional Area Metal (in.²)	Inside Sectional Area (ft²)
$\frac{1}{8}$	0.405	40	0.068	0.269	0.072	0.00040
		80	0.095	0.215	0.093	0.00025
$\frac{1}{4}$	0.540	40	0.088	0.364	0.125	0.00072
		80	0.119	0.302	0.157	0.00050
$\frac{3}{8}$	0.675	40	0.091	0.493	0.167	0.00133
		80	0.126	0.423	0.217	0.00098
$\frac{1}{2}$	0.840	40	0.109	0.622	0.250	0.00211
		80	0.147	0.546	0.320	0.00163
		160	0.187	0.466	0.384	0.00118
$\frac{3}{4}$	1.050	40	0.113	0.824	0.333	0.00371
		80	0.154	0.742	0.433	0.00300
		160	0.218	0.614	0.570	0.00206

(continued through page 452)

Nominal Pipe Size (in.)	Outside Diam. (in.)	Schedule No.	Wall Thickness (in.)	Inside Diam. (in.)	Cross-Sectional Area Metal (in.^{2})	Inside Sectional Area (ft^2)
1	1.315	40	0.133	1.049	0.494	0.00600
		80	0.179	0.957	0.639	0.00499
		160	0.250	0.815	0.837	0.00362
1½	1.900	40	0.145	1.610	0.799	0.01414
		80	0.200	1.500	1.068	0.01225
		160	0.281	1.338	1.429	0.00976
2	2.375	40	0.154	2.067	1.075	0.02330
		80	0.218	1.939	1.477	0.02050
		160	0.343	1.689	2.190	0.01556
2½	2.875	40	0.203	2.469	1.704	0.03322
		80	0.276	2.323	2.254	0.02942
		160	0.375	2.125	2.945	0.02463
3	3.500	40	0.216	3.068	2.228	0.05130
		80	0.300	2.900	3.016	0.04587
		160	0.437	2.626	4.205	0.03761
4	4.500	40	0.237	4.026	3.173	0.08840
		80	0.337	3.826	4.407	0.07986
		120	0.437	3.626	5.578	0.07170
		160	0.531	3.438	6.621	0.06447
5	5.563	40	0.258	5.047	4.304	0.1390
		80	0.375	4.813	6.112	0.1263
		120	0.500	4.563	7.953	0.1136
		160	0.625	4.313	9.696	0.1015
6	6.625	40	0.280	6.065	5.584	0.2006
		80	0.432	5.761	8.405	0.1810
		120	0.562	5.501	10.71	0.1650
		160	0.718	5.189	13.32	0.1469

Nominal Pipe Size (in.)	Outside Diam. (in.)	Schedule No.	Wall Thickness (in.)	Inside Diam (in.)	Cross-Sectional Area Metal (in.^{2})	Inside Sectional Area (ft^2)
8	8.625	20	0.250	8.125	6.570	0.3601
		30	0.277	8.071	7.260	0.3553
		40	0.322	7.981	8.396	0.3474
		60	0.406	7.813	10.48	0.3329
		80	0.500	7.625	12.76	0.3171
		100	0.593	7.439	14.96	0.3018
		120	0.718	7.189	17.84	0.2819
		140	0.812	7.001	19.93	0.2673
		160	0.906	6.813	21.97	0.2532
10	10.75	20	0.250	10.250	8.24	0.5731
		30	0.307	10.136	10.07	0.5603
		40	0.365	10.020	11.90	0.5475
		60	0.500	9.750	16.10	0.5158
		80	0.593	9.564	18.92	0.4989
		100	0.718	9.314	22.63	0.4732
		120	0.843	9.064	26.24	0.4481
		140	1.000	8.750	30.63	0.4176
		160	1.125	8.500	34.02	0.3941
12	12.75	20	0.250	12.250	9.82	0.8185
		30	0.330	12.090	12.87	0.7972
		40	0.406	11.938	15.77	0.7773
		60	0.562	11.626	21.52	0.7372
		80	0.687	11.376	26.03	0.7058
		100	0.843	11.064	31.53	0.6677
		120	1.000	10.750	36.91	0.6303
		140	1.125	10.500	41.08	0.6013
		160	1.312	10.126	47.14	0.5592

F-2   Standard Tubing Gages

Outside Diam. (in.)	Wall Thickness B.W.G. and Stubs' Gage	(in.)	Inside Diam. (in.)	Cross-Sectional Area (ft²)	Inside Sectional Area (ft²)
½	12	0.109	0.282	0.1338	0.000433
	14	0.083	0.334	0.1087	0.000608
	16	0.065	0.370	0.0888	0.000747
	18	0.049	0.402	0.0694	0.000882
	20	0.035	0.430	0.0511	0.001009
¾	12	0.109	0.532	0.2195	0.00154
	13	0.095	0.560	0.1955	0.00171
	14	0.083	0.584	0.1739	0.00186
	15	0.072	0.606	0.1534	0.00200
	16	0.065	0.620	0.1398	0.00210
	17	0.058	0.634	0.1261	0.00219
	18	0.049	0.652	0.1079	0.00232
1	12	0.109	0.782	0.3051	0.00334
	13	0.095	0.810	0.2701	0.00358
	14	0.083	0.834	0.2391	0.00379
	15	0.072	0.856	0.2099	0.00400
	16	0.065	0.870	0.1909	0.00413
	17	0.058	0.884	0.1716	0.00426
	18	0.049	0.902	0.1463	0.00444
1¼	12	0.109	1.032	0.3907	0.00581
	13	0.095	1.060	0.3447	0.00613
	14	0.083	1.084	0.3042	0.00641
	15	0.072	1.106	0.2665	0.00677
	16	0.065	1.120	0.2419	0.00684
	17	0.058	1.134	0.2172	0.00701
	18	0.049	1.152	0.1848	0.00724
1½	12	0.109	1.282	0.4763	0.00896
	13	0.095	1.310	0.4193	0.00936
	14	0.083	1.334	0.3694	0.00971
	15	0.072	1.358	0.3187	0.0100
	16	0.065	1.370	0.2930	0.0102
	17	0.058	1.384	0.2627	0.0107
	18	0.049	1.402	0.2234	0.0109

(continued)

Outside Diam. (in.)	B.W.G. and Stubs' Gage	Wall Thickness (in.)	Inside Diam (in.)	Cross-Sectional Area (ft²)	Inside Sectional Area (ft²)
1¾	10	0.134	1.482	0.6803	0.0120
	11	0.120	1.510	0.6145	0.0124
	12	0.109	1.532	0.5620	0.0128
	13	0.095	1.560	0.4939	0.0133
	14	0.083	1.584	0.4346	0.0137
	15	0.072	1.606	0.3796	0.0141
	16	0.065	1.620	0.3441	0.0143
2	10	0.134	1.732	0.7855	0.0164
	11	0.120	1.760	0.7084	0.0169
	12	0.109	1.782	0.6475	0.0173
	13	0.095	1.810	0.5686	0.0179
	14	0.083	1.834	0.4998	0.0183
	15	0.072	1.856	0.4359	0.0188
	16	0.065	1.870	0.3951	0.0191

PROBLEMS_____

1.1 Air is enclosed in the space between the outside and inside walls of a house. This space is $3\frac{5}{8}$ in. wide and sufficiently large in the other dimensions that heat transfer may be considered one-dimensional. If we may assume conduction to be the dominant mode of heat transfer between the walls at temperatures 122°F and 73°F, respectively, what steady-state heat flux would occur?

If, somehow, this space were evacuated to a pressure of 0.05 atmospheres, what heat flux would exist?

Compare the heat flux calculated for air with that when the space between walls is filled with rock wool insulation.

1.2 Glass storm windows are used to decrease heat transfer between the inside of a house and its surroundings. On a cold day, when the outside air temperature

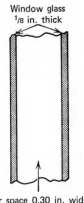

Window glass
$^1/_8$ in. thick

Air space 0.30 in. wide

is $-10°F$ and the air inside the house is maintained at $71°F$, how much heat will be lost through a picture window measuring 6 ft by 12 ft? A section through the storm window is shown in the figure.

The convective heat transfer coefficients at the inside and outside window surfaces are 3.5 Btu/hr-ft²-°F and 2.5 Btu/hr-ft²-°F, respectively.

What will be the temperature drop through each of the glass panes?

At what average temperature will the enclosed air be?

1.3 Compare the heat loss through the storm window under the conditions described in the previous problem with that through a regular glass window using a $\frac{1}{8}$-inch-thick pane. The values of h given in problem 1.2 may be used.

1.4 The walls of a house are constructed with a 4-inch-thick layer of brick, $\frac{1}{2}$ in. of Celotex, an air space $3\frac{5}{8}$ in. thick, another $\frac{1}{2}$-in. layer of Celotex, and $\frac{1}{4}$ in. of wood paneling. The temperature at the outside surface of the brick is $30°F$ and that at the inside of the wood paneling (exposed to room air) is $74°F$.

Find the heat flux through this wall if
  (a) the air space transfers heat by conduction only.
  (b) there are convection currents in the air space making the equivalent conductance of this space 2.0 Btu/hr-ft²-°F.
  (c) the space between inside and outside walls is filled with glass wool.

$$k_{\text{brick}} = 0.38 \text{ Btu/hr-ft-}°F$$
$$k_{\text{Celotex}} = 0.027 \text{ Btu/hr-ft-}°F$$
$$k_{\text{wool}} = 0.12 \text{ Btu/hr-ft-}°F$$
$$k_{\text{glass wool}} = 0.027 \text{ Btu/hr-ft-}°F$$

1.5 Given the same conditions and dimensions on the composite wall described in the previous problem, determine the heat flux for inside and outside air temperatures of $74°F$ and $20°F$, respectively, with convective heat transfer coefficients of 3 Btu/hr-ft²-°F and 8.5 Btu/hr-ft²-°F at inside and outside wall surfaces, respectively. Solve this problem for each of the three cases specified in problem 1.4.

1.6 Consider the case of steady-state one-dimensional conduction through a medium having a cross-sectional area decreasing linearly from the value A_0 at $x = 0$ to A_L at $x = L$. Evaluate the heat flow rate. The temperatures at $x = 0$ and $x = L$ are T_0 and T_L, respectively.

1.7 A boiler is to be insulated such that the heat loss will not exceed 750 Btu per hour per square foot of wall area. What thickness of asbestos is required if the inner and outer surfaces of the insulation are to be at $1500°F$ and $400°F$, respectively?

1.8 If the boiler in the previous problem is to be insulated with asbestos having one surface at $1500°F$, what thickness of insulation will be required if the adjacent air is to be at $80°F$ with an h value of 11 Btu/hr-ft²-°F? What will be the surface temperature of the insulation? The heat flux is to be no more than 750 Btu/hr-ft².

1.9 A composite wall is to be made of $\frac{3}{8}$-inch stainless steel, 3 in. of corkboard, and $\frac{3}{4}$ in. of plastic ($k = 1.3$ Btu/hr-ft-°F). The outside surface of the stainless steel is to be 250°F and the temperature of the plastic surface will be 60°F.

(a) Determine the thermal resistance of each layer of material.

(b) Determine the heat flux.

(c) Find the temperature at each surface of the corkboard.

1.10 In the previous problem the same composite wall is exposed to gases at 250°F and 70°F with convective heat transfer coefficients of 45 Btu/hr-ft²-°F and 6 Btu/hr-ft²-°F at the steel and plastic surfaces, respectively. Find

(a) the heat flux.

(b) the surface and interfacial temperatures for the wall.

(c) which individual thermal resistance is controlling.

1.11 A steel plate 1/2-inch thick and with surfaces measuring 4 ft by 6 ft lies on the ground on a summer day when the surrounding air is at 95°F. The heat transfer coefficient between the steel surface and surrounding air is 1.0 Btu/hr-ft²-°F. Solar energy is incident on the steel with a flux of 360 Btu/hr-ft²-°F. If the steel surface is considered black, what will be its surface temperature under the conditions described?

1.12 For the conditions of the previous problem, estimate the surface temperature of the steel if it is "gray" with an emissivity of 0.85. (85% of the radiant energy incident on the surface is absorbed; the other 15% is reflected; 85% of the black-body emission is actually emitted from the surface.) The sky may be assumed to absorb all radiant energy leaving the steel surface. What would be the temperature of the steel surface if convection were neglected?

1.13 Solve problem 1.11 for the temperature of the steel plate if it is in a rack with air at 95°F adjacent to both surfaces. Solar energy will still be incident on one side.

1.14 A 2-inch-thick aluminum plate measuring 10 inches in diameter is heated from below by a hot plate with its top surface and edges exposed to air at 75°F. The surface conductance (including radiation) may be taken as 8 Btu/hr-ft²-°F. How much heat must be supplied by the hot plate if the upper surface of the aluminum is to be 160°F?

1.15 Compare the solution to problem 1.14 with the case of a circular disk of the same dimension made of

(a) copper.

(b) stainless steel.

(c) asbestos.

(d) lead.

1.16 A 1-1/2-inch schedule-80 steel pipe with its outside surface at 400°F is to be located in air at 90°F with the convective heat transfer coefficient between the steel surface and surrounding air equal to 1.6 Btu/hr-ft²-°F. It is proposed to add 85% magnesia insulation to the pipe surface to reduce the heat loss to one-half of that for the bare pipe. What thickness of insulation should be used if the heat transfer coefficient, h_0, and outside steel surface temperature retain the same values given above?

1.17 Solve the previous problem with all conditions as specified except that the surface conductance varies according to $h_0 = 1.1/D_o^{\frac{1}{4}}$, where D_0 is the outside diameter of the insulation in feet.

1.18 Saturated steam at 400 psia enters a 2-inch schedule-40 mild steel pipe. The pipe is insulated with 3 in. of 85% magnesia and is surrounded by 55°F air. The convective heat transfer coefficients at the inside pipe surface and at the outside surface of the insulation are 270 Btu/hr-ft²-°F and 4.5 Btu/hr-ft²-°F, respectively. Determine

    (a) the thermal resistance of each portion of the heat flow path.
    (b) the total heat loss per foot of insulated pipe.
    (c) the heat flux based upon the outside surface area of the steel pipe.
    (d) the temperatures at the inside pipe surface, the steel-magnesia interface, and outside surface of the insulation.
    (e) the amount of steam condensed in a 20-foot length of pipe.
    (f) the overall heat transfer coefficient for the insulated pipe.

1.19 A plane wall has thickness L and its two surfaces maintained at temperatures T_0 and T_L, respectively. If the thermal conductivity of the wall material varies according to $k = k_0(1 + \alpha T + \beta T^2)$, develop an expression for the steady-state one-directional heat flux.

1.20 An asbestos pad, square in cross section, measures 4 in. on a side at its large end; the sides decrease linearly to 2 in. at the small end. The pad is 6 in. high with the temperatures at the large and small ends held at 800°F and 150°F, respectively. If one-dimensional heat conduction is assumed and convective losses from the sides are neglected, what will be the rate of heat conduction through the asbestos?

1.21 Work the previous problem for the case where the temperature at the small end is 800°F and that at the large end is 150°F.

2.1 Starting with the cylindrical form of the heat equation, under steady-state conditions Laplace's equation applies and appears in the form of equation (2-56).

    (a) To what form does the expression reduce if conduction is in the radial direction only?
    (b) Solve for the temperature variation $T(r)$ in the case of radial conduction with the boundary conditions

$$T(r_i) = T_i$$
$$T(r_o) = T_o$$

    (c) Express the heat flow rate q_r, which follows from the expression generated in (b).
    (d) What is the shape factor for this configuration?

2.2 For steady-state conduction in spherical coordinates, Laplace's equation is of the form given by equation (2-57).

    (a) What form does this equation take for conduction in the radial direction only?

    (b) Solve for $T(r)$ in the case of radial conduction with boundary conditions

$$T(r_i) = T_i$$
$$T(r_o) = T_o$$

    (c) Using the temperature profile from (b), evaluate the heat flow rate q_r.

    (d) What is the shape factor for this configuration?

2.3 The sketch below shows a cylindrical wall with the surfaces at $r = r_o$ and $r = r_i$ insulated such that heat flow is in the θ direction.

    (a) Explain (or show) why, under these conditions, isotherms are radial lines and heat flow is in the θ direction.

    (b) Reduce equation (2-56) to the applicable form for the conditions specified.

    (c) Solve for $T(\theta)$ with boundary conditions $T(o) = T_h$, $T(\pi/2) = T_c$.

    (d) Using the result of (c), express q_θ, the heat flow rate in the θ direction.

    (e) What is the shape factor for this configuration?

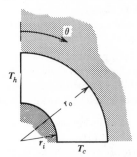

2.4 Solve (d) and (e) of problem 2.3, using the result of problem 2.1(d).

2.5 An electric household iron with a stainless steel sole plate is initially at room temperature T_0. When the switch is turned on, electrical energy generates heat uniformly at a rate of $\dot{q}$ Btu/hr-ft^3 in the steel sole plate. For a volume V of stainless steel, and for surface area A, exposed to air with a convective coefficient between the iron and room air of h, it is desired to find the temperature of the steel sole plate as a function of time.

    (a) Formulate this problem, using a lumped parameter approach, assuming T as a function of time only.

    (b) Solve the expression from (a) for $T(t)$.

    (c) Formulate this problem, using a differential approach. Model the steel sole plate as a rectangular solid, with uniform internal generation of heat. Convection will occur from all surfaces but one.

2.6 A nuclear fuel rod may be modeled as a long cylinder with uniform volumetric generation of energy $\dot{q}$, and convection at the cylindrical surface to coolant with the convective heat transfer coefficient h applying at the surface. For an initial fuel rod temperature T_0, coolant temperature T, rod diameter D, and all properties of the fuel rod material considered constant,

(a) formulate the problem from a lumped parameter approach to find rod temperature as a function of time.

(b) solve the equation derived in (a) to express $T(t)$.

2.7 If the cylindrical fuel rod of problem 2.6 is very long, it may be safely assumed that $T = T(r, t)$.

(a) Formulate this problem, starting with the general differential expression for conduction, equation (2-50).

(b) Solve the equation generated in (a) for the steady-state temperature distribution in the fuel rod with the conditions specified.

2.8 If the cylindrical fuel rod described in problem 2.6 is sufficiently slender and its conductivity is good enough, we may safely assume that $T = T(x)$ only, i.e., that temperature is lumped in the r direction. For such a case, with a fuel rod of length L, and with internal generation of energy a function of x, i.e., $\dot{q} = \dot{q}(x)$,

(a) formulate the problem to find $T(x, t)$ starting with the more complete expression, equation (2-50).

(b) solve for the steady-state temperature distribution, $T(x)$, under the conditions specified.

2.9 A flat plate, initially at temperature T_0, is placed in a bath at temperature T_∞ with a heat transfer coefficient h applying at the surface-liquid interface. The plate thickness is $2L$ and its properties may be considered constant. The bath temperature may also be taken as constant.

(a) Formulate the problem in lumped parameter form, i.e., for $T(t)$.

(b) Solve the equation from (a).

(c) Formulate the problem in differential form, i.e., for $T(x, t)$.

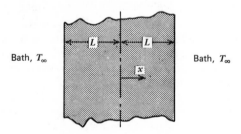

(d) Sketch the temperature variation in x that you would expect from $t = 0$ to $t = \infty$.

2.10 A solid sphere of radius R initially at uniform temperature T_0 is suddenly dropped into a bath at temperature T_∞. For T_∞ and solid properties constant,

(a) formulate the problem in lumped parameter form, i.e., $T = T(t)$.

(b) solve for $T(t)$ in the lumped parameter case.

(c) formulate the problem in differential form, i.e., for $T(r, t)$.

2.11 A liquid at temperature T_e enters a channel as shown between two hot plates with a uniform energy flux imposed at both walls. The liquid flows at constant velocity v, and its properties may be taken to be constant. Under these conditions,

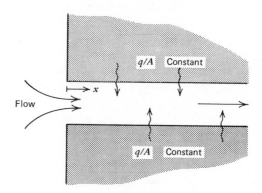

(a) formulate the problem for finding the lumped fluid temperature as a function of x, the distance down the channel from the entrance.

(b) solve for $T(x)$.

2.12 Solve the previous problem for $T(x)$ in the case of a fluid flowing in the channel between two parallel walls but with the heat flux from both walls given by the expression

$$\frac{q}{A} = \alpha + \beta \sin \frac{\pi x}{L}$$

where α and β are constants and L is the length of the plate assembly in the x direction.

2.13 A liquid at temperature T_e enters a tube of diameter D, which is uniformly wrapped with a heating element such that a constant heat flux is imposed at the tube wall. The liquid flows at constant velocity v, and its properties may be assumed constant. For these conditions,

(a) formulate the problem in which the lumped fluid temperature is expressed as a function of x, the distance downstream from the tube entrance.

(b) solve for $T(x)$.

2.14 Given the conditions of the previous problem with the exception that wall heat flux is a function of x, i.e., $q/A = (q/A)(x)$,

(a) formulate the problem for finding $T(x)$.

(b) solve for $T(x)$, given that $q/A = \alpha + \beta \sin \pi x/L$.

(c) sketch the variation of q/A and T with x. How will these sketches change if α and/or β increase?

2.15 Nuclear reactor fuel elements may be considered as flat plates in which thermal energy is generated uniformly at a rate $\dot{q}$ Btu/hr-ft^3. The plates are clad with material having thickness δ, in which no energy generation occurs. Coolant at temperature T_∞ flows between the plates, and the heat transfer coefficient h may be considered constant between the plate surface and the coolant. It is desired that the temperature variation within a fuel element be found. Dimensions are shown in the sketch below.

(a) Formulate the problem for finding $T(x)$, in the fuel elements and cladding.
(b) Solve for $T_p(x)$ in the fuel elements, and $T_c(x)$ in the cladding.
(c) Sketch the temperature variation in the fuel elements and cladding.
(d) Show how the temperature profile sketched in (c) will change if h is increased.

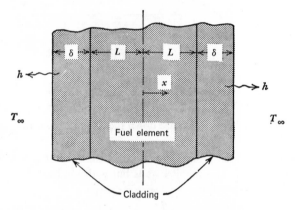

2.16 Solve problem 2.14 for the case of a fuel element being a cylinder of diameter D, clad with material of uniform thickness δ.

2.17 A plate of thickness L has its surfaces exposed to fluids at temperatures T_1 and T_2, respectively, where $T_1 > T_2$.

The heat transfer coefficients h_1 and h_2 may be used to express the convective heat fluxes at surfaces 1 and 2, respectively. For a plate initially at the uniform temperature T_0, where $T_0 < T_2$, formulate the problem in lumped parameter form and solve for $T(t)$.

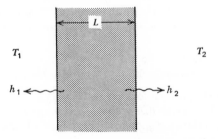

2.18 For the plate and conditions described in the previous problem, formulate the problem in differential form.
How will your formulation change if $T_1 > T_0 > T_2$?

2.19 For problem 2.17, reduce the differential formulation to steady-state conditions. Solve for $T(x)$ and sketch the steady-state temperature profile.

2.20 Approximating the earth's surface as that of a semi-infinite medium, formulate the problem in which the temperature within the earth, given as $T(x, t)$, is to be found when the initial temperature of the earth was uniform at T_0, and a constant solar flux is incident on the surface at $x = 0$.

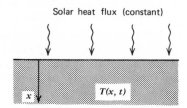

Solar heat flux (constant)

$T(x, t)$

Sketch the temperature profile you expect within the earth after a short time t_1 has elapsed and after a longer period of elapsed time t_2.

2.21 Repeat the previous problem but include the consideration of convection a the surface with surface conductance h and air temperature T_0.

3.1 A finned surface is shown in the accompanying sketch. The temperature at the fin root is T_0, the temperature of the surrounding air is T_∞, and the surface conductance is h. For the node arrangement shown, formulate the problem in difference form for finding the steady-state temperature distribution, using the heat balance approach.

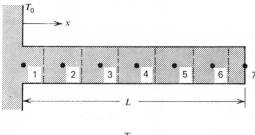

T_0

x

1 2 3 4 5 6 7

L

T_∞

3.2 If, in the previous problem, the material is stainless steel, the cross-sectional area of the fin is 0.004 ft², the fin perimeter is 0.035 ft, $h = 100$ Btu/hr-ft²-°F, $T_0 = 850$°F, and $T_\infty = 1800$°F, estimate

$w = .5'$
$L = .6'$
$t = .02'$

(a) the steady-state temperature distribution in the fin.
(b) the rate of heat transfer to the fin root.

3.3 Formulate numerically the transient problem in which the finned surface described in problem 3.1 is initially at temperature T_0 and is subjected to air at T_∞ for $t > 0$.

Develop formulations in which the solution will be

(a) explicit (forward difference for the time derivative).
(b) implicit (backward difference for the time derivative).
(c) implicit (Crank-Nicholson method for considering time derivatives).

Comment on stability criteria for each approach used.

3.4 Formulate the situation described in problem 2.3 numerically. Given $r_o = 5$ in., $r_i = 4$ in., $T_h = 400°F$, and $T_c = 250°F$, solve numerically for the steady-state circumferential temperature distribution.

3.5 Given the hollow rectangular section shown in the sketch with known surface temperatures, formulate the problem for numerical determination of the steady-state temperatures at the 20 nodes. Note that symmetry can substantially decrease the work in solving for temperature distribution.

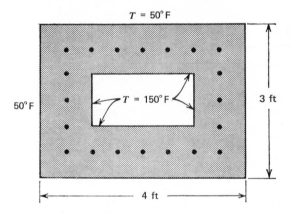

$T = 50°F$

$50°F$

$T = 150°F$

3 ft

4 ft

3.6 For the situation depicted in the previous problem, solve for the temperatures at each internal node, using Gauss elimination. The wall material is masonry brick.

3.7 Solve for the nodal temperatures in problem 3.4, using Gauss-Seidel iteration. The wall material is masonry brick.

3.8 Formulate the transient heat conduction problem in which the system described in problem 3.5 is initially at a uniform temperature of $50°F$. Accomplish this formulation in

(a) explicit form.
(b) implicit form, using a first backward difference for $\partial T / \partial t$.
(c) implicit form, using the Crank-Nicholson approach.

3.9 Problem 2.9 describes a one-dimensional transient situation. Formulate this problem numerically with the node arrangement as sown in the sketch in

(a) explicit form.
(b) implicit form, using a first backward difference for $\partial T/\partial t$.
(c) implicit form, using the Crank-Nicholson approach.

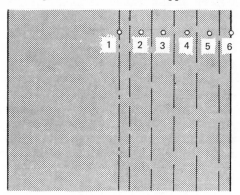

3.10 Develop a numerical formulation for the steady-state situation described in problem 2.15 for a clad nuclear reactor fuel element.

3.11 For the previous problem, develop the numerical formulation for the transient problem with the fuel element initially at uniform temperature T_∞. What time increments would be recommended in solving this problem? Develop your schemes in

(a) explicit form.
(b) implicit form, using a first backward difference for $\partial T/\partial t$.
(c) implicit form, using the Crank-Nicholson approach.

4.1 A composite wall is constructed of 1/2 in. of aluminum, 1/2 in. of corkboard, and 1/8 in. of plastic ($k = 1.3$ Btu/hr-ft^2-°F). The outside temperature of the aluminum is 250°F and the plastic surface is held at 80°F. How much heat will be transferred per square foot of wall surface under these conditions? What will be the temperatures on either side of the corkboard?

4.2 For the composite wall described in problem 4.1, what will be the heat flux if 1/4-inch aluminum rivets are used to hold the wall together? The rivets are in a square array and are located on 3-inch centers.

4.3 The composite wall described in problem 4.1 is exposed to 325°F air on the aluminum side ($h = 60$ Btu/hr-ft^2-°F) and to 60°F air on the plastic side ($h = 6$ Btu/hr-ft^2-°F). What will be the heat flux and the temperatures at the aluminum and plastic wall surfaces?

4.4 For the temperatures and values of h given in the previous problem, determine the heat flux and surface temperatures for the composite wall with aluminum rivets as described in problem 4.2. Does any one thermal resistance control heat flow in this case? If so, which one?

4.5   A pipe carrying 300 psi saturated steam passes through a room in which the air temperature is 80°F. If the values of h on the inside and outside pipe walls are 1700 Btu/hr-ft²-°F and 7 Btu/hr-ft²-°F, respectively, what will be the heat loss per foot of pipe? The pipe is 2-1/2-inch, schedule-80 mild steel.

4.6   If the steam line described in the previous problem is insulated with 2 inches of 85% magnesia, what will be the decrease in heat loss from the situation where the pipe is bare? All conditions stated for the pipe in problem 4.5 may be considered to remain the same.

4.7   If it is desired to reduce the heat loss from the bare pipe considered in problem 4.5 by 50%, what thickness of 85% magnesia insulation will be required?

4.8   A 3-inch schedule-80 mild steel steam line is to carry 250°F steam through a tunnel where the temperature may get as high as 100°F. In order that the surface temperature never exceed 140°F, how much 85% magnesia insulation must be added to the pipe wall? The values of h applying at inside and outside surfaces are 850 Btu/hr-ft²-°F and 12 Btu/hr-ft²-°F, respectively.

4.9   Solve for the required thickness of insulation when the pipe of problem 4.8 transports steam at 500°F.

4.10   A 2-inch schedule-40 steel pipe transports 60 psi saturated steam 40 ft through a room with air temperature of 70°F. The pipe is insulated with 1-1/2 in. of 85% magnesia which costs $0.66 per foot. If the cost of heating steam is $0.28 per 10^5 Btu, how long must the line be in service to justify the cost of insulation? The value of h at the outside surface of the insulation is 6 Btu/hr-ft²-°F.

4.11   Solve for the required length of service for the insulated steam pipe of problem 4.10 if the line is exposed to 0°F air for 60 ft of its length with h on the outside surface equal to 16 Btu/hr-ft²-°F.

4.12   Liquid nitrogen is stored in a spherical container measuring 14 inches in diameter. At one atmosphere total pressure the saturation temperature of nitrogen is 139.3°R, the latent heat of evaporation is 85.7 Btu/lb$_m$, and the specific volume of saturated liquid and saturated vapor are 0.020 and 3.48 ft³/lb$_m$, respectively.

It is desired to insulate the outside of the spherical stainless steel container with asbestos such that some liquid nitrogen will remain 36 hours after the vessel is filled. How thick should the asbestos insulation layer be? The vessel is vented to the atmosphere at 75°F, and the outside surface coefficient is 8 Btu/hr-ft²-°F.

4.13   How fast will one-half of the liquid nitrogen in the spherical vessel described in the previous problem be lost if the 1/4-inch-thick stainless steel container is not insulated? The surface coefficient may be taken to be 8 Btu/hr-ft²-°F.

4.14   Liquid nitrogen, properties for which are given in problem 4.12, is stored in a Dewar flask, which may be considered a spherical container with inside diameter of 10 inches. The flask walls are made of 0.040-inch-thick Pyrex glass with 1/4 in. of evacuated space between. Repeated observations indicate that

0.5 lb_m of the liquid nitrogen is lost by evaporation each 48-hour period. What is the effective conductance of the gas at low pressure in the flask walls under these conditions if h outside is 6 Btu/hr-ft²-°F and the room air surrounding the flask is at 72°F?

4.15   The steady-state rate of heat conduction through a plane wall is, according to equation (4-5), $q = (kA/L)\,\Delta T$. For steady-state conduction through a cylindrical wall, the heat flow rate is given by equation (4-10) as

$$q = \frac{2\pi kL}{\ln \dfrac{r_0}{r_i}} \Delta T$$

(a) Show that equation (4-10) can be put into the form of equation (4-5), where the effective area $\bar{A}$ in the expression

$$q = \frac{k\bar{A}}{r_o - r_i} \Delta T$$

is the logarithmic mean area given by

$$A_{lm} = \frac{2\pi L(r_o - r_i)}{\ln \dfrac{r_o}{r_i}}$$

(b) If the arithmetic mean area $\pi L(r_o + r_i)$, rather than the logarithmic mean, is used for a cylindrical wall, what will be the percent error for values of r_o/r_i of 1.5, 3, and 5?

4.16   Find the appropriate value of $\bar{A}$ for a spherical wall when heat flow is expressed in the form

$$q = \frac{k\bar{A}}{r_o - r_i} \Delta T$$

Repeat (b) of problem 4.15 for the spherical case.

4.17   Nichrome, having a resistivity of 110 $\mu\Omega$-cm, is to be used as a heating element in a 10-kw heater. The nichrome surface temperature should not exceed 2400°F. Other design features include

input power is available at 12 volts.
surrounding air temperature is 200°F.
outside surface coefficient is 200 Btu/hr-ft²-°F.

(a) What diameter nichrome is necessary for a 3-foot-long heater under these conditions?
(b) What length of 14-gage wire will be required to satisfy the given criteria?
(c) Are the answers to (a) and (b) extremely sensitive to h? What will be the answers to (a) and (b) if h is 200 Btu/hr-ft²-°F?

4.18 A sandwich heater is constructed of nichrome wire wound back-and-forth, covered by a 1/8-inch layer of asbestos on both sides, with 1/8 in. of stainless steel on the outside of each piece of asbestos. For an outside surface co-efficient of 3 Btu/hr-ft²-°F and a maximum temperature of nichrome limited to 1200°F, how much power in watts/ft² must be supplied to the heater? What will be the stainless steel surface temperature?

4.19 A 1-inch nominal copper tube has its outside surface maintained at 230°F. It is proposed to add fins made of 3/32-inch-thick copper 3/4 in. high to the tube surface to increase the heat transfer rate. If 12 equally spaced longitudinal fins of the dimensions given are added to the tube surface, what will be the percent increase in heat transfer from the tube? The surrounding air is at 80°F and the surface coefficient may be taken as 1.8 Btu/hr-ft²-°F.

4.20 To establish the importance of thermal conductivity on finned-tube heat transfer rates, rework problem 4.19 for materials with the following values of thermal conductivity:

$$k = 132 \text{ Btu/hr-ft-°F (aluminum)}$$
$$k = 70 \text{ Btu/hr-ft-°F (brass)}$$
$$k = 40 \text{ Btu/hr-ft-°F (iron)}$$
$$k = 24.8 \text{ Btu/hr-ft-°F (mild steel)}$$
$$k = 9.8 \text{ Btu/hr-ft-°F (stainless steel)}$$

Plot percent increase in heat transfer for fins of the dimensions prescribed versus thermal conductivity.

4.21 Construct a plot of percent increase in heat transfer of a finned tube over a bare tube as a function of outside surface coefficient with all other values as given in problem 4.19. Consider values of h ranging from 1–150 Btu/hr-ft²-°F.

4.22 Circular fins measuring 1/32 in. thick and 3/4 in. high are to be added to the 1-inch tube of problem 4.19 whose surface is at 230°F. If 12 copper fins are added per foot, find the percent increase in heat transfer compared to the bare tube. As in problem 4.19, the surface coefficient and air temperature are 1.8 Btu/hr-ft²-°F and 80°F, respectively.

4.23 A metal rod 1 inch in diameter has one end in a furnace and the remainder exposed to 75°F air with a surface coefficient of 4 Btu/hr-ft²-°F applying. If, on the rod, temperatures measured 3 inches apart are 258°F and 197°F, estimate the thermal conductivity of the rod material.

4.24 A steel I-beam with dimensions as shown has its upper and lower flanges adjacent to materials at 1000°F and 250°F, respectively. The surrounding air is at 80°F, and a value of $h = 12$ Btu/hr-ft²-°F applies at the metal surfaces.

(a) Assuming a negligible temperature change through each flange, generate an expression for the web temperature T, as a function of x, the distance away from the 1000°F flange.
(b) Plot $T(x)$.
(c) Evaluate the rate of heat transfer per foot into or out of the web at both flanges.

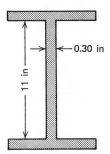

4.25 Repeat problem 4.24 but with the I-beam made of aluminum. All other information can be considered the same as for the previous problem.

4.26 An iron bar 2 ft long is used as a support in a stack through which combustion gases at 800°F pass. The bar is 1/2 inch in diameter; the surface heat transfer coefficient is 150 Btu/hr-ft²-°F. Determine the maximum temperature along the bar if

(a) the chimney walls where the bar is attached are maintained at 350°F.

(b) the bar ends are held at temperatures of 300°F and 400°F, respectively.

4.27 Saturated steam at atmospheric pressure is to be condensed on the outside surface of a tube whose temperature is controlled by 50°F water running through the inside. The tube is 1-inch 16-BWG nickel; the effective value of h at the inside tube surface is 300 Btu/hr-ft²-°F, and on the condensing vapor side, $h = 2200$ Btu/hr-ft²-°F. It is proposed to add circular nickel fins measuring 1/16 in. thick and 7/8 in. high, with 1 in. between centers, to the tube surface. What increase in condensation rate should be expected with the fins added?

4.28 How much heat must be added to solder two very long pieces of 1/8-inch-diameter copper wire together? The surrounding air temperature is 70°F and the surface coefficient may be taken as 3 Btu/hr-ft²-°F. The solder temperature must be at least 450°F before it can melt.

4.29 A copper rod, 3/8 inch in diameter and 3 ft long, runs between two copper bus bars whose temperatures remain at 60°F. The surrounding air temperature is 60°F and the surface coefficient is 5 Btu/hr-ft²-°F. How much current can the copper rod carry if its temperature is not to exceed 150°F? The resistivity of copper is 1.72×10^{-6} Ω-cm.

4.30 A cylindrical rod is added to the surface of a hot container so that it can be carried. It is desired that the center temperature of the handle not be above 120°F. For the dimensions of the handle as shown (p. 470), 70°F surrounding air temperature, and surface coefficient equal to 5 Btu/hr-ft²-°F, how hot may be the surface to which a 3/8-inch mild steel handle is attached?

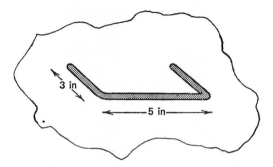

4.31 Example 4.6 dealt with a two-dimensional fin, the solution for the case of surface temperatures given according to $T(0, y) = F(y)$, $T(x, \pm l) = T_\infty$, $T(\infty, y) = T_\infty$; the solution is expressed by equation (4-55). Evaluate $T(x, y)$, using equation (4-55) as the starting point when the function $F(y)$ is $T_{max} \cos \pi y/2l$.

4.32 Solve for $T(x, y)$ as in the previous problem with the function $F(y)$ given as $F(y) = T_{max} l^2 [1 - (y/l)^2]$.

4.33 A 2-inch-OD pipe with its surface maintained at 200°F is placed eccentrically within a 6-inch-ID pipe whose surface is at 60°F. The center of the 2-inch pipe is 3/4 in. away from the center of the 6-inch pipe. The space between the pipes is filled with asbestos. Find the rate of heat transfer between pipes by (a) flux plotting and (b) using the shape factor from Table 4.3.

4.34 Hot flue gases at 450°F are transported through the 6-inch-ID circular opening in a concrete chimney structure with the dimensions shown. Inside and outside temperatures of the chimney surfaces are maintained at 420°F and 100°F, respectively. Find the heat loss per foot of chimney using (a) flux plot techniques and (b) Table 4.3.

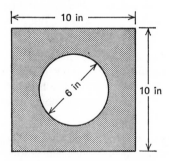

4.35 A rectangular tunnel 3 ft wide and 6 ft high is dug in permafrost ($k = 0.06$ Btu/hr-ft-°F) with the top at a depth of 2 ft below the surface, which is at

~10°F. If the tunnel walls are at 40°F, find the heat loss to the surface, using flux plotting techniques.

4.36 A pipe line carrying fuel oil is laid in permafrost with the pipe walls maintained at 35°F. For ground surface temperature of −30°F, find the heat loss if the center of the 8-inch-OD pipe is 4 ft below the surface. Compare the results obtained, using flux plotting techniques, with those using the shape factor determined from Table 4.3. For permafrost, $k = 0.06$ Btu/hr-ft²-°F.

4.37 What will be the rate of heat transfer from the surface of the earth to a spherical container 3 ft in diameter with the container center at a depth of 6 ft below the earth's surface? The container holds liquid oxygen at −297°F, and the earth's surface is 60°F. Thermal conductivity of soil may be taken as 0.4 Btu/hr-ft-°F.

4.38 Two pipes are buried well below the surface of the earth ($k = 0.4$ Btu/hr-ft-°F), one containing steam at 240°F, the other water at 55°F. What will be the rate of heat transfer between the two fluids per foot of pipe if the steam line is 8 inches in diameter, the water line has a diameter of 3 in., and they are 12 in. apart center to center? Solve with a flux plot and with a shape factor from Table 4.3.

4.39 A calrod heating element 0.5 inch in diameter is encased in the center of a 3-inch-square aluminum block. The calrod-aluminum interface is at a temperature of 600°F and the outer aluminum surface is at 250°F. What will be the heat loss per foot of this composite system?

4.40 For conditions given in the previous problem, find the heat loss if construction errors result in the calrod heater being in the positions shown.

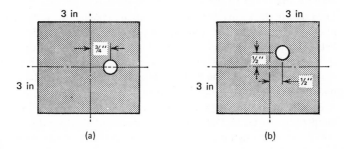

(a)   (b)

4.41 How much steam will condense in 30 ft of 3-inch-OD pipe buried 1 ft below ground level as shown (p. 472)? The surface of the ground is 60°F and the steam enters saturated at 450°F. The thermal conductivity of dry earth may be taken as 0.20 Btu/hr-ft-°F.

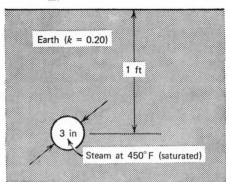

4.42 Most of the preceding 11 problems involving two-dimensional steady-state heat transfer can be solved, using the numerical procedures similar to those employed in Examples 4.10 and 4.11. Starting with the two-dimensional form of Laplace's equation, develop a numerical technique, flow chart, and FORTRAN program for solving typical two-dimensional steady-state conduction problems, which is compatible with the computer system available to you. Using your program, solve numerically

    (a) problem 4.31.
    (b) problem 4.32.
    (c) problem 4.34.
    (d) problem 4.39.
    (e) problem 4.40.

4.43 A copper bar, initially at 400°F, measures 0.2 ft by 0.5 ft by 10 ft in length. How long after being exposed to 80°F air with $h = 12$ Btu/hr-ft²-°F will it take for the center of the copper bar to reach a temperature of 200°F?

4.44 Lead shot in the form of spheres 0.2 inch in diameter are quenched in 90°F oil after being heated to a uniform temperature of 420°F. For a value of h between the lead spheres and the oil of 30 Btu/hr/ft²-°F, evaluate the time-temperature history of a shot after it is dropped into the oil bath. If a shot reaches the bottom of the bath 12 seconds after its first contact, what will be its temperature when it touches bottom?

4.45 A cylinder measuring 2 ft in height with a diameter of 3 in. is initially at a temperature of 60°F. Determine the amount of time for the center to reach 600°F when the cylinder is placed in a 1200°F medium with $h = 16$ Btu/hr-ft²-°F at the surface, for cylinders made of the following materials:

    (a) stainless steel.
    (b) aluminum.
    (c) copper.
    (d) mild steel.
    (e) asbestos.

4.46 Given a 3-inch-diameter aluminum cylinder initially at 60°F placed in a 1200°F medium with a surface coefficient of 16 Btu/hr-ft²-°F, determine the time required for the center temperature to reach 600°F for cylinder heights of

    (a) 3 inches.
    (b) 6 inches.
    (c) 1 foot.
    (d) 2 feet.
    (e) 5 feet.

4.47 A tire carcass is placed in a jig with 320°F steam admitted all around with an extremely large value for the effective surface coefficient. The thermal diffusivity of rubber is approximately 0.0025 ft²/hr. For an initial temperature of 70°F, an effective tire thickness of 1.25 in, and a required temperature at the central layer of 275°F, determine the time required for vulcanization to be completed.

4.48 How long will the heat treatment take for a slab of glass initially at a uniform temperature of 70°F to have its temperature everywhere reach at least 730°F? The glass blank is exposed to 810°F air with a surface coefficient of 3 Btu/hr-ft²-°F applying. The glass is cylindrical, measuring 16 inches in diameter and 1.75 in. thick.

4.49 A carrot whose shape may be approximated as a cylinder 8 in. long and 3/4 inch in diameter is initially at room temperature, 70°F, and then dropped into boiling water at atmospheric pressure. Properties of the system and the carrot may be taken as follows:

$$h = 350 \text{ Btu/hr-ft}^2\text{-°F}$$
$$k = 0.28 \text{ Btu/hr-ft-°F}$$
$$c_p = 0.95 \text{ Btu/lb}_m\text{-°F}$$
$$\rho = 64 \text{ lb}_m/\text{ft}^3$$

How long must the carrot cook if the requirement is that the minimum temperature reached be 195°F?

4.50 Determine the required cooking time for a 6-ounce turnip exposed to the same conditions as described for the carrot in the preceding problem. Properties of carrots and turnips may be considered equal. Assume the turnip to be spherical.

4.51 Find the temperature at the center of a solid concrete block measuring 1 ft by 1 ft by 4 in. thick after 2 hours' exposure to saturated steam at 210°F. The concrete was initially at a uniform temperature of 80°F; the surface coefficient may be taken as 300 Btu/hr-ft²-°F.

4.52 A stainless steel billet, in the shape of a cylinder 10 in. long and 4 inches in diameter, is heated to a uniform temperature of 1200°F preparatory to forming. How long may the billet sit in still 80°F air, with a surface coefficient of 8 Btu/hr-ft²-°F applying, before its surfaces reach the limiting temperature of 900°F? What will be the center temperature in the billet at this time?

4.53 Two sheets of plastic, initially at 70°F, each 3/4 in. thick, are to be bonded together with glue which sets at 300°F. To heat the plastic sheets such that the glued layer reaches the required temperature, two heated steel plates, held at 320°F, are placed and held next to the plastic surfaces. Determine (a) the required time for the glue to set and (b) the temperatures 1/4 in. and 1/2 in. from the steel at the time the setting temperature of the glue is reached. The properties of plastic which apply are

$$k = 0.090 \text{ Btu/hr-ft-}°F$$

$$\alpha = 0.0030 \text{ ft}^2/\text{hr}$$

4.54 Water, initially at 50°F, is contained in a thin-walled cylindrical vessel 24 inches in diameter. The water and its container are placed in an oil bath held at a uniform temperature of 280°F. The cylinder is immersed to a depth of 2 ft, the water is well stirred and is 2 ft deep in the container, and the heat transfer coefficient at the cylinder-oil interface is 35 Btu/hr-ft²-°F. Plot the water temperature as a function of time up to 1 hr.

4.55 Lord Kelvin estimated the age of the earth to be 9.8×10^7 years, assuming the original temperature of the earth to be 7000°F and using the following properties of the earth's crust:

$$\alpha = 0.0456 \text{ ft}^2/\text{hr}$$

$$T = 0°F$$

$$\left.\frac{\partial T}{\partial y}\right|_{y=0} = 0.02°F/\text{ft (observed)}$$

Using the expression for unsteady-state conduction in a semi-infinite medium, determine whether or not Lord Kelvin's estimate of the earth's age is reasonable.

4.56 Referring to the previous problem, determine the depth below the surface of the earth where the rate of cooling is greatest.

4.57 A brick wall, initially at 100°F, is exposed to 50°F air with a surface coefficient of 6 Btu/hr-ft²-°F. If the wall is taken as semi-infinite, what will its surface temperature be after 1 hr? 2 hr? 10 hr? Find the temperature at a depth of 1 ft corresponding to each of the elapsed times stated.

4.58 Using the integral solution for a semi-infinite wall developed in Section 4.2–3.2, find the penetration depth for the brick wall and conditions specified in the previous problem.

4.59 A concrete cylinder, 6 inches in diameter and 16 in. long, has its ends capped, which essentially insulates them against heat flow. For steam at atmospheric pressure surrounding the cylinder with a surface coefficient of 22 Btu/hr-ft²-°F, show graphically the temperature profile in the cylinder after 5 minutes, 30 minutes, 1 hour, 2 hours, and 6 hours of elapsed time. The cylinder was initially at a uniform temperature of 80°F.

4.60 Find the location and magnitude of the minimum temperature in the concrete cylinder described in the previous problem at each of the times specified. Work this problem for the condition of (a) one end uncapped and (b) both ends uncapped with heat transfer through both ends.

4.61 A large concrete wall 18 in. thick may be considered infinite in the other two directions. From an initial condition of uniform temperature at 65°F throughout, the wall is exposed to hot gas at 350°F on one surface with a heat transfer coefficient at the surface of 45 Btu/hr-ft²-°F. Air at 65°F and a surface coefficient of 1.5 is adjacent to the other wall. (a) What will be the temperatures at each surface after 1 hour of exposure to these conditions? (b) How long will it take for the center of the wall to reach a temperature of 100°F? (c) When the center temperature reaches 100°F, how much energy will the wall have absorbed per square foot?

4.62 A 2-inch-thick plate, considered infinite in the other directions, is initially at a uniform temperature of 400°F. The plate is then immersed in 120°F oil with a value of h equal to 12 Btu/hr-ft²-°F applying at both surfaces. Plot the temperature profile across the plate after 15 min of elapsed time for materials with the following values of thermal conductivity:

(a) 132 Btu/hr-ft²-°F
(b) 23 Btu/hr-ft²-°F
(c) 8.2 Btu/hr-ft²-°F
(d) 0.73 Btu/hr-ft²-°F
(e) 0.20 Btu/hr-ft²-°F
(f) 0.10 Btu/hr-ft²-°F

4.63 Given the 2-inch plane surface and other conditions from the previous problem, evaluate the time required for the center temperature to reach 100°F if it is extracted from the oil bath after 15 minutes of immersion. When taken out of the oil, 65°F air is adjacent to both surfaces and an h value of 3 Btu/hr-ft²-°F applies. Solve for each of the materials specified in problem 4.62.

4.64 An asbestos wall 6 in. thick is initially at 70°F uniformly. One surface of the wall is suddenly exposed to hot gases at 600°F with a surface coefficient given by $h = 0.10(\Delta T)^{1/3}$, where ΔT is the temperature difference between the hot gas and the wall surface in °F and h has units of Btu/hr-ft²-°F. If the other wall is assumed insulated, how long will it be until the insulated wall reaches a temperature of 250°F? What will be the temperature of the hot wall at this time?

4.65 Given an asbestos wall 3 in. thick with a temperature profile initially linear, with values of 350°F and 150°F at each surface, respectively, determine the time it would take for the center to reach a temperature of 150°F if both surfaces are exposed to 60°F air with h values of 0.90 applying.

4.66 Determine the temperature at each surface of a plane wall after steady-state conditions have been reached under the following conditions:

wall thickness:	2 in.
initial uniform temperature through wall:	70°F
medium adjacent to cool wall:	air at 70°F
surface coefficient at cool wall, Btu/hr-ft²-°F:	$h = 0.19(\Delta T)^{1/3}$
wall material:	masonry brick

The hot wall is subjected to a constant heat flux of 100 Btu/hr-ft²-°F.

4.67 Determine the time required for the center temperature of the wall described in the previous problem to change by 95 % of the difference between initial and steady-state values.

4.68 After steady-state conditions are reached in the brick wall described in problem 4.66, the heat source is removed from the hot wall and cooling occurs from both surfaces to air at 70°F with h at both surfaces given by the expression $h = 0.19(\Delta T)^{1/3}$. How long will it take for
(a) the hot surface to reach 150°F?
(b) the center to reach 150°F?
(c) seventy-five percent of the absorbed energy while heated, to be released to the surrounding air?

4.69 Given a linear temperature profile in the earth near the surface, increasing by an amount 0.5°F per foot of depth from a value of 35°F at the surface, determine the time required for a pipe buried at a depth of 8 ft to reach a temperature of 32°F if the surface is suddenly exposed to 0°F air with a surface coefficient of 1.2 Btu/hr-ft²-°F. Properties of soil are $k = 0.8$ Btu/hr-ft-°F and $\alpha = 0.02$ ft²/hr.

4.70 For soil having the properties $k = 0.8$ Btu/hr-ft-°F and $\alpha = 0.02$ ft²/hr, determine the temperature at a depth of 1 ft below the surface after 5 hours' elapsed time when the surface is held at 1200°F. The soil temperature was initially uniform at 50°F. Compare answers to this problem, using
(a) semi-infinite wall analysis.
(b) integral analysis, assuming a parabolic temperature profile.
(c) graphical analysis.

4.71 Solve for the temperature 1/2 ft below ground surface after 5 hours, as in problem 4.70, except for the surface condition of a constant heat flux of 150 Btu/hr-ft². As in the previous problem, compare answers obtained by using
(a) semi-infinite wall (exact) analysis.
(b) integral analysis, assuming the temperature profile in soil to be parabolic.
(c) graphical analysis.

4.72 In Example 4.19 a one-dimensional transient conduction problem was solved numerically, using an explicit solution. Starting with the applicable partial differential equation, develop an explicit numerical solution technique, show the flow chart, and include a program listing compatible with your local computer system which will yield an explicit solution for one-dimensional transient conduction problems. Using your program, solve
   (a) problem 4.47.
   (b) problem 4.48.
   (c) problem 4.53.
   (d) problem 4.61.
   (e) problem 4.62.
   (f) problem 4.63.
   (g) problem 4.64.
   (h) problem 4.65.
   (i) problem 4.66.

4.73 In Example 4.20 a one-dimensional transient heat conduction problem was solved numerically, using implicit techniques. Starting with the governing partial differential equation, develop an implicit numerical solution scheme, show the flow chart, and include a program listing compatible with your local computer system for one-dimensional transient heat conduction problems. Using your program, obtain numerical solutions for
   (a) problem 4.47.
   (b) problem 4.48.
   (c) problem 4.53.
   (d) problem 4.61.
   (e) problem 4.62.
   (f) problem 4.63.
   (g) problem 4.64.
   (h) problem 4.65.
   (i) problem 4.66.

4.74 In Example 4.21 a numerical solution for multidimensional transient conduction was illustrated. Starting with the governing differential equation, develop a solution scheme, flow chart, and program listing compatible with your local computer system for solving multidimensional transient conduction problems. Using your program, obtain a numerical solution to
   (a) problem 4.45 (cylindrical coordinates).
   (b) problem 4.49 (cylindrical coordinates).
   (c) problem 4.50 (spherical coordinates).
   (d) problem 4.51 (rectangular coordinates).
   (e) problem 4.52 (cylindrical coordinates).

5.1 At what distance from the leading edge of a flat plane surface will transition from laminar to turbulent flow occur if the fluid is
   (a) air at 60°F, atmospheric pressure?
   (b) water at 80°F?

(c) glycerin at 60°F?
(d) mercury at 80°F?
Free stream velocity is 50 ft/sec.

5.2 Estimate the force on a 10-foot-diameter circular sign in a hurricane wind with velocity of 120 mph. Estimate the moment exerted at ground level if the sign is mounted on top of a 6-inch-diameter cylindrical supporting column 15 ft tall.

5.3 Estimate the steady-state rotational speed in rpm of the anemometer shown when local wind velocity is 40 mph. Friction may be assumed negligible; standard atmospheric conditions may be used.

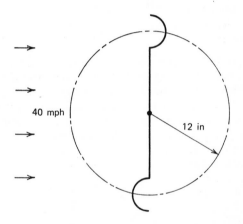

40 mph

12 in

5.4 Find a velocity profile for the laminar boundary layer of the form

$$\frac{v_x}{v_\infty} = a + by + cy^2 + dy^3$$

with a zero pressure gradient.

5.5 Develop dimensionless expressions for boundary layer thickness and local and mean skin friction coefficients for the case of a laminar boundary layer over a plane surface, assuming the velocity profile within the boundary layer to be of the form
(a) $v_x = a + by$
(b) $v_x = a + by + cy^2$
(c) $v_x = a + b \sin cy$

5.6 Compare the boundary layer thickness and the local skin friction coefficient with flow over a plane surface at a point where $\mathrm{Re}_x = 10^6$, for
(a) laminar flow
(b) turbulent flow

assuming each boundary layer begins at the leading edge of the plate. Compare the total drag exerted on the plate for each flow condition.

5.7 Find the pressure drop for hydraulic fluid at 80°F flowing in a 0.20-inch-diameter tube, 30 ft long, at a rate of 10 gal/hr.

5.8 Determine the pressure drop which must be overcome for air at 70°F and atmospheric pressure to flow through a rectangular duct measuring 10 in. by 18 in. with an average velocity of 20 ft/sec. The duct is 120 ft long.

5.9 With water at 60°F flowing through a piping system at 0.5 ft³/sec, the pressure drop is measured as 14 psi. What flow rate will occur with this same pressure drop if the fluid is
(a) air at 60°F, atmospheric pressure?
(b) Freon-12?
(c) hydraulic fluid?
(d) kerosene?
Assume flow to be fully turbulent.

5.10 Find the flow rate of 100°F hydraulic fluid which will be achieved with a pressure drop of 15 psi through a line 40 in. long, having a diameter of 0.15 in.

5.11 Equation (5-86) was developed from a dimensional analysis, taking the wall temperature as constant. Another possibility for a boundary condition at the wall is constant *heat flux*.

Using dimensional analysis, show that the modified Grashof number for the case of a specified wall heat flux is expressed as

$$Gr^* = \frac{\beta g(q/A)L^4}{\nu^2 k}$$

5.12 Using dimensional analysis, show that for unsteady-state conduction in a plane wall, the following parameters apply:

$$\frac{T - T_\infty}{T_0 - T_\infty}, \quad \frac{\alpha t}{L^2}, \quad \frac{hL}{k}, \quad \frac{x}{L}$$

5.13 A plane surface 8 in. wide and 2 ft long is maintained at a temperature of 200°F. Evaluate the local heat transfer coefficient at the end of the plate, and find the mean value of h and the total heat transferred if the fluid flowing parallel to the surface with a velocity of 10 ft/sec is
(a) air at 80°F, atmospheric pressure.
(b) water at 80°F.
(c) hydraulic fluid at 80°F.
(d) kerosene at 80°F.

5.14 Develop dimensionless expressions for thermal boundary layer thickness and local and mean Nusselt numbers for laminar boundary layer flow over a plane isothermal surface, using integral analysis. The velocity and temperature profiles within the boundary layer may be assumed of the form

(a) $\dfrac{v_x}{v_\infty} = \dfrac{T - T_s}{T_\infty - T_s} = a + by$

(b) $\dfrac{v_x}{v_\infty} = \dfrac{T - T_s}{T_\infty - T_s} = a + by + cy^2$

(c) $\dfrac{v_x}{v_\infty} = \dfrac{T - T_s}{T_\infty - T_s} = a + by + cy^2 + dy^3$

(d) $\dfrac{v_x}{v_\infty} = \dfrac{T - T_s}{T_\infty - T_s} = a + b \sin cy$

5.15 The figure shows the case of flow parallel to a plane surface where, for $0 \le x < X$, the plane surface and fluid are both at temperature T_∞. For $x \ge X$, the plate temperature T_s is greater than T_∞ and held constant. Assuming velocity and temperature profiles both of the form

$$\frac{v_x}{v_\infty} = \frac{T - T_s}{T_\infty - T_s} = a + by + cy^2 + dy^3$$

show that the ratio of boundary layer thickness is represented by

$$\xi = \frac{\delta_t}{\delta} \simeq \frac{1}{\mathrm{Pr}^{1/3}}\left[1 - \left(\frac{X}{x}\right)^{3/4}\right]^{1/4}$$

and that the local Nusselt number may be expressed as

$$\mathrm{Nu}_x \simeq 0.33\left[\frac{\mathrm{Pr}}{1 - (X/x)^{3/4}}\right]^{1/3}\mathrm{Re}^{1/2}$$

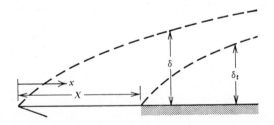

5.16 For natural convection adjacent to a plane vertical isothermal wall, show that the integral equations applying to the hydrodynamic and thermal boundary layers are

$$-\nu\left.\frac{\partial v_x}{\partial y}\right|_{y=0} + \beta g\int_0^{\delta_t}(T - T_\infty)\,dy = \frac{d}{dx}\int_0^{\delta}v_x^2\,dx$$

$$\alpha\left.\frac{\partial T}{\partial y}\right|_{y=0} = \frac{d}{dx}\int_0^{\delta_t}v_x(T_\infty - T)\,dy$$

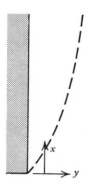

5.17 Assuming parabolic velocity and temperature profiles of the form

$$\frac{v}{v_x} = \left(\frac{y}{\delta}\right)\left(1 - \frac{y}{\delta}\right)^2$$

$$\frac{T - T_\infty}{T_s - T_\infty} = \left(1 - \frac{y}{\delta}\right)^2$$

and the integral relations given in problem 5.16, show that the following expressions result:

$$\frac{2\alpha}{\delta} - \frac{d}{dx}\left(\frac{\delta v_x}{30}\right)$$

$$-\frac{v v_x}{\delta} + \beta g \Delta T \frac{\delta}{3} = \frac{d}{dx}\left(\frac{\delta v_x^2}{105}\right)$$

Next assume that, for δ and v_x varying with x according to

$$\delta = Ax^a \qquad\qquad v_x = Bx^b$$

the boundary layer thickness and local Nusselt number become

$$\frac{\delta}{x} = 3.94(\text{Pr} + 0.953)^{1/4}\text{Pr}^{-1/2}\text{Gr}_x^{-1/4}$$

$$\text{Nu}_x = 0.508(\text{Pr} + 0.953)^{-1/4}\text{Pr}^{1/2}\text{Gr}^{1/4}$$

5.18 Starting with the equations for δ and Nu_x in the previous problem, valuate δ and h_x at $x = 1/2$ ft, 1 ft, and 5 ft for
  (a) air at 60°F, atmospheric pressure.
  (b) water at 60°F.
  (c) hydraulic fluid at 60°F.
  (d) kerosene at 60°F.
The wall temperature may be considered constant at 160°F.

5.19 Find expressions for the local and mean Nusselt numbers applying for a flat plate using the integral approach for a turbulent boundary layer with velocity and temperature profiles both given by the 1/7 power law expression according to

$$\frac{v_x}{v_\infty} = \frac{T - T_s}{T_\infty - T_s} = \left(\frac{y}{\delta}\right)^{1/7}$$

Example 5.2 may be used as a guide in your solution.

5.20 Air at 100°F, 60 psi enters the space between two adjacent fuel plates in a nuclear reactor core. Air flow is 6000 lb_m/hr-ft^2, and the fuel plates are 4 ft long with 1/2 in. of space between them. Heat flux along the plate surfaces varies according to

$$\frac{q}{A} = \alpha + \beta \sin \frac{\pi x}{L}$$

where α = 250 Btu/hr-ft^2
β = 4000 Btu/hr-ft^2
x = distance from leading edge
L = total plate length

For these conditions, prepare plots for
(a) heat flux versus x.
(b) mean air temperature versus x.

5.21 Water at 50°F enters a 1-1/2-inch 16-BWG tube 8 ft long with a velocity of 35 ft/sec. Steam at 240°F is condensing on the outside tube wall. Estimate the mean value of the heat transfer coefficient at the tubing-water interface, using
(a) the Reynolds analogy.
(b) the Prandtl analogy.
(c) the von Kármán analogy.
(d) the Colburn analogy.
Which analogy do you believe most accurate? Most inaccurate? Why?

5.22 For the conditions described in problem 5.21, compare the results obtained by using the Reynolds analogy and the Colburn analogy with regard to total heat transfer and the exiting water temperature.

5.23 Solve problem 5.21 for the case of air at 50°F entering the 1-1/2-inch tube with all other conditions being the same.

5.24 Evaluate the total heat transfer and exit air temperature for the situation considered in problem 5.23. Compare results, using the Reynolds analogy and Colburn analogy.

5.25 Problems 5.22 and 5.24 relate to two possible approaches to condenser design. If a condenser is to be used in a process where the major concern is the maximum condensation rate in minimum space, would you specify water or air as the heat transfer medium inside the condenser tubes? Justify your answer quantitatively.

5.26 Water at 50°F enters a 1-1/4-inch 16-BWG tube, 5 ft long, with a velocity of 15 ft/sec. The outside surface of the tube is wrapped uniformly with nichrome ribbon; in total, 80 kw of power is supplied to the heater, and it is determined that 94% of this energy is transferred to the water flowing in the tube. Determine

(a) the heat flux at the wall.

(b) the mean temperature of the water at the tube exit.

5.27 For the conditions described in the previous problem, evaluate and plot the variation in tube surface temperature along its axis; compare the profiles obtained by using both the Reynolds analogy and the Colburn analogy.

5.28 The heat flux applied at the surface of a 1-1/4-inch 16-BWG tube varies sinusoidally along its 5-foot length.

If the total power applied, with air axial distribution as described, is 80 kw, with 94% transferred to water flowing through the inside of the tube, evaluate and plot the tube surface temperature, using both the Reynolds and Colburn analogies. The water enters the tube at 50°F, flowing with a velocity of 15 ft/sec.

5.29 Problems 5.27 and 5.28 involve the variation in surface temperature of a tube inside which water flows subjected to a specified wall heat flux. In both cases, the total energy transmitted to the water is the same; however, the axial distribution of heat flux is constant in one case and sinusoidal in the other. How do the maximum wall temperatures compare for these two types of heat flux variation? Use the Colburn analogy for purposes of making this comparison.

5.30 Air at 50°F and a velocity of 15 ft/sec enters a 1-1/4-inch 16-BWG tube. The heat flux at the wall is uniform over the 5-foot length of copper tubing, with 80 kw of energy being supplied to the wall and 94% of this energy being transferred to the air. Under these conditions of operation, determine

(a) the wall heat flux.

(b) the mean air temperature at the tube exit.

5.31 Evaluate and plot the surface temperature variation along the tube axis for the situation described in problem 5.30. Compare results obtained by using the Reynolds and Colburn analogies.

5.32 With 50°F air entering at 15 ft/sec, evaluate and plot the variation in surface temperatures along the axis of a 1-1/4-inch 16-BWG copper tube subjected to a heat flux distribution which is sinusoidal. The total energy transmitted to the air is 94% of that supplied at the tube surface, which is 80 kw. Compare wall temperature profiles obtained by using both the Reynolds and Colburn analogies.

5.33 The fuel plates in a nuclear reactor are 4 ft long and stacked such that a 1/2-inch gap exists between each plate and those adjacent to it. The heat flux along each plate surface varies sinusoidally, according to the equation

$$\frac{q}{A} = \alpha + \beta \sin \frac{\pi x}{L}$$

where $\alpha = 250$ Btu/hr-ft^2
        $\beta = 8000$ Btu/hr-ft^2
        $x =$ distance from leading edge
        $L =$ total plate length

If air at 120°F, 80 psi, flowing at a mass velocity of 6000 lb$_m$/hr-ft^2 is used to cool the plates, evaluate and plot
(a) the heat flux versus x.
(b) mean air temperature versus x.
(c) plate surface temperature versus x, using the Colburn analogy.

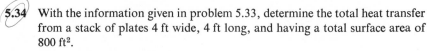

5.34 With the information given in problem 5.33, determine the total heat transfer from a stack of plates 4 ft wide, 4 ft long, and having a total surface area of 800 ft^2.

5.35 Solve problem 5.33 with all information as given except that the coolant is water entering at 120°F, with a mass velocity of 4.2×10^7 lb$_m$/hr-ft^2.

5.36 Determine the surface temperature of a 1-kw immersion heater in 80°F stagnant water. The heater is cylindrical, with a length of 6 in. and a diameter of 3/4 in. Consider both the cases when the heater axis is (a) horizontal and (b) vertical.

5.37 Solve problem 5.36 for the same conditions, except that the surrounding stagnant fluid is
(a) hydraulic fluid at 40°F.
(b) mercury at 120°F.

5.38 A steel ingot 1 ft in diameter and 20 in. high, heated to the uniform temperature of 800°F, is placed on a concrete floor and allowed to cool in 100°F ambient air. Estimate the length of time required for the surface of the ingot to cool to 150°F. What will be the maximum temperature in the ingot at this time?

5.39 A 2-inch 18-BWG aluminum tube transports steam such that its outside surface remains at 250°F. If this tube is surrounded by still air at 40°F, what heat flux will result if the tube orientation is
(a) horizontal?
(b) vertical?

5.40 Solve for the heat flux for the tube of problem 5.39, except that the surrounding fluid is 40°F water.

5.41 Liquid oxygen at −320°F is contained within a tank which is spherical with an outside diameter of 30 in. The tank is constructed of stainless steel, 1/8-inch thick. Determine the rate of heat transfer to the tank if it is surrounded by 60°F still air.

5.42 Determine the required thickness of glass wool to be applied to the spherical tank containing liquid oxygen described in the previous problem for the transfer to be reduced to one-tenth of its value for the bare tank. All specifications given in problem 5.41 still apply.

5.43 A transformer mounted on a power pole must dissipate 350 watts of power to the surrounding air which, on a summer day, may reach 95°F. For such a situation with no wind, estimate the surface temperature of the transformer. Consider the transformer to be a cylinder 20 inches in diameter and 40 in. high.

5.44 An electrical transmission line carries 6000 amps of current through a conductor with a resistance of 1.6×10^{-6} ohms per foot of length. The outside diameter of the line is 7/8 in. Estimate the surface temperature of this line if it is surrounded by still air at 85°F.

5.45 A "swimming pool" nuclear reactor consists of 30 rectangular plates, each measuring 1 ft wide and 3 ft high. These plates, oriented vertically, are spaced such that a 2-inch gap exists between each vertical surface. If the coolant is 70°F water and the maximum allowable plate temperature is 205°F, what is the maximum power level at which the reactor might operate?

5.46 A thin sheet of iron measuring 3 ft by 3 ft is supported vertically and surrounded by 60°F still air. Taking the iron to be a perfectly emitting surface and considering all emission from the plate to be absorbed, determine the surface temperature of the sheet if it is subject to a solar heat flux of 160 Btu/hr-ft².

5.47 A solar energy collector measuring 20 ft by 20 ft is installed on a rooftop in a horizontal position. The collector surface temperature is 160°F, and the incident solar heat flux is 200 Btu/hr-ft². If the surrounding still air is at 60°F, what is the heat loss from the collector by convection? What would be the convective loss if the collector were criss-crossed with ridges spaced 1 ft apart?

5.48 For flow inside a closed conduit with the conduit surface at a temperature T_0 and the "mixed-mean" temperature of the fluid expressed as T_m, the convective heat transfer coefficient, or the "uuit convective conductance," is defined as

$$ h \equiv \frac{q/A}{T_0 - T_m} $$

For a "fully developed" temperature profile, the requirement is that

$$ \frac{\partial}{\partial x}\left(\frac{T_0 - T}{T_0 - T_m}\right) = 0 $$

Show that, for a fully developed temperature profile, the convective heat transfer coefficient h, defined above, is constant.

5.49 Pursuing the discussion in Section 5.3-1.1 related to fully developed laminar flow inside a circular conduit with a fully developed temperature profile, show that, for a uniform wall heat flux, the Nusselt number has a value of 4.364.

5.50 Using the values of local Nusselt number with laminar flow in a constant heat flux pipe listed in Table 5.4, calculate and plot mean fluid temperature and pipe surface temperature for the case where an organic liquid is heated from

60°F to 150°F in a 1/4-inch-ID tube, 4 ft long, heated electrically. The liquid flow rate is 9 lb_m/hr, and its properties, which may be assumed constant, are as follows:

$$\rho = 47 \ lb_m/ft^3$$
$$c_p = 0.5 \ Btu/lb_m\text{-}°F$$
$$k = 0.079 \ Btu/hr\text{-}ft\text{-}°F$$
$$\mu = 1.6 \ lb_m/ft\text{-}hr$$

5.51 Referring to the organic fluid and conditions described in the previous problem, determine and plot the mean fluid temperature and local heat flux as a function of axial position along the 4-foot length of tube where the tube surface is maintained at the constant temperature of 210°F.

5.52 A valve on a hot water line is opened such that a flow of 0.07 ft/sec of water is allowed. The water pipe surface is maintained at 80°F, and the hot water leaves the valve at 180°F. Determine the temperature of the water leaving a 5-foot section of 1/2-inch schedule-40 pipe under these conditions. What is the heat loss?

5.53 Oil, with properties given in the table below, enters a bank of six 3/4-inch schedule-40 steel tubes at 80°F. Steam at 212°F is condensed on the outside surface of the tubes. The oil flow rate is 90 lb_m/hr, and each tube is 9 ft long. Determine the oil temperature at the heater exit and the total heat transferred.

T (°F)	ρ(lb_m/ft³)	c_p(Btu/lb_m-°F)	k(Btu/hr-ft-°F)	μ(lb_m/sec-ft)
80	56.8	0.44	0.077	27.80×10^{-3}
100	56.0	0.46	0.076	15.30×10^{-3}
150	54.3	0.48	0.075	5.30×10^{-3}
200	54.0	0.51	0.074	2.50×10^{-3}

5.54 What is the maximum velocity at which the oil whose properties are given in the preceding problem may flow in the 3/4-inch schedule-40 steel tubes and have the flow remain laminar? At this maximum flow rate, determine the total heat transfer and exiting oil temperature if all other conditions are the same as described in problem 5.53.

5.55 For the oil whose properties are given in problem 5.53, flowing at the maximum allowable velocity such that laminar flow is maintained, how many 3/4-inch schedule-40 steel tubes are required to accommodate 90 lb_m/hr of total flow? With this number of tubes, what total heat transfer to the oil will be achieved and what will be its temperature as it exits the heater?

5.56 Determine the total heat transfer from an insulated steam pipe with outside surface temperature of 140°F which passes through a room where the air temperature is 70°F. The pipe is 28 ft long and its outside diameter, including insulation, is 8-1/2 in.

5.57 Determine the heat loss from the insulated steam pipe with dimensions and surface temperature as given in problem 5.56, if the pipe extends between two buildings 28 ft apart on a day when the outside air is at 70°F and the wind velocity normal to the pipe is 50 mph.

5.58 A 1-1/2-inch 16-BWG copper tube 10 ft long has its outside surface maintained at 230°F. Air at 50°F and atmospheric pressure surrounds the tube. Compare the heat flux from the tube to air under the following conditions:
(a) horizontal tube, still air.
(b) vertical tube, still air.
(c) horizontal tube, air flows normal to tube at 45 ft/sec.
(d) horizontal tube, air flows parallel to tube axis at 45 ft/sec.

5.59 A copper wire, 1/8 inch in diameter, is insulated with a 1/4-inch-thick layer of packed glass wool ($k = 0.020$ Btu/hr-ft-°F) and surrounded by 60°F still air. The wire carries 60 amps of current; the resistivity of copper is 1.72 × 10^{-6} ohm-cm. Determine
(a) the convective heat transfer coefficient between the air and insulation surface.
(b) the temperature at either surface of the glass insulation.

5.60 Determine the heat transfer coefficient and the temperature at each surface of the packed glass wool insulation for the conditions described in the preceding problem when the air flows across the insulated wire with a velocity of 40 ft/sec.

5.61 Water is used as a coolant in a condenser in which steam at 300 psia enters a 2-1/2-inch schedule-80 steel pipe as saturated vapor and leaves as saturated liquid. The condensation process occurs in an 8-foot length of pipe with a steam rate of 1000 lb_m/hr. The cooling water flows in the annular space between the smaller pipe and a 4-inch schedule-80 pipe which is concentric with it. Water is available at 55°F. To avoid local boiling, the mean water temperature should not exceed 160°F. Is this arrangement capable of condensing all of the steam which enters? If not, how would you suggest changing the design to make it so?

5.62 Saturated liquid nitrogen at atmospheric pressure ($T_{sat} = 139°R$) is transferred between a storage cylinder and a flask through a 1/2-inch 18-BWG copper tube which is 20 in. long. If this transfer occurs when 80°F air is blowing across the copper tube at 30 ft/sec, what is the rate of heat gain during the period of time that these conditions exist?

5.63 It is determined to insulate the copper line described in the previous problem with a 1-inch thickness of 85% magnesia. If other conditions remain as specified in problem 5.62, what will be the rate of heat transfer with the insulation in place?

5.64 Water is to be used as coolant in a heat exchanger where steam condenses on the outside surface of the tubes. Design considerations establish the steam pressure, water temperatures entering and leaving the exchanger, water flow rate, and pressure drop on the water side. How does the heat transfer area for this situation vary with the inside diameter of the condenser tubes?

5.65 A water droplet, which may be analyzed as a sphere with diameter of 0.05 in., has reached its terminal velocity in 75°F air. Taking the water droplet temperature to be 145°F, what will be the heat transfer rate due to convection only?

5.66 A droplet of fuel oil at 180°F is injected at 280 ft/sec into a diesel engine combustion chamber. The droplet is 0.001 inch in diameter, and the gas in the cylinder is at 1300°F and 680 psia. Gas properties may be assumed the same as air under the same conditions of state; fuel oil properties may be taken as constant, with the following values:

$$\rho = 51.8 \; lb_m/ft^3$$
$$c_p = 0.54 \; Btu/lb_m\text{-}°F$$
$$k = 0.073 \; Btu/hr\text{-}ft\text{-}°F$$
$$\mu = 83 \times 10^{-5} \; lb_m/sec\text{-}ft$$
$$Pr = 22$$

How long will it take for a fuel droplet to reach 600°F, its auto-ignition temperature?

5.67 A cylindrical pipe whose surface temperature is 300°F has an outside diameter of 4 in. Air at 40°F blows across the pipe, normal to the pipe axis at 25 ft/sec. Estimate the heat transfer rate per foot of pipe under these conditions.

5.68 Hydraulic fluid at 175°F is pumped at a velocity of 18 ft/sec across a bundle of tubes arranged according to model number 1 in Figure 5.40. If the 3/8-inch tubes have a surface temperature of 50°F, what value will the convective heat transfer coefficient have?

5.69 Compare the heat transfer capability of 3/8-inch-OD tubes arranged in rows oriented according to model 2 in Figure 5.40 with that of the same configuration except wider spacing, as in model 5 of the same figure. Hydraulic fluid flowing with conditions specified in problem 5.68 may be used for comparison.

5.70 Compare the heat transfer capability of 3/8-inch-OD tubes arranged in banks oriented according to model number 1 in Figure 5.40 with that of the same configuration except wider spacing, as in model 4 of the same figure. Hydraulic fluid flowing with conditions specified in problem 5.68 is the fluid flowing through the tube banks.

5.71 A refrigerated truck, with its outside surface maintained at 50°F, travels along a road on a hot day at 50 mph. If the air temperature is 95°F, estimate the rate of heat transfer from the top and sides of the truck. The refrigerated portion of the truck is 8 ft wide, 10 ft high, and 12 ft long.

5.72 Estimate the rate of heat transfer by convection, on a winter day when the air temperature is 20°F, from the flat roof of a large building with inside air temperature of 72°F. The roof measures 100 ft in length by 125 ft in width and is constructed of 3/4 in. of wood ($k = 0.14$ Btu/hr-ft-°F), 3/8 in. of fibreboard ($k = 0.023$ Btu/hr-ft-°F), and 1/4 in. of tar ($k = 0.09$ Btu/hr-ft-°F). Consider the case where (a) the air is still and (b) wind is blowing with a velocity of 15 mph.

5.73 Estimate the required velocity of 80°F air to be blown across a pane of glass, initially at 350°F uniformly, such that its surface temperature will reach

100°F within 30 min. The glass is 3/8 in. thick and lies on a support such that it loses heat from its top surface only.

5.74 What is the required velocity of air to cool the glass described in problem 5.73 if cooling occurs on both sides of the glass surface?

5.75 Water at 120°F enters the tubes shown with a velocity of 5 ft/sec. The tubes are 1/2-inch 16-BWG copper and are imbedded in concrete with centers 1 in. below the concrete surface, as shown. Estimate the concrete surface temperature if still air at 65°F is adjacent to the surface. What will be the heat flux under these conditions? How much will the water temperature change per foot of pipe?

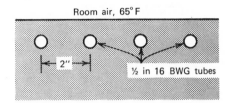

Room air, 65° F

2″

½ in 16 BWG tubes

5.76 Estimate the concrete surface temperature for the conditions of the preceding problem, except that the air flow is across the concrete surface with a velocity of 15 ft/sec. The boundary layer flow may be taken to be turbulent everywhere, with a total characteristic length of concrete of 20 ft.

The surface tension of water is temperature dependent; it may be calculated from the expression $\sigma = (5.28 \times 10^{-3})(1 - 0.0013\ T)$, where T is in °F and σ is in lb /ft. This expression may be used in determining σ, where needed, in each of the following problems.

5.77 A cylindrical copper heating element measuring 2 ft in length with a diameter of 3/4 in. is immersed in water. For a surface temperature of 250°F and a system pressure of 1 atm, estimate the value of h which applies and the rate at which heat is transferred between the heating element and the water, which is at 212°F.

5.78 Given a cylinder with the dimensions specified in problem 5.77 initially at a uniform temperature of 550°F, suddenly immersed in water at its saturation temperature and atmospheric pressure, show roughly, the time-temperature history of the cylinder. How long will it be until the cylinder temperature reaches 250°F?

5.79 Estimate the equilibrium surface temperature of a cylindrical immersion heater 5 in. long and 1 inch in diameter, rated at 500 watts. Assume the heater to be immersed in
  (a) water at 60°F, heater horizontal.
  (b) water at 212°F, atmospheric pressure.

5.80 Nichrome wire, 0.02 inch in diameter, is immersed in saturated water at 240°F. For a wire surface temperature of 2100°F, estimate the heat transfer rate per foot.

5.81   Twelve thousand watts of electrical energy are to be dissipated through copper plates measuring 4 in. by 8 in. by 1/4 in., immersed in water at its saturation temperature of 220°F. How many plates would you suggest using? What is your estimate of the surface temperature of the plates under these conditions?

5.82   A steel plate is removed from a heat treating operation at 700°F, then immediately immersed into a water bath at 212°F.
   (a) Construct a plot showing heat flux versus plate temperature for this system.
   (b) Construct a plot of h versus plate temperature.
   (c) Plot plate temperature versus time if the mild steel plate is 1 in. thick and 18 in. square.

5.83   A pan of water at room temperature is placed in an oven preheated to 450°F. Will the water boil? Explain your answer.

5.84   The heat flux between a hot pipe wall and water flowing through the pipe is to be 3×10^6 Btu/hr-ft^2. The pipe ID is 1 in. and it is 8 ft long. The water enters at 212°F. What is the recommended flow rate of water for safe operation under these conditions? Explain all design criteria leading to your recommendation.

5.85   A cylindrical piece of metal heated initially to 500°F is immersed in saturated. water at 1 atm pressure. The dimensions are 6 in. high by 3 inches in diameter. Estimate the time required for the maximum temperature in the cylinder to fall below 300°F if it is composed of
   (a) copper.
   (b) nickel.
   (c) brass.
   Assume the cylinder to be resting on one of its bases.

5.86   Saturated 200°F steam is condensed on a 3/4-inch-diameter tube with a surface temperature of 160°F. Find the value of h which applies if the tube is 5 ft long and oriented
   (a) horizontally.
   (b) vertically.
   What will be the condensation rate in each of these cases?

5.87   What will be the total heat transfer rate and rate of condensation if there are 6 tubes in a vertical bank with the tube axes horizontal, with conditions being those described in the previous problem?

5.88   A plane surface, oriented vertically, is 2 ft high and has a temperature of 75°F. What will be the heat transfer coefficient and rate of condensation if saturated steam at 1 atm pressure surrounds the surface?

5.89   A horizontal 1-inch-OD tube, surrounded by saturated steam at 200°F, has its surface temperature maintained at 140°F. What is the heat transfer coefficient on the outside tube surface for this condition? What is the total rate of heat transfer for 3 ft of tube?

5.90 Under the conditions of the previous problem, what must be the mean heat transfer coefficient on the inside tube wall if water at 55°F enters the 1-inch 15-BWG copper tube? What must be the rate of flow of water to achieve this condition? What is the corresponding pressure drop for the water?

5.91 Solve problem 5.89 if the orientation of the tube is vertical.

5.92 Saturated steam at atmospheric pressure is enclosed within a 1-inch-ID tube oriented vertically, with its surface maintained at 160°F. Construct a plot showing the fraction of pipe cross section filled with condensate as a function of distance from the top of the pipe. What happens as the fraction of cross section occupied by condensate approaches unity?

5.93 Saturated 212°F steam flows between two vertical surfaces held at 140°F which are 1/2-inch apart. The steam flow rate is 100 lb_m/hr per foot of width. How tall may this system be in order that the steam velocity not exceed 25 ft/sec?

5.94 If a circular pan has its bottom surface maintained at 200°F and is adjacent to saturated steam at 230°F, show with a plot the depth of condensate accumulated versus time up to 1 hour. Neglect any effect of the sides of the pan.

5.95 A pan, 18 in. square, has a 3/4-inch lip on all sides, and its bottom surface is maintained at 180°F. How long after being exposed to 212°F saturated steam will it take for condensate to spill over the lip if the pan orientation is
   (a) horizontal?
   (b) inclined at 5° with the horizontal?
   (c) inclined at 10° with the horizontal?

5.96 A window pane 16 in. wide by 30 in. high has air at 70°F adjacent to one surface. If saturated steam at 210°F is adjacent to the other surface, estimate the amount of steam which will be condensed in 1 hour.

6.1 It is desired to estimate surface temperatures of black emitters, using a detector which absorbs radiant energy in the wavelength band between 0.6 and 4.6 μ. What correction factor must be used with this detector to correct its readings for surfaces at
   (a) 1000°R?
   (b) 3000°R?
   (c) 5000°R?
Characterize the errors inherent in this detector at each of these temperatures, i.e., the fraction of emission falling below and above the instrument's range.

6.2 For the radiation detector described in problem 6.1, generate and plot, as a function of absolute temperature, the fraction of total surface emission that will be absorbed by the instrument. On this same plot show the fraction of emission occurring below 0.6μ and that which lies above 4.6μ.

6.3 Using Appendix F-1, generate and plot the spectral black body emissive power as a function of wavelength for a surface at
   (a) 800°R.
   (b) 2400°R.
   (c) 5000°R.

6.4 The temperature of the sun is approximately 10,000°R. If the visible light range is taken as that for which $0.4\mu < \lambda < 0.7\mu$, what fraction of solar emission is visible? What fraction of solar emission is in the ultraviolet range? the infrared range? At what wavelength is the spectral emissive power of the sun a maximum?

6.5 The "greenhouse" effect is a manifestation of transmission of low wavelength energy and absorption or reflection of higher wavelength emission. If silica glass transmits 92 % of the radiation incident in the 0.33μ to 2.6μ range and is opaque to radiation at other wavelengths, what fraction of incident solar energy will be transmitted? What fraction of the energy emitted from surfaces whose average temperature is 95°F will be transmitted?

6.6 A satellite, in a circular orbit 500 miles above the earth, has a diameter of 60 in., and its surface is made of polished aluminum. The earth may be considered to be at a constant temperature of 55°F, with an effective emissivity of 0.95. Estimate the temperature of the satellite surface.

6.7 A water tank, with its bottom and sides insulated, is covered by a glass plate having radiant transmission characteristics given as

$$\tau_\lambda = \begin{cases} 0 & 0 < \lambda < 0.33\mu \\ 0.92 & 0.33\mu < \lambda < 2.6\mu \\ 0 & 2.6\mu < \lambda \end{cases}$$

$$\rho_\lambda = 0.08 \quad \text{all } \lambda$$

The glass is 1/4 in. thick and its bottom surface is 1/2 in. above the water surface. The water is 6 in. deep. For an average incident solar irradiation of 220 Btu/hr-ft²,
   (a) determine the average water temperature if radiation effects only are considered.
   (b) determine the average water temperature, taking into account conduction and convection as well as radiation; the surrounding air temperature may be taken as 70°F.
   (c) solve (b) if the water and glass cover are in contact.

6.8 A cylindrical gasoline storage tank 25 ft high and 40 ft in diameter is painted white ($\rho = 0.82$). For a solar irradiation of 350 Btu/hr-ft², determine the temperature of the tank surface if the sun's rays are
   (a) straight down on the tank top.
   (b) at an angle of 45° from the plane of the tank top.
The air temperature adjacent to the tank surface is 60°F; be sure to include natural convection effects when the air is still.

6.9 Solve problem 6.9 for the tank surface temperature if there is a wind of 30 mph blowing across the tank.

6.10 Solve for the temperature of the tank specified in problem 6.8 when the sun's rays are directed at an angle of 45° with the plane of the top surface on a still day, with

(a) the tank top painted white ($\rho = 0.82$) and the sides gray ($\rho = 0.12$).

(b) the entire tank painted gray.

6.11 Evaluate the view factor F_{d1-d2} between differential areas dA_1 and dA_2, oriented on plane surfaces which are parallel to each other and a distance D apart.

6.12 Using the result of the previous problem, express, but do not attempt to solve, the view factor between the differential plane area $d1$ and area A_2 with dimensions shown.

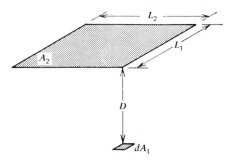

6.13 Using the result of the previous problem, express, but do not attempt to solve for, the view factor between finite areas A_1 and A_2, oriented on parallel planes with dimensions as shown in the figure.

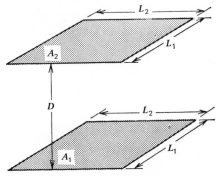

6.14 Determine the view factor between a rectangular surface A_1, measuring 20 ft by 20 ft, and another rectangular surface A_2, measuring 1 ft by 20 ft, centered above A_1 and parallel to A_1 at a distance of 6 ft.

6.15 Determine the radiant exchange between a black surface measuring 1 ft by 1 ft, at 700°F, and a black rectangular area measuring 20 ft by 20 ft, maintained at 75°F, if the two surfaces are oriented in the fashion shown (p. 494).

6.16 Determine the view factor F_{1-2} for surfaces oriented as shown in the sketches (p. 494).

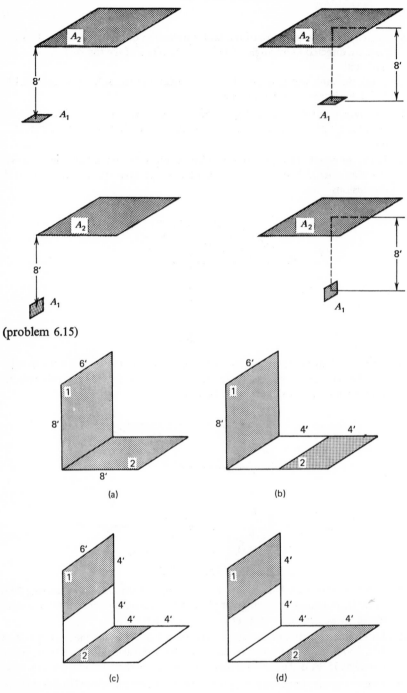

(problem 6.15)

(a)

(b)

(c)

(d)

(problem 6.16)

6.17 Determine the view factor F_{1-2} for surfaces oriented as shown.

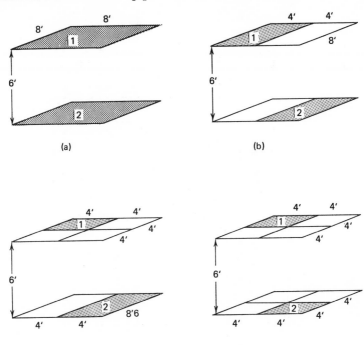

(a)

(b)

(c)

(d)

6.18 Determine the view factor F_{1-2} for surfaces oriented as shown (p. 496).

6.19 How much of the energy emitted from a floor measuring 8 ft by 10 ft escapes directly through a 1 ft by 1 ft window centered on one wall of the room measuring 8 ft by 8 ft?

6.20 How much of the energy emitted from one wall measuring 8 ft by 8 ft escapes directly through a 1 ft by 1 ft window which is centered on the opposite wall of the room and is 10 ft away?

6.21 Two cylindrical surfaces of infinite length are oriented with configuration and dimensions shown (p. 496).

(a) Using the crossed-strings method, show that F_{1-2} becomes

$$F_{1-2} = \frac{4L(L/4 + R)^{1/2} + 2R \sin^{-1}[R/(L/2 + R)] - L - 2R}{\pi R}$$

or, letting $X = L/2R + 1$,

$$F_{1-2} = \frac{2}{\pi}\left[(X^2 + 1)^{1/2} + \sin^{-1}\left(\frac{1}{X}\right) - X\right]$$

(b) Show that this expression approaches the proper limit for infinite parallel planes.

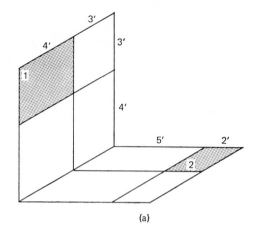

(a)

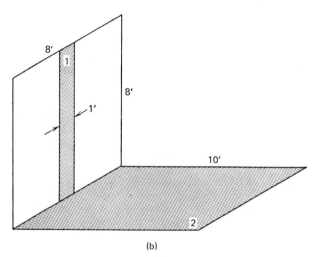

(b)

(problem 6.18)

(problem 6.21)

6.22 Compare the result of problem 6.21 for infinite cylindrical areas with the answer obtained in example 6.10 for infinite opposed parallel planes. Show that these results are compatible in the appropriate limiting condition.

6.23 Given an infinite cylinder and an infinite plane oriented as shown, use the crossed-strings approach to generate the view factor F_{1-2}.

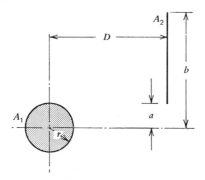

6.24 Two infinitely long parallel planes are 5 ft apart, with widths of 1 ft and 20 ft, respectively. Evaluate the view factor between these two planes, using Hottel's crossed-strings approach. How does F_{1-2} for infinite planes compare with that for the case of finite planes? (See problem 6.14.)

6.25 A long cylindrical surface of radius $R = 2$ ft is to be covered by a plane with a slit of width $R/3$ running along its center, as shown. The plane has its lower surface maintained at 400°F and is black. The cylindrical surface is well insulated. Find
    (a) the temperature of the cylindrical surface.
    (b) the net heat that must be supplied to maintain the temperature of the plane surface at 400°F.

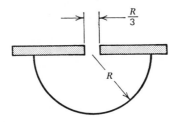

6.26 A hemispherical surface with radius $R = 2$ ft is covered by a plate with a hole of diameter $R/3$ drilled in its center. The lower surface of the plane is black and is maintained at 400°F by an imbedded heater. The hemispherical surface is well insulated. Find
    (a) the temperature of the hemispherical surface.
    (b) the net rate which heat must be supplied to the heater to maintain the temperature of the plane surface at 400°F.

6.27 The cylindrical enclosure shown consists of side walls maintained at 540°F with $\varepsilon = 0.70$, one base maintained at 40°F with $\varepsilon = 0.35$, and the other base reradiating. Determine

    (a) the net heat exchange between the hot and cold surfaces.

    (b) the temperature of the reradiating surface.

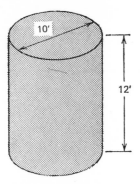

6.28 The circular base of the cylindrical tank shown may be considered a reradiating surface. The cylindrical side walls have an emissivity of 0.75 and are maintained at 540°F. The top of the tank is open to the surroundings, which are at an effective temperature of 50°F and may be considered black. Determine the net heat loss to the surroundings.

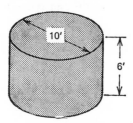

6.29 Determine the net heat exchange between the floor and ceiling in a 12 ft by 24 ft room 8 ft high if the side walls are considered reradiating. For floor and ceiling temperatures of 80°F and 61°F, respectively, determine the net heat exchange. The floor surface is wood and the ceiling is painted white.

6.30 Electronic equipment is housed in a rectangular box measuring 2 ft by 2 ft by 3 ft high. The energy generated by the electrical equipment is 1 kw. The box surface is oxidized aluminum. The energy loss is from five sides to surroundings at an effective temperature of 40°F. Still air at 40°F surrounds the box. Find the temperature of the box surface if

    (a) all heat loss is by radiation.

    (b) all heat loss is by natural convection.

    (c) heat is lost by both radiation and natural convection.

6.31 Two circular plates are parallel to each other, spaced 2 ft apart; both are 5 ft in diameter. One surface is made of commercial aluminum and is maintained at 1400°F; the other surface is polished stainless steel at 200°F. The surroundings may be assumed to absorb all energy which escapes this system. Find

    (a) the net energy loss from the hot surface.

    (b) the net energy gain (or loss) by (from) the stainless steel surface.

    (c) the net energy exchange between the two circular surfaces.

    (d) the total energy lost to the surroundings.

6.32 If another 5-foot-diameter plate is placed midway between the two surfaces described in the previous problem, what will be the net energy loss from the hot surface? The new plate is polished brass. Draw the thermal circuit for this problem.

6.33 A 3-inch-OD rusted iron pipe passes horizontally through a room whose walls are at 72°F and have an effective emissivity of 0.92. Room dimensions are 10 ft high by 20 ft by 35 ft; 20 ft of pipe is exposed. The pipe surface is at 210°F, and air at 70°F surrounds the pipe. Determine

    (a) the energy loss from the pipe due to radiation.

    (b) the energy loss from the pipe due to convection.

6.34 Two surfaces, maintained at 1040°F and 140°F, respectively, both having an emissivity of 0.6, may be considered parallel infinite planes. What will be the net energy exchange between these surfaces?

6.35 Determine the heat flux and the equilibrium temperature of the shields if radiation shields are placed between the surfaces specified in the previous problem. The shields also have $\varepsilon = 0.6$. Solve for the case of

    (a) one shield.

    (b) two shields.

    (c) three shields.

6.36 A long (effectively infinite) cylinder has a surface temperature of 540°F. This cylinder, which has a diameter of 2 in., is enclosed within a concentric, 12-inch-ID cylinder, whose temperature is maintained at 80°F. The surfaces of inner and outer cylinders have emissivities of 0.60 and 0.28, respectively. Find the net heat transfer between the two cylinders.

6.37 If a thin-walled aluminum cylinder ($\varepsilon = 0.20$) 6 inches in diameter is placed between the two surfaces described in problem 6.36, find

    (a) the net heat exchange.

    (b) the temperature of the aluminum shield.

6.38 Liquid nitrogen, at 130°R, is to be stored in a spherical container 10 inches in diameter. A concentric sphere with 12-inch diameter surrounds the storage vessel, and the intervening space is evacuated. This storage vessel is in a room with air temperature of 76°F. The viewing surfaces are flashed with aluminum, having an effective emissivity of 0.05. Assuming the temperature of the inner sphere to be 139°F and neglecting conduction through the outer spherical shell, find:

(a) the temperature at the outer spherical surface.

(b) the evaporation rate of liquid nitrogen under these conditions in pounds per hour. (h_{f_g} for nitrogen at 1 atm pressure is 85.7 Btu/lb$_m$.)

6.39 A thermocouple with surface area of 0.3 in.^{2} is placed in the center of a 3-inch-diameter duct whose walls are at 200°F and through which hot gas is flowing. The duct walls have an effective emissivity of 0.8, and that of the thermocouple is 0.6. The convective heat transfer coefficient between hot gases and the thermocouple junction (assumed to be spherical) is 25 Btu/hr-ft^2-°F. If the thermocouple reading is 340°F, what is your estimate of the actual temperature of the flowing gas?

6.40 Solve for the gas temperature in problem 6.39 if a cylindrical radiation shield 2 inches in diameter, made of commercial aluminum sheet, is placed around the thermocouple. The shield is long enough so that it may be safely assumed to completely obscure the view between the thermocouple and the pipe walls. What will be the temperature of the radiation shield?

6.41 A cubical enclosure is shown in the figure. The bottom surface is heated to 1200°F, and one of the sides is maintained at 700°F. The remaining sides and top may be considered reradiating surfaces. Emissivities of the floor and side wall are 0.6. If all sides are 3 ft in length, determine the heat transfer from the floor to the 700°F side wall. What will be the temperature of the reradiating surfaces? Solve this problem, using an electrical analogy.

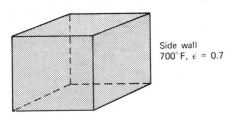

Side wall
700°F, $\epsilon = 0.7$

Floor, 1200°F, $\epsilon = 0.7$

6.42 In Example 6.13 an iterative solution was achieved for a cubical enclosure, using equations (6-101), (6-115), and (6-117). Solve problem 6.41, using this same iterative approach.

6.43 In Example 6.14 a 3-zone radiation problem was solved by means of the digital computer, with Gauss-Seidel iteration used to obtain the final result.

Starting with the same equations used in the example, generate a computer program which will solve these kinds of problems. Present the flow diagram and a program listing which is compatible with your local computer system.

Using your program, solve problem 6.41 for

(a) 3 zones.

(b) 4 zones.

(c) 5 zones.

6.44 The figure shows, schematically, a duct with square cross section which extends to infinity along its axis. Determine the heat transfer at each surface for the following conditions, using whatever solution technique you wish:

   (a) surface 1; $T = 1200°F$, $\varepsilon = 0.6$
      surface 2; $T = 800°F$, $\varepsilon = 0.44$
      surface 3; $T = 400°F$, $\varepsilon = 1.0$
      surface 4 reradiating
   (b) surface 1; $T = 1200°F$, $\varepsilon = 0.6$
      surface 2; $T = 800°F$, $\varepsilon = 0.44$
      surface 3; $T = 400°F$, $\varepsilon = 1.0$
      surface 4; $T = 600°F$, $\varepsilon = 0.8$

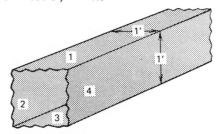

6.45 Solve for the heat transfer at each surface in an infinite rectangular duct, with temperatures and emissivities of the walls as given in (a) and (b) of the preceding problem. The duct dimensions are 1 ft by 3 ft, with walls 2 and 4 being 1 ft wide; walls 1 and 3 have a width of 3 ft.

6.46 A square duct measuring 10 in. by 10 in. has water vapor at 1 atm and 640°F flowing through it. The duct walls are at 300°F and have an emissivity of 0.8. Determine the rate of radiant energy transfer between the water vapor and duct walls under these conditions.

6.47 A gas mixture at 1000°F and 3 atm pressure enters a circular duct measuring 8 inches in diameter, with an inside surface which may be considered black. The gas is 18% CO_2 and 24% H_2O vapor by volume, the remainder being nonradiating. Determine the rate of heat ransfer between this gas mixture and the pipe walls if the pipe is maintained at 300°F. Neglect convective effects.

6.48 If the gas mixture specified in problem 6.47 enters the 8-inch-ID black pipe at 1000°F, flowing at 120 cu ft/min, estimate the length of the pipe required for the mean gas temperature to decrease to 900°F. The nonabsorbing fraction of the gas mixture may be assumed to be nitrogen.

6.49 A gas mixture at 1200°F and 8 atm pressure is introduced into a 10-foot-diameter spherical cavity with black walls which was initially evacuated. At the time of introducing the gas, the cavity walls are at 600°F. What is the rate of heat transfer under these conditions? The gas is composed of 20% CO_2 and the remainder nonradiating constituents.

6.50 A gas consisting of 22% CO_2 and 78% oxygen and nitrogen leaves a lime kiln at 2000°F and enters a 6-inch-square duct with a mass velocity of 0.4 lb_m/ft^2

per second. The specific heat of the gas is 0.28 Btu/lb$_m$-°F. It is to be cooled to 1000°F by interacting with the duct walls, having an emissivity of 0.9 and an inside surface temperature maintained at 800°F. The convective heat transfer coefficient between the gas and duct walls is 1.5 Btu/hr-ft^2-°F.

(a) Determine the length of duct required to cool the gas to 1000°F.

(b) Determine the fractions of total heat transfer due to radiation and to convection.

(c) If the duct length is increased to twice that evaluated in (a), what will be the exiting gas temperature?

7.1 One hundred thousand pounds per hour of water are to pass through a heat exchanger with a resulting temperature increase from 140°F to 200°F. Combustion gas at 800°F with specific heat of 0.24 Btu/lb$_m$-°F is available at a rate of 100,000 lb$_m$/hr to heat the water. The overall heat transfer coefficient for this system is 16 Btu/hr-ft^2-°F. Determine (a) the exit temperature of the gas and (b) the required heat transfer area of a single-pass counterflow heat exchanger to accomplish the specified energy transfer.

7.2 An oil having a specific heat of 0.45 Btu/lb$_m$-°F enters a single-pass counterflow heat exchanger at 260°F with a flow rate of 18000 lb$_m$/hr. The oil must be cooled to 180°F, using water as the other fluid; water at 50°F is available at a rate of 15,000 lb$_m$/hr. Find the required heat transfer area for an overall heat transfer coefficient of 42 Btu/hr-ft^2-°F.

7.3 Given the initial fluid temperatures, flow rates, overall heat transfer coefficient, and required heat transfer specified in the previous problem, what area will be required if the exchanger configuration is

(a) shell-and-tube with oil in the tubes; one shell pass, four tube passes.

(b) crossflow; water mixed, oil unmixed.

7.4 Water enters a counterflow double-pipe heat exchanger at 60°F and a flow rate of 180 lb$_m$/min. The water is to be heated to 140°F, using an oil, having a specific heat of 0.44 Btu/lb$_m$-°F, which enters the exchanger at 220°F and leaves at 90°F. The exchanger has an overall heat transfer coefficient of 50 Btu/hr-ft^2-°F.

(a) What heat transfer area is required?

(b) What area is required if the configuration is to be shell-and-tube with water making one shell pass and the oil two tube passes?

(c) What exit water temperature will result if, for the area determined in (a), the water flow rate is decreased to 140 lb$_m$/min?

7.5 A house, using a warm water heating system, requires 95,000 Btu/hr for heating purposes. Water is heated from 80°F to 125°F for this purpose in a heat exchanger through which hot air enters at 210°F and leaves at 120°F. Of the heat exchangers specified below, choose the one which will be most compact for this use:

(a) a counterflow unit with U = 30 Btu/hr-ft^2-°F, surface-to-volume ratio of 130 ft^2/ft^3.

(b) a crossflow unit, water unmixed and air mixed; U = 40 Btu/hr-ft^2-°F, surface-to-volume ratio of 100 ft^2/ft^3.

(c) a crossflow unit with both fluids unmixed; $U = 50$ Btu/hr-ft²-°F, surface-to-volume ratio of 90 ft²/ft³.

7.6 Determine the required heat transfer surface area for a heat exchanger constructed from 1-inch, 14-BWG mild steel tubing to cool 55,000 lb/hr of benzene from 150°F to 103°F, using 50,000 lb/hr of water available at 50°F. The overall heat transfer coefficient, based upon the outer-tube area, is 110 Btu/hr-ft²-°F. Consider each of the following arrangements:

  (a) parallel flow, single pass.
  (b) counterflow, single pass.
  (c) shell and tube with benzene in the shell, two shell passes, 48 tube passes.
  (d) crossflow with one tube pass and one shell pass, shell-side fluid (benzene) mixed.

7.7 Hydraulic fluid (MIL-M-5606) flows in the tubes of a single-shell, two-tube-pass heat exchanger entering at 160°F and leaving at 100°F, with a flow rate of 5000 lb/hr; the oil is in turbulent flow. Water enters the shell at 60°F and leaves at 80°F. If the same exchanger is used with the same water flow rate and inlet temperature, but with the oil flowing at 3500 lb/hr and entering at 200°F, what will be the exit temperatures of the two fluids?

*7.8 A crossflow heat exchanger is diagrammed on p. 504. This unit is to be used to heat air which enters at 60°F, flowing at 6500 lb/hr, by means of turbine exhaust gases which are available at 1550°F, flowing at 5000 lb/hr. It is desired to find the exit temperature of the air.

*7.9 A heater is needed to heat water from 60°F to 160°F, flowing at 6000 lb/hr, using 130 psi saturated steam condensing on the outside of 1-1/4-inch 16-BWG copper tubes. A steam-side surface coefficient of 1400 Btu/hr-ft²-°F may be assumed. The tubes are to be 10 ft long, and the desired water velocity is 5 ft/sec.

Determine the required number of tubes per pass in a shell-and-tube heat exchanger, the required heat transfer area, and the arrangement which will accomplish the necessary heat transfer.

*7.10 Solve for the required information in the previous problem when a fouling resistance of 0.002 hr/ft²-°F/Btu is considered for the water side and one of 0.0005 hr-ft²-°F/Btu is considered on the steam side.

*7.11 Aniline is cooled from 125°F to 90°F in the inner pipe of a double-pipe single-pass heat exchanger. Cooling water flows countercurrently to the aniline, entering at 60°F and leaving at 82°F. The exchanger is constructed of a 7/8-inch, 16-BWG copper tube jacketed with a 1-1/2-inch schedule-40 steel pipe. The velocity of the aniline is to be 5 ft/sec and that of the water is 4 ft/sec. Determine

  (a) the overall coefficient based on the outside area of the inner pipe.
  (b) the required length of exchanger if the water flow rate is 3000 lb/hr.

---

* In this and subsequent problems with an (*) sign, the overall heat transfer coefficient is not given. It will thus be necessary to determine this or related quantities in order to solve the problem.

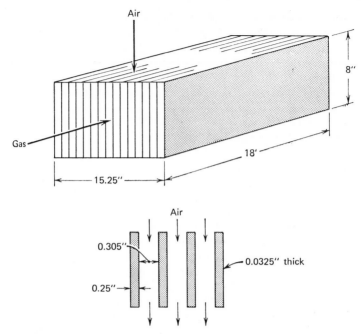

(problem 7.8)

*7.12 Hydraulic fluid is being cooled from 200°F to 110°F in the annulus of a double-pipe heat exchanger by water which enters the inside pipe at 60°F and leaves at 95°F. The proposal is made that the double-pipe unit be lengthened to cool the hydraulic fluid to a lower temperature. If the fluid streams flow concurrently, determine
    (a) the minimum temperature to which the hydraulic fluid may be cooled.
    (b) the exiting hydraulic fluid temperature as a function of the fractional increase in heat exchanger length.

*7.13 Given the same double-pipe heat exchanger as in the previous problem, with the same entering temperatures and flow rates of the two fluid streams except that the streams flow countercurrently, determine
    (a) the exit temperature of each stream.
    (b) the lowest temperature to which the oil might be cooled.

*7.14 Forty thousand standard cubic feet per hour of nitrogen are to be heated from 75°F to 145°F by condensing saturated steam at 35 psia. Nitrogen is to flow through 1-inch 14-BWG tubes with a mass velocity of 8500 lb/hr-ft^2.
    (a) Determine the required number of 8-foot tubes and the number of passes for this nitrogen preheating to be accomplished. One shell pass of condensing steam is to be used.
    (b) What flow rate of carbon dioxide in lb/hr would be anticipated were this exchanger used to preheat CO_2 rather than nitrogen? Steam at the same conditions would still be condensed in the shell.

*7.15 Air flows across a bank of tubes through which hot gas is introduced. Both gases are at atmospheric pressure; the air enters at 80°F, and hot gas enters the tubes at 1450°F. Air flows at a rate of 2500 lb/hr, and the hot gas flow rate is 4000 lb/hr. Forty 1-inch 18-BWG mild steel tubes are used; the arrangement and dimensions are shown in the accompanying figure. Determine the heat transfer acomplished per hour and the exiting temperature of each fluid.

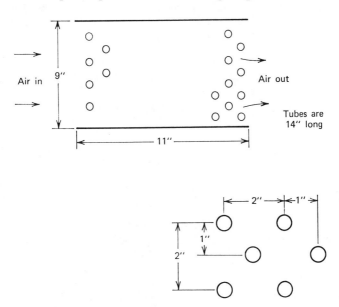

7.16 A shell-and-tube heat exchanger with one tube pass is to be designed in which water in the tubes is to be heated by condensing steam in the shell. The steam is to condense filmwise and the water is to be in turbulent flow in the smooth tubes. Specified quantities include water flow rate, entering and exiting water temperatures, steam condensing temperature, and available pressure drop (entering and exiting losses neglected). Assuming that the water flow remains turbulent, and that the thermal resistance of the tube wall and condensate film are negligible, evaluate the effect of tube diameter on the total required heat transfer area in the exchanger.

7.17 Two identical heat exchangers are constructed, each 12 ft long. These units consist of a 2-inch schedule-40 steel pipe concentric with a 3-inch schedule-40 pipe. Water, at 65°F, enters the inner pipe of one unit and flows through both exchangers in series at a rate of 50 gal/min. Two hot water streams are available for supplying heat to the cold stream, one at 140°F and 30 gpm, the other at 210°F and 30 gpm. If the hot streams may be mixed arbitrarily at any location before entering this system, specify the arrangement which will yield maximum heat transfer. For this optimum arrangement, what will be the exiting temperature of the cold stream?

7.18 A condenser in an automobile air conditioner has a design capacity to remove 50,000 Btu/hr when the car is traveling at 40 mph when the outside air temperature is 95°F. The condensing system fluid is Freon-12, which remains at 150°F, and the air temperature rise across the condenser is 12°F. The overall heat transfer coefficient, for design purposes, is 30 Btu/hr-ft²-°F. When the automobile velocity changes, so, of course, does the performance of the air conditioning unit. Based upon the assumption that the mass flow rate of air varies directly as velocity and that V varies as velocity to the 0.7 power, generate a plot of condenser capacity in Btu/hr as a function of automobile speed over the range from 10 to 60 mph.

7.19 An oil cooler in an engine utilizes jacket water at 150°F to cool lubricating oil from 210°F to 165°F, flowing at 2600 lb/hr. A small shell-and-tube heat exchanger consisting of one shell pass and two tube passes, with oil in the shell, is to be used for this purpose. Fifty 1/2-inch 18-BWG copper tubes are involved, 25 per pass. The oil has a specific heat of 0.42 Btu/lb-°F, and the oil-side heat transfer coefficient may be taken as 45 Btu/hr-ft²-°F. If the cooling water rise is to be 5°F, how long must the heat exchanger tubes be?

7.20 Exhaust steam from a turbine is to be used to heat 50,000 lb/hr of water from 60°F to 115°F. Steam is to be in the shell of a 2-shell-pass, 8-tube-pass heat exchanger in which each tube pass employs 20 1-1/4-inch 16-BWG copper tubes. The heat transfer coefficients on steam and water sides are 700 Btu/hr-ft²-°F and 250 Btu/hr-ft²-°F, respectively. What is the required length of the tubes in this exchanger? Neglect surface fouling.

7.21 Solve for the length of tubes for conditions specified in the previous problem if fouling resistances of 0.001 hr-ft²-°F/Btu are assumed for both the steam and water sides of the heat exchanger tubes.

*7.22 A single-pass shell-and-tube heat exchanger with steam condensing in the shell is to be used to heat 5000 gpm of water from 60°F to 100°F. The steam-side heat transfer coefficient is 1800 Btu/hr-ft²-°F. Pumping capacity limits the pressure drop through the tubes to 14 psi. Calculate the length and number of 1-inch schedule-40 steel tubes which will accomplish the necessary heat transfer with the pumping capacity available.

*7.23 A shell-and-tube heat exchanger, with 50 psi saturated steam condensing in the shell, is to heat 550 lb/min of N-butyl alcohol from 80°F to 180°F. The exchanger employs 3/4-inch 16-BWG copper tubes, and the allowable pressure drop in the tubes is 1.5 psi. Neglecting the steam-side resistance as well as the conductive resistance through the thin-walled copper tubes, determine
   (a) the velocity of alcohol in the tubes.
   (b) the number of tubes in the bundle.
   (c) the length of the tubes.

*7.24 Benzene with a flow rate of 80,000 lb/hr is to be cooled from 160°F to 100°F by 80,000 lb/hr of water available at 70°F. Assuming an overall heat transfer coefficient from Table 7.2, estimate the surface area required for the following exchanger configurations:
    (a) crossflow, single pass, both fluids unmixed.
    (b) crossflow, two passes.
    (c) crossflow, single pass, with water mixed, benzene unmixed.

INDEX